DATE DUE

Demco, Inc. 38-293

VITAMINS AND HORMONES
VOLUME 48

Editorial Board

FRANK CHYTIL
ANTHONY R. MEANS
BERT W. O'MALLEY
MICHAEL SPORN
ARMEN H. TASHJIAN, JR.

VITAMINS AND HORMONES
ADVANCES IN RESEARCH AND APPLICATIONS

Editor-in-Chief

GERALD LITWACK

Department of Pharmacology
Jefferson Cancer Institute
Thomas Jefferson University Medical College
Philadelphia, Pennsylvania

Volume 48

ACADEMIC PRESS

San Diego New York Boston
London Sydney Tokyo Toronto

This book is printed on acid-free paper. ∞

Copyright © 1994 by ACADEMIC PRESS, INC.
All Rights Reserved.
No part of this publication may be reproduced or transmitted in any form or by any means, electronic or mechanical, including photocopy, recording, or any information storage and retrieval system, without permission in writing from the publisher.

Academic Press, Inc.
525 B Street, Suite 1900, San Diego, CA 92101-4495

United Kingdom Edition published by
Academic Press Limited
24–28 Oval Road, London NW1 7DX

International Standard Serial Number: 0083-6729

International Standard Book Number: 0-12-709848-8

PRINTED IN THE UNITED STATES OF AMERICA
94 95 96 97 98 99   QW   9 8 7 6 5 4 3 2 1

Former Editors

ROBERT S. HARRIS
Newton, Massachusetts

JOHN A. LORRAINE
*University of Edinburgh
Edinburgh, Scotland*

PAUL L. MUNSON
*University of North Carolina
Chapel Hill, North Carolina*

JOHN GLOVER
*University of Liverpool
Liverpool, England*

GERALD D. AURBACH
*Metabolic Diseases Branch
National Institute of Diabetes
and Digestive and Kidney Diseases
National Institutes of Health
Bethesda, Maryland*

KENNETH V. THIMANN
*University of California
Santa Cruz, California*

IRA G. WOOL
*University of Chicago
Chicago, Illinois*

EGON DICZFALUSY
*Karolinska Sjukhuset
Stockholm, Sweden*

ROBERT OLSON
*School of Medicine
State University of New York at Stony Brook
Stony Brook, New York*

DONALD B. MCCORMICK
*Department of Biochemistry
Emory University School of Medicine
Atlanta, Georgia*

Contents

PREFACE xi

Molecular and Cellular Aspects of Insulin-like Growth Factor Action
HAIM WERNER, MARTIN ADAMO, CHARLES T. ROBERTS, JR., AND DEREK LEROITH

I.	General Introduction	1
II.	Ligands	2
III.	Receptors	6
IV.	IGF Binding Proteins	15
V.	Physiological Roles of the IGF System	20
VI.	Organ/System- Involvement of IGFs	27
VII.	Conclusions	38
	References	39

Heterologous Expression of G Protein-Linked Receptors in Pituitary and Fibroblast Cell Lines
PAUL R. ALBERT

I.	Introduction	59
II.	Heterologous Expression Systems	61
III.	GH Pituitary Cells as Models for Signal Transduction	64
IV.	Fibroblast Cells as Models for Signal Transduction	85
V.	Conclusion and Future Prospects	96
	References	98

Receptors for the TGF-β Ligands Family
CRAIG H. BASSING, JONATHAN M. YINGLING, AND XIAO-FAN WANG

| I. | Introduction | 111 |
| II. | TGF-β Ligands | 113 |

III.	TGF-β Receptors	117
IV.	TGF-β Signal Transduction	131
V.	Perspectives and Future Directions	151
	References	152

Biological Actions of Endothelin

KATHERINE STEPHENSON, CHANDRASHEKHAR R. GANDHI, AND MERLE S. OLSON

I.	Introduction	157
II.	Discovery of Endothelin	158
III.	Endothelin Structure	159
IV.	Endothelin Structure–Activity Relationships	161
V.	Processing of Endothelin	162
VI.	Genes of the Endothelin Family	164
VII.	Structural Organization of the Endothelin Gene	164
VIII.	Factors That Stimulate Endothelin Production	166
IX.	Endothelin Receptors	166
X.	Localization of Endothelin	169
XI.	Endothelin-Activated Transmembrane Signaling Systems	170
XII.	Biological Actions of Endothelin	172
XIII.	Pathophysiology of Endothelin	175
XIV.	Endothelin and the Liver	177
XV.	Conclusions	183
	References	183

Cyclic ADP–Ribose: Metabolism and Calcium Mobilizing Function

HON CHEUNG LEE, ANTONY GALIONE, AND TIMOTHY F. WALSETH

I.	Introduction	199
II.	Cyclic ADP–Ribose	201
III.	Enzymes Involved in the Metabolism of Cyclic ADP–Ribose	207
IV.	Cyclic ADP–Ribose-Dependent Ca^{2+} Release	221
V.	Cyclic ADP–Ribose Receptor	236
VI.	Physiological Roles of Cyclic ADP–Ribose	244
VII.	Conclusion	249
	References	251

A Critical Review of Minimal Vitamin B_6 Requirements for Growth in Various Species with a Proposed Method of Calculation

Stephen P. Coburn

I.	Characteristics of Vitamin B_6 Metabolism	259
II.	Vitamin B_6 Requirements for Growth	266
III.	Discussion	287
IV.	Summary	290
	References	291

Preface

This volume of *Vitamins and Hormones* marks the first under my editorship. I appreciate the long history of the serial and the in-depth coverage of related areas of research found in previous volumes that have established this important source as a major reference work crossing several fields.

The past decade has emphasized that the progress of science is unifying and that disciplines are merging closer than ever before. In consequence, it seems an appropriate time to broaden the scope of these volumes.

The Editorial Board has been partially reconstituted and new names may be added to future listings of the Board. The members of the Board have assisted me greatly in pointing out potential authors of cutting edge research and I expect to rely heavily on the suggestions of Board members in the future in addition to my own ideas. The contributing authors have been very cooperative in completing their manuscripts in a timely fashion and Academic Press has helped in many ways.

In this eclectic volume, there are three reports dealing with receptorology: one on the mechanism of action of insulin-like growth factor from the Derek LeRoith laboratory; another is on the heterologous expression of G protein-linked receptors by Paul R. Albert and the third is on receptors of the TGF-β ligands family from the Xiao-Fan Wang laboratory. These contributions are followed by chapters on the action of endothelins from the Merle S. Olson laboratory with collaborators, functions of cyclic ADP-ribose from the Hon Cheung Lee laboratory, and finally the determination of vitamin B_6 requirements in various species by Stephen P. Coburn.

I thank Dr. Charles Crumly and Richard Van Frank at Academic Press for their guidance and interest.

GERALD LITWACK

Molecular and Cellular Aspects of Insulin-like Growth Factor Action

HAIM WERNER, MARTIN ADAMO, CHARLES T. ROBERTS, JR., AND DEREK LeROITH

National Institute of Diabetes and Digestive and Kidney Diseases
National Institutes of Health
Bethesda, Maryland 20892

I. General Introduction
II. Ligands
    A. Introduction
    B. IGF-I
    C. IGF-II
III. Receptors
    A. Introduction
    B. IGF-I Receptor
    C. IGF-II/M-6 Receptor-P
IV. IGF Binding Proteins
    A. Structure
    B. IGF Binding
    C. Regulation of Expression
    D. Function
V. Physiological Roles of the IGF System
    A. Cell Cycle
    B. *In Vivo* Biological Actions
    C. Development
    D. *In Vitro* Effects
VI. Organ/System-Involvement of IGFs
    A. Reproductive System
    B. Nervous Tissue
    C. Cancer
VII. Conclusions
    References

I. General Introduction

The insulin-like growth factors (IGF-I and IGF-II) are pluripotent factors that regulate growth, differentiation, and the maintenance of differentiated function in numerous tissues and in specific cell types. Both IGFs are produced in largest amounts by the liver and are secreted into the circulation, where they function as classical endocrine agents by interacting with specific cell-surface receptors present on target tissues. The widespread distribution of IGF receptors, and the production and secretion of the IGFs themselves by almost all extrahepatic tissues, strongly suggests that these factors employ autocrine and paracrine modes of action in addition to their endocrine effects. The actions of the IGFs are also influenced by a family of IGF binding proteins that are found in the circulation and in extracellular fluids; these proteins may have positive or negative effects on IGF action through different mechanisms. The overall actions of these important molecules thus are governed by a complex interplay between ligands, receptors, and binding proteins; the levels and the regulation of each component contribute significantly to the ultimate biological effects manifested in a given situation. In this chapter, we describe in some detail the structure, expression, and regulation of the different components of the IGF system as well as their roles.

II. Ligands

A. Introduction

IGF-I, IGF-II, and insulin (Fig. 1) constitute a family of structurally related hormones that exhibit 40–50% amino acid homology with each other (Blundell *et al.*, 1983; Daughaday and Rotwein, 1989; Rechler and Nissley, 1990; Sussenbach, 1989). The mature IGF-I and IGF-II peptides are similar to proinsulin because both contain B and A domains that are analogous to the B and A chains of insulin. In the IGFs, however, these domains are linked by a C domain that is slightly smaller than the C peptide of proinsulin, and, in contrast to the proinsulin C peptide, is not removed during processing of the prohormones (Fig. 1). Mature IGF-I and IGF-II contain an extension to the A domain, the D domain, that is not found in insulin. In addition, both IGF prohormones contain C-terminal E peptides that are cleaved during processing of the precursors into the mature circulating peptides (Van den Brande, 1990; Bell *et al.*, 1986).

B. IGF-I

The IGF-I gene has been mapped to the long arm of chromosome 12 in humans (Tricoli *et al.*, 1984) and to the central region of the homologous chromosome 10 in mice (Taylor and Grieco, 1991). The mammalian gene consists of at least six exons and, in rodent and humans, encompasses at least 90 kb of chromosomal DNA (Rotwein *et al.*, 1986;

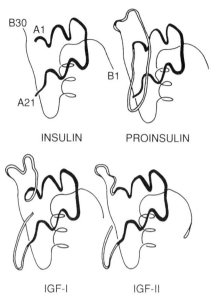

FIG. 1. Predicted tertiary structures of the insulin-like growth factor (IGF) family of peptides.

Shimatsu and Rotwein, 1987a). Exons 1 and 2 encode distinct, mutually exclusive 5' untranslated regions (UTRs) as well as distinct N termini of the IGF-I signal peptide because of several in-frame translation initiation codons (Fig. 2). The common C-terminal amino acids of the signal peptide are encoded by the 5' end of exon 3. All of the mature peptide coding sequence is present in the remainder of exon 3 and in part of exon 4. The E-peptide coding sequences are contained in exons 4, 5, and 6; the remainder of exon 6 (and part of exon 5 in the human gene) encodes 3'-UTRs (Rotwein, 1986; Roberts et al., 1987a,b).

Transcription of the mammalian gene and processing of the primary transcript are extremely complex processes. Alternative leader exons 1 and 2 are transcribed from distinct start sites and appear to be regulated by separate promoters. The exon 1 promoter lacks core regulatory elements such as TATA and CCAAT boxes and, possibly as a consequence, transcription of this exon is initiated from at least four distinct sites dispersed over a 350-bp region (Adamo et al., 1991a,b; Jansen et al., 1991; Kim et al., 1991; Hall et al., 1992; Shemer et al., 1992). The exon 2 promoter, which contains both TATA- and CCAAT-like motifs at appropriate positions, directs transcription initiation from a cluster of sites located 50–70 bp upstream of the 3' end of this

FIG. 2. Structure of mammalian IGF-I and IGF-II genes. Exons are numbered. Known promoter sites in the IGF-II gene are labeled P1–P4.

exon (Adamo et al., 1991a,b; Jansen et al., 1991; Kim et al., 1991; Hall et al., 1992).

IGF-I mRNAs resulting from the use of the alternative promoters and transcription start sites are differentially expressed and regulated. Exon 1-containing transcripts are expressed ubiquitously and, in fact, constitute the major mRNA species in every rat tissue examined (Lowe et al., 1987; Adamo et al., 1991a). Transcription of exon 2 is more limited, occurring in the liver and in a few extrahepatic tissues. The liver (the major source of circulating IGF-I) is particularly enriched in exon 2 transcripts. During development, hepatic exon 1 transcripts appear earlier than exon 2 transcripts; the expression of the latter increases markedly at the onset of growth hormone (GH)-dependent linear growth (Hoyt et al., 1988; Adamo et al., 1989,1991b; Kikuchi et al., 1992). GH is a major regulator of liver IGF-I gene transcription (Mathews et al., 1986; Bichell et al., 1992) and, under most experimental conditions, GH increases the levels of exon 2 transcripts more than the levels of exon 1 transcripts (Lowe et al., 1987; Foyt et al., 1992).

In addition to the transcriptional regulation of IGF-I gene expression, two facets of IGF-I gene expression suggest the possibility of post transcriptional regulation. (1) In rats, a 186-bp region of exon 1 is spliced out of ~20% of liver IGF-I mRNA (Shimatsu and Rotwein, 1987b; Shemer et al., 1992), potentially influencing the translation of IGF-I mRNAs (see subsequent discussion). (2) Alternative E peptide-encoding IGF-I mRNAs result from alternative splicing of exon 5. Although the splicing pattern is generally constitutive (i.e., ~95% of IGF-I mRNAs lack exon 5 and encode the A form of the E peptide; ~5% contain exon 5 and encode the B form), evidence exists that, in the liver, GH treatment favors the retention of exon 5 (Lowe et al., 1988).

Currently, however, no function for the two divergent E peptides has been defined.

Northern blot hybridization reveals multiple IGF-I mRNA transcripts varying from ~1 kb to more than 7 kb in length. The difference in size primarily reflects different lengths of 3'-UTRs resulting from differential polyadenylation site usage (Lund et al., 1989). The 7.6-kb species, for example, contains almost 6.5 kb of 3'-UTR encoded by exon 6 (Heppler et al., 1990; Steenbergh et al., 1991). These transcripts are less stable than the smaller ones, possibly because of a series of AT-rich destabilizing elements.

In addition to regulation at the level of transcription, mRNA stability, and processing, IGF-I gene expression is also potentially regulated at translational and posttranslational levels. *In vivo* studies suggest that the 5'-UTRs in IGF-I mRNAs affect translatability. For example, the fully spliced exon 1 variant is enriched in rat liver polysomes relative to the full-length mRNA (Foyt et al., 1991,1992). Further, the presence of multiple in-frame translation initiation codons predicts the expression of preprohormones with different N-terminal extensions of the core IGF-I signal peptide (Roberts et al., 1987b; Rotwein et al., 1987), which could result in variations in intracellular processing, targeting, and action.

C. IGF-II

The human IGF-II gene spans about 30 kb of chromosomal DNA on the distal end of the short arm of chromosome 11, adjacent to the 3' end of the insulin gene (Tricoli et al., 1984). The IGF-II gene consists of nine exons. The mature IGF-II peptide is encoded by exons 7, 8, and 9; the long 3'-UTR is also contained within exon 9 (Fig. 2; Ueno et al., 1989). Expression of the human IGF-II gene is controlled by four different promoters (P1–P4) that precede some of the leader exons. As in the IGF-I gene, these leader exons also encode multiple 5'-UTRs (Dull et al., 1984).

In humans, the promoters are differentially activated in a tissue- and development-specific manner (de Pagter-Holthuizen et al., 1987,1988; Holthuizen et al., 1990; Soares et al., 1985). Promoters P2, P3, and P4 are active in fetal and most adult nonhepatic tissues. In adult liver, these promoters are turned off and promoter P1 is activated. Promoter P1, which is only found in the human gene, is a TATA-less GC-rich promoter; these two features are compatible with the heterogeneous transcription initiation observed. P3 and P4, on the other hand, contain a TATA box; in addition, P3 has a CCAAT box. As a result, transcription from these promoters occurs from specific sites. Potentially correlated with these findings, the P3 and P4 promoters are more actively regulated than the P1 promoter (Matsuguchi et al., 1990).

As for IGF-I, multiple IGF-II mRNA transcripts (ranging from ~2.2 kb to ~ 6.0 kb) are produced but, unlike IGF-I, the size heterogeneity is a function of both promoter usage (and, therefore, length of 5'-UTR; Irminger et al., 1987) and different lengths of 3'-UTR resulting from the use of multiple polyadenylation sites (Rechler, 1991).

Regulation of IGF-II gene expression also occurs at the translational level. mRNAs transcribed from the P2 and P4 promoters (5.0 kb and 4.8 kb) have shorter 5'-UTRs and are preferentially associated with polysomes, relative to P3-derived transcripts (6.0 kb), which have longer 5'-UTRs (Nielsen et al., 1990; Meinsma et al., 1991). Unlike IGF-I mRNAs, however, IGF-II mRNAs encode a single E peptide and signal peptide (Rechler, 1991).

III. Receptors

A. Introduction

The IGFs exert their biological actions by binding to a family of specific membrane-associated glycoprotein receptors that include the insulin, IGF-I, and IGF-II receptors (Czech, 1989). Molecular cloning of the receptor cDNAs and genes confirmed earlier conclusions that each receptor is the product of a distinct gene. The insulin and IGF-I receptors are, however, closely related structures (Fig. 3). The IGF-II receptor, on the other hand, is identical to the cation-independent mannose-6-phosphate (M-6-P) receptor. Interestingly, this bifunctional property of the IGF-II receptor appears to be unique to mammalian species. As discussed later, most, if not all, of the physiological effects of the IGFs are mediated by the IGF-I receptor. Thus, we concentrate in this chapter on that receptor and include comparisons to the closely related insulin receptor when appropriate. We do, however, describe certain relevant aspects of the IGF-II/M-6-P receptor.

B. IGF-I Receptor

The human IGF-I receptor is the product of a single-copy gene located at the distal end of the long arm of chromosome 15. Like the insulin receptor, this gene consists of 22 exons spanning a minimum

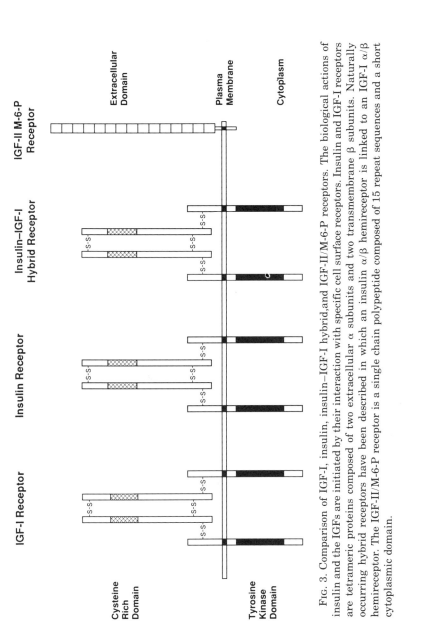

FIG. 3. Comparison of IGF-I, insulin, insulin–IGF-I hybrid,and IGF-II/M-6-P receptors. The biological actions of insulin and the IGFs are initiated by their interaction with specific cell surface receptors. Insulin and IGF-I receptors are tetrameric proteins composed of two extracellular α subunits and two transmembrane β subunits. Naturally occurring hybrid receptors have been described in which an insulin α/β hemireceptor is linked to an IGF-I α/β hemireceptor. The IGF-II/M-6-P receptor is a single chain polypeptide composed of 15 repeat sequences and a short cytoplasmic domain.

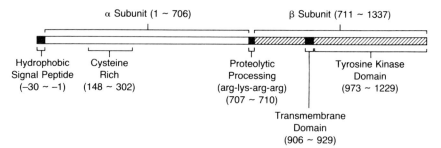

FIG. 4. Human IGF-I receptor precursor. Molecular closing of human IGF-I receptor cDNAs showed the presence of an open reading frame of 4101 nucleotides, starting with an ATG translation initiation codon and flanked by a TGA codon for translation termination.

of 100 kb of chromosomal DNA (Ullrich et al., 1986; Abbott et al., 1992).

The mature receptor is composed of two α subunits, which are entirely extracellular and are mainly involved in ligand binding, and two transmembrane β subunits, which include a tyrosine kinase domain in their cytoplasmic portion. The α and β subunits are synthesized colinearly as part of a 1367-amino-acid precursor molecule (Fig. 4). The precursor is N-glycosylated at several sites, principally in the α subunit. Following cleavage of a basic tetrapeptide (Arg–Lys–Arg–Arg) located after residue 706, the subunits are produced and subsequently linked by disulfide bonds to form the mature heterotetrameric receptor in an $(\alpha\beta)_2$ configuration.

The IGF-I receptor resembles the insulin receptor in overall primary and tertiary structure. Both have cysteine-rich domains in the α subunits and very homologous tyrosine kinase domains in the β subunits. Amino acids homologies are variable, being highest (84%) in the kinase domain and lowest (40%) in the cysteine-rich domain.

1. Expression

The IGF-I receptor is expressed in virtually every tissue and cell type. Expression is regulated developmentally as well as by many other physiological and pathological conditions. Fetal expression of the receptor is very high in most tissues, strongly supporting the suggestion that this receptor plays an essential role in organogenesis. During postnatal development, expression in most tissues falls dramatically to adult levels. Since circulating IGF-I levels rise postnatally, the decrease in IGF-I receptor may be the result of down-regulation (Fig. 5). A similar inverse correlation between IGF-I levels and IGF-I receptor

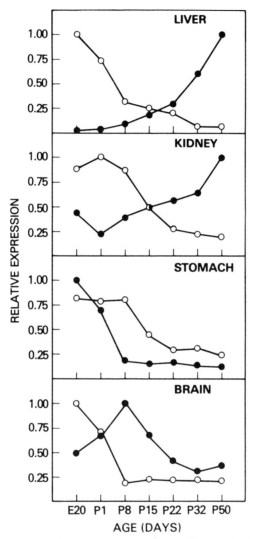

FIG. 5. Tissue-specific developmental changes in IGF-I and IGF-I receptor gene expression in the rat. The relative changes in levels of mRNA encoding IGF-I (●) and IGF-I receptor (○) have been plotted as a function of age. E, embryo/fetus; P, postnatal.

expression has been noted in fasting and insulinopenic diabetes, in which circulating IGF-I levels are reduced and IGF-I receptor expression is increased (Lowe et al., 1989a). However, this up-regulation is not universally observed in every tissue; thus, the expression of IGF-I receptors in organs that are separated from the general circulation by an endothelial barrier, for example, in brain and testes, is less affected by fluctuations in the levels of IGF-I. In addition to the blood-borne factors involved in the regulation of expression of IGF-I receptors, tissue-specific factors are implicated in this regulation (Werner et al., 1990a). These factors include, among others, follicle stimulating hormone (FSH) in the ovary, platelet-derived growth factor (PDGF) in fibroblasts, and estrogen in breast tumors.

The characterization of the IGF-I receptor promoter has revealed a number of interesting features (Werner et al., 1990b,1992a; Cooke et al., 1991). Despite the absence of classic TATA and CCAAT boxes, the IGF-I receptor gene contains a single major transcription initiation site contained within an "initiator" sequence. This unique transcription start site defines a very long 5'-UTR of ~1 kb. The 5' flanking sequence and the 5'-UTR are very GC rich. These features are highly conserved in mammals and are currently being studied to elucidate the factors involved in regulating the expression of this gene.

2. Binding Characteristics

The α subunits of the insulin and IGF-I receptors lie entirely extracellularly and are responsible for binding the ligands. The ligand binding site(s) on the IGF-II receptor is also present in its long extracellular domain. The IGF-I receptor binds IGF-I and IGF-II with equally high affinity ($\sim 10^{-10}$ M), whereas the insulin receptor demonstrates high affinity for insulin but a lower affinity for IGF-I and IGF-II ($\sim 10^{-8}$ M). The IGF-II/M-6-P receptor, on the other hand, binds IGF-II better than IGF-I and does not bind insulin (Fig. 6). The differences in binding affinities of the insulin and IGF-I receptors for the various ligands are presumably the result of structural differences in the α subunits. Indeed, using chimeric receptors, investigators have determined the essential role of the cysteine-rich domain of the IGF-I receptor (encoded by exon 3) for IGF-I binding. In contrast, in the insulin receptor, regions outside the cysteine-rich domain, both N- and C-terminal, are responsible for high-affinity insulin binding.

Binding studies reveal another important difference between insulin and IGF-I receptors. IGF-I receptor binding studies generally yield linear Scatchard plots, suggesting a single class of receptors. In contrast, insulin receptors classically generate curvilinear plots, suggesting either multiple classes of receptors or negative cooperativity. The structural basis for these phenomena has not been determined.

3. Signal Transduction

As for the insulin receptor, the initial activation of the IGF-I receptor after ligand binding involves autophosphorylation of the receptor β

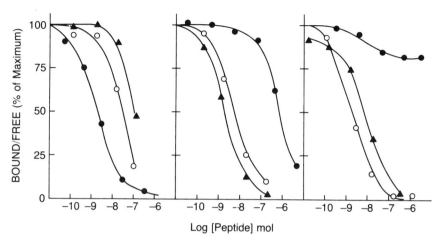

FIG. 6. Competition–inhibition curves for ^{125}I-ligand binding using ^{125}I-insulin (left), ^{125}I-IGF-I (middle), and ^{125}I-IGF-II (right), with increasing concentrations of unlabeled insulin (●), IGF-I (△), and IGF-II (○).

subunit. Whereas the epidermal growth factor (EGF) receptor, a single transmembrane protein, requires dimerization prior to autophosphorylation (Spaargaren et al., 1991), the insulin and IGF-I receptors already exist as tetrameric complexes. Thus, IGF-I binding to the α subunit of the receptor results in rapid activation of tyrosine kinase activity intrinsic to the β subunit (Jacobs and Cuatrecasas, 1986; Ikari et al., 1988). Receptor aggregation following IGF-I binding or treatment with antibody αIR3, a specific antibody against the human IGF-I receptor, may play a role in enhancing the signal through the IGF-I receptor (Ikari et al., 1988).

As for the insulin receptor, autophosphorylation of the IGF-I receptor occurs initially on tyrosine residues 1131, 1135, and 1136 in the tyrosine kinase domain. Other tyrosine residues in the β subunit are subsequently phosphorylated. The importance of the activation of the tyrosine kinase activity in receptor signaling was demonstrated by experiments in which mutations were introduced at the lysine residue at position 1003 (in the ATP binding site) and at the triple tyrosines. In each mutant, receptor autophosphorylation is absent or markedly reduced; also, all the biological actions studied, including glucose uptake, thymidine incorporation, c-*fos* and c-*jun* induction, and so on, were reduced (Kato, et al., 1993).

Autophosphorylation of the β subunit activates the tyrosine kinase activity toward endogenous substrates. The earliest substrate to be phosphorylated on tyrosine residues is a cytosolic 185-kDa phosphoprotein (pp185 or IRS-I). IRS-I contains six tyrosine residues in a Tyr–Met–X–Met (YMXM, where X is any amino acid) motif. This motif, which is also present in the PDGF receptor, mediates binding of phosphatidyl-inositol 3' (PI3) kinase to the PDGF receptor. Indeed, tyrosine-phosphorylated IRS-I binds the 85-kDa subunit of PI3 kinase via its two SH_2 domains, which are presumed to bind to phosphorylated tyrosines within the YMXM sequence (Rothenberg et al., 1991; Sun et al., 1991; Yamamoto et al., 1992). The IRS-I-bound 85-kDa subunit, in turn, complexes with the 110-kDa catalytic subunit, thereby increasing the activity of PI3 kinase. IRS-I is fairly widespread, being expressed in cells derived from liver, neural (N18), and epidermal (KB) tissues (White et al., 1987; Kadowaki et al., 1987; Shemer et al., 1987).

In addition to IRS-I, several other proteins are phosphorylated after IGF-I stimulation. During the proliferative myoblast stage of L6 skeletal muscle cells, IGF-I stimulates tyrosine phosphorylation of a 175-kDa protein, pp175. In contrast to IRS-I, pp175 is a cytoskeleton-associated protein, since it is only solubilized by sodium dodecylsulfate (SDS) and not by Triton X-100 (Beguinot et al., 1989). When myoblasts differentiate into myocytes, pp175 is not detected. pp175 is also found in rat liver and FRTL-5 (thyroid-derived) cells, in which its phosphorylation is enhanced by IGF-I (Condorelli et al., 1989). The specific role of this protein in mediating IGF-I action in these cells has yet to be elucidated.

In cultured fetal chick neurons, pp70, a unique tyrosine-phosphorylated protein unrelated to neurofilaments, is an endogenous substrate for the IGF-I receptor (Kenner and Heidenreich, 1991). On the other hand, in cultured kidney mesangial cells, IGF-I induces tyrosine phosphorylation of a number of nuclear proteins of varying sizes. One of these proteins has been identified as the transcription factor c-*jun* (Oemar et al., 1991). Phosphorylation of c-*jun* may result in increased overall transcriptional activity. In bovine chromaffin cells, IGF-I rapidly activates mitogen-activated protein 2 (MAP-2) kinase by tyrosine phosphorylation. In turn, MAP-2 kinase phosphorylates myelin basic protein but not histone or ribosomal S6 protein (Cahill and Perlman, 1991).

In addition to tyrosine phosphorylation of the IGF-I receptor β subunit, autophosphorylation may occur on serine and threonine residues. Stimulation of NIH-3T3 fibroblasts transfected with the human IGF-I receptor cDNA with IGF-I or phorbol ester, an agent that enhances protein kinase C (PKC) activity, resulted in serine phosphorylation. Phorbol ester was more effective than IGF-I in this respect (Pillay et al., 1991). In certain cell lines, serine phosphorylation of the receptor has been shown to decrease the activation of agonist-induced receptor tyrosine kinase activity. In addition to inducing serine phosphorylation of the receptor itself, IGF-I also enhances the serine/threonine phosphorylation of some proteins such as casein kinase II in BALB/c3T3 cells (Klarlund and Czech, 1988). Several other second messenger pathways are also activated by the IGF-I receptor.

Since many cellular events are modulated by the level of phosphorylation of various proteins, kinases and phosphatases play essential roles in the regulation of cellular functions. Several phosphatases have been identified, some of which are plasma membrane-bound (receptor-like) and others of which are cytosolic. IGF-I receptor autophosphorylation in the C127 mouse mammary tumor cell line was inhibited by coexpression of the transmembrane tyrosine phosphatase CD45. This inhibition resulted in inhibition of IGF-I-induced thymidine incorporation in these cells (Mooney et al., 1992). Similarly, dephosphorylation of the IGF-I receptor by membrane-associated phosphatases was demonstrated *in vitro* using extracts of rat liver and placenta (Peraldi et al., 1992). The actions of IGF-I, like those of insulin, are mimicked by vanadate and zinc, two compounds that inhibit phosphotyrosine phosphatases. Thus, accumulating evidence suggests that the phosphorylation state of the IGF-I receptor is important for its biological function and, further, that this process is tightly regulated by the opposing actions of kinases and phosphatases.

4. Receptor Heterogeneity

An increasing amount of immunological and biochemical evidence suggests the presence of naturally occurring tetrameric hybrid receptors (Moxham et al., 1989) composed of an insulin receptor $\alpha\beta$ hemireceptor joined to an IGF receptor $\alpha\beta$ hemireceptor by disulfide bonds. Since the insulin and IGF-I $\alpha\beta$ hemireceptor precursors undergo identical processing, nonidentical $\alpha\beta$ hemireceptors may be randomly sorted to form hybrid receptors *in vivo* (Fig. 3). Physiological concentrations of IGF-I bind to these hybrid receptors and stimulate autophosphorylation of both the IGF-I and the insulin receptor β subunit components by intramolecular phosphorylation. In contrast, these hybrid receptors exhibit a markedly reduced affinity for insulin.

Functional characterization of these hybrids, as well as study of their tissue distribution, is currently underway (Treadway et al., 1989). The physiological role of hybrid receptors is unknown, although a number of interesting possibilities can be proposed. For example, the binding characteristics of hybrid receptors could explain the ability of IGF-I to stimulate typical insulin-like metabolic functions as the result of transphosphorylation of the insulin hemireceptor. Thus, activation of hybrid receptors by IGF-I may add a new dimension to IGF-I signaling properties, in addition to its characteristic responses via the typical IGF-I receptor.

C. IGF-II/M-6-P RECEPTOR

Competitive ligand binding studies originally suggested that, in addition to the IGF-I receptor, which shows approximately equal affinities for IGF-I and IGF-II, a second distinct IGF receptor with a high affinity for IGF-II is expressed by various tissues. This receptor, a large 250kDa protein, does not bind insulin. The deduced amino acid sequence based on molecular cloning of the IGF-II receptor cDNA showed it to be identical to the cation-independent M-6-P receptor (Morgan et al., 1987; Lobel et al., 1988; MacDonald et al., 1988). This bifunctional IGF-II/M-6-P receptor is largely extracellular. The extracellular portion contains 15 contiguous repeats, each with a similar pattern of eight cysteine residues. One of these repeats is similar to the type II fibronectin repeat. The cytoplasmic tail is relatively short and does not contain a tyrosine kinase domain, although, as described subsequently, it may be linked to a G-protein signaling pathway.

1. Ligand Binding

IGF-II binds to the IGF-II/M-6-P receptor with an affinity of approximately 10^{-9} M, whereas IGF-I binds with lower affinity. Phosphomannosyl residues of lysosomal enzymes bind at a separate site. Interestingly, IGF-II and the lysosomal enzymes may show reciprocal inhibition of binding, presumably by inducing a conformational change in the receptor (Kiess et al., 1989,1990).

2. Expression

IGF-II/M-6-P receptor levels in the rat are high in fetal tissues, but decline dramatically in late gestation and in the early postnatal period. This developmental pattern parallels that of IGF-II. Thus, IGF-II and the IGF-II/M-6-P receptor may play important roles in fetal development, possibly in the process of tissue remodeling. IGF-II could modulate the cycling of lysosomal enzymes by the receptor during development. A circulating form of the IGF-II/M-6-P receptor has been identified. This molecule lacks the cytoplasmic portion of the receptor, suggesting that the serum IGF-II/M-6-P protein arises by proteolytic cleavage of the membrane-bound receptor. Despite the fact that significant levels of this receptor are found in serum, no known function has been ascribed to this form, although conceivably the circulating form of the IGF-II/M-6-P receptor represents another, essentially IGF-II-specific, IGF binding protein.

3. Function

IGF-II/M-6-P receptors target recently synthesized lysosomal enzymes from the *trans*-Golgi network to lysosomes. In addition, lysosomal enzymes that have escaped the cell via constitutive bulk flow secretion are internalized into the cell via a small number of IGF-II/M-6-P receptors that are found on the cell surface. The IGF-II/M-6-P receptor also internalizes surface-bound IGF-II, delivering it to the lysosomal compartment where it is degraded.

Compared with the extensive evidence supporting an important role for the IGF-I receptor in mediating biological responses, only a few reports have suggested a role for the IGF-II/M-6-P receptor in signal transduction. Nishimoto and co-workers have presented evidence that binding of IGF-II to the IGF-II/M-6-P receptor results in reduced pertussis toxin-induced ATP-ribosylation of a heterotrimeric G protein (G_{i2}) Nishimoto *et al.*, 1987,1989; Okamoto *et al.*, 1990). These studies also demonstrated that a 14-amino-acid peptide, representing residues 2420–2423 of the cytoplasmic domain of the receptor, bound and activated G_{i2} for binding to GTPγS. These interesting studies may explain reports that IGF-II/M-6-P receptors can transduce signals and mediate biological responses, including Ca^{2+} influx into BALB/c-3T3 cells, clonal growth of K562 erythroleukemic cells, and generation of inositol phosphate (IP_3) and diacyl glycerol (DAG) by kidney proximal tubules (Hari *et al.*, 1987; Matsmaga *et al.*, 1988; Rogers and Hammerman, 1988).

IV. IGF-Binding Proteins

A. Structure

IGF-I and IGF-II, unlike insulin, are present in the circulation and in extracellular fluids tightly bound to a family of specific IGF binding proteins (IGFBP). To date, six different IGFBPs have been isolated and cloned (Baxter and Martin, 1989; Table I). The six IGFBPs contain regions with strong homology (Mohan *et al.*, 1989; Shimasaki *et al.*, 1991a,b; Drop *et al.*, 1992). Specifically, the hydrophobic cysteine-rich N-terminal region, the C-terminal region, and the alignment of 18 cysteines in these regions are highly conserved. Sequence similarity

TABLE I
HUMAN IGF BINDING PROTEINS

Protein	Mass (kDa)	Source for purification	Relative binding affinities for IGFs	RGD sequence	Glycosylation
BP1	25	Amniotic fluid, placenta	IGF-I = IGF-II	+	
BP2	31.3	BRL-3A and MDBK cells, human serum	IGF-II > IGF-I	+	
BP3	28.7	Plasma	IGF-I = IGF-II	−	N-linked
BP4	25.9	Human osteosarcomas, prostatic carcinoma, colon carcinoma, glioblastoma	IGF-I = IGF-II	−	N-linked
BP5	28.5	C2 myoblast conditioned media, human bone	IGF-I = IGF-II	−	
BP6	22.8	Cerebrospinal fluid, human serum	IGF-II >> IGF-I	−	O-linked

between the IGFBPs ranges from 47 to 60%, suggesting that these proteins evolved by successive gene duplications. The IGFBP-1, -2, and -3 genes are composed of four protein-encoding exons; however, IGFBP-3 has an additional exon that encodes the 3'-UTR. Biochemical analysis suggests that the cysteines are involved in disulfide bond formation between residues in the N-terminal and C-terminal domains. IGFBP-1 and IGFBP-2 each contain an Arg–Gly–Asp (RGD) sequence at the C terminus that is also found in extracellular matrix proteins that bind to integrin receptors.

IGFBP-1 undergoes phosphorylation on serine residues, resulting in at least four isoforms that migrate differently on gel electrophoresis. IGFBP-3 and IGFBP-4 contain asparagine-linked glycosylation sites and clusters of serine/threonine residues that are potential O-linked glycosylation sites. Plasma IGFBP-3 migrates on gels as two heavily glycosylated proteins. IGFBP-6 contains O-linked oligosaccharide chains, whereas IGFBP-1, -2, and -5 do not contain any potential glycosylation sites.

B. IGF BINDING

All six binding proteins bind IGF-I and IGF-II with apparent association constants (K_a) in the range of $0.1–1.0 \times 10^{-9}$ M. IGFBP-1, -3, -4, and -5 bind IGF-I and IGF-II equally, whereas IGFBP-2 and -6 have a higher affinity for IGF-II than for IGF-I. Generally, the binding protein affinities for the IGFs are greater than those of the IGF receptors.

The B chain of the IGF-I molecule seems to play a role in binding to the IGFBPs, since the N-terminally truncated des(1–3) and QAYL (Gln 3–Ala 4–Tyr 15–Leu 16) IGF analogs have markedly reduced affinity for all IGFBPs tested but retain near normal affinity for IGF receptors (Szabo et al., 1988). The region(s) of the IGFBPs involved in IGF binding has not yet been clearly determined.

C. REGULATION OF EXPRESSION

1. *IGFBP-1*

IGFBP-1 is widely expressed; the highest levels are seen in the liver. IGFBP-1 mRNA levels are high at fetal and postnatal stages and fall dramatically at adult stages (Ooi et al., 1990). IGFBP-1 liver mRNA levels are regulated by insulin, GH, and dexamethasone (Luo et al., 1990; Powell et al., 1991). Insulinopenic states, including diabetes and fasting, result in markedly increased mRNA and protein levels. Dexamethasone also increases IGFBP-1 levels, whereas GH suppresses IGFBP-1 levels (Murphy et al., 1991). Thus, IGFBP-1 expression is positively correlated with states of growth retardation and insulin resistance. The promoter region of the IGFBP-1 gene contains an insulin-response element (IRE) similar to the putative IRE found in the phosphoenol pyruvate carboxykinase (PEPCK) gene promoter.

2. IGFBP-2

IGFBP-2 expression and regulation is similar to that of IGFBP-1, that is, IGFBP-2 is also widely expressed in fetal tissues, with levels falling during development. Highest levels are found in liver. Both hypophysectomy and diabetes cause an increase in mRNA and protein levels (Ooi et al., 1990). Expression in adult brain is restricted to the choroid plexus where IGFBP-2 colocalizes with IGF-II mRNA (Tseng et al., 1989).

3. IGFBP-3

IGFBP-3 is expressed by a number of tissues and cells. In contrast to IGFBP-1 and IGFBP-2, increased levels of IGFBP-3 are most closely correlated with conditions of growth stimulation. For example, levels are highest in adult liver and are stimulated primarily by GH, directly or via IGF-I. Thus, hypophysectomy results in a marked decrease in circulating IGFBP-3 levels, which are restored by GH or IGF-I administration.

4. IGFBP-4

IGFBP-4 was cloned from a human osteosarcoma cell line in which parathyroid hormone (PTH) regulated the levels of its mRNA through a cAMP-dependent mechanism (LaTour et al., 1990). IGFBP-4, as well as IGFBP-3 and IGFBP-5, is secreted by fibroblasts in response to increased intracellular cAMP (Camacho-Hubner et al., 1992). In addition, IGF-I itself may regulate IGFBP-4 and IGFBP-5 secretion from fibroblasts by posttranscriptional mechanisms. Studies on the tissue distribution of IGFBP-5 have revealed that the highest expression occurs in adult kidney, in marked contrast to observations for other IGFBPs (Shimasaki et al., 1991b). The potential significance of tissue-specific expression of IGFBPs lies in the possibility that these molecules modulate the tissue-specific actions of the IGFs.

D. FUNCTION

The biological actions of the IGFs, although mediated by their receptors on the surface of target cells, are modulated by the IGFBPs. This important regulatory function of the IGFBPs is primarily determined by certain structural characteristics.

The majority of serum IGF circulates as a large 150-kDa ternary complex composed of three subunits: IGFBP-3, an acid-labile subunit (ALS), and the hormone (IGF-I or IGF-II). The remaining IGFs are bound by circulating IGFBP-1, -2, and -4, forming complexes of 30–40 kDa. The large IGFBP-3 complex probably acts to protect the IGF peptides from degradation and to prolong their half-life in the circulation. The half-life of IGFs in the 150-kDa complex is 12–16 hr, whereas the IGFs bound to the lower molecular weight (LMW) fraction have a half-life of 30 min and the half-life of the free peptide is only 10 min. In addition, bound IGFs are unable to interact with insulin receptors, despite their very high levels, and therefore do not induce hypoglycemia.

In contrast to the high molecular weight complex, which cannot leave the circulation, IGFs may leave the circulation bound to IGFBP-1. This transfer is enhanced by insulin (Bar et al., 1990). At the cellular level, IGFBPs have been shown to modulate the biological actions of the IGFs. These modulatory actions have been studied primarily using cell cultures. Thus, their relevance to IGF action in vivo has not yet been clarified. IGFBPs may accumulate in the extracellular matrix and bind all the local IGF peptide, thereby inhibiting IGF action (Ross et al., 1989). Alternatively, these proteins may be induced to release this "store" of IGF slowly in close proximity to the cell-surface receptor, thereby enhancing IGF action (Elgin et al., 1987; De Mellow and Baxter, 1988; Blum et al., 1989). Several mechanisms appear to regulate this process. (1) Cell-free IGFBPs have an affinity for the IGFs that is almost 10-fold greater than the affinity of membrane-bound IGFBPs (2) Phosphorylation of IGFBP-1, -2, and -3 on serine residues enhances the affinity for IGFs (Frost and Tseng, 1991; Jones et al., 1991). Thus, affinity alterations secondary to tissue partitioning, or covalent modifications of IGFBPs, could explain the ability of IGFBPs to regulate the availability and, therefore, the actions of free IGFs at the receptor level. (3) Specific proteases have been described that degrade the IGFBPs, thereby releasing the IGFs (Hossenlopp et al., 1990). Differential regulation of this degradative process could also regulate the level of free IGFs at the receptor. (4) The IGFBPs may have IGF-independent effects by associating with the cell surface, possibly via their RGD sequences or some hydrophobic interaction (Ruoslathi and Pierschlaber, 1987).

V. Physiological Roles of the IGF System

A. Cell Cycle

The cell cycle consists of several phases: the presynthetic phase (G_1), the phase of DNA synthesis (S), the premitotic phase (G_2), and mitosis (M); quiescent cells are considered to be in the noncycling G_0 phase. The G_1 phase of the cell cycle forms the gap between mitosis and readily observed DNA synthesis and is, therefore, the phase marked by initial synthesis of DNA, histones, and some enzymes (Surmacz et al., 1987). In fibroblasts, G_1 requires many hours (~12), during which a number of necessary events appear to occur sequentially. Thus, G_1 can be divided into subphases referred to as competence, entry, progression, and assembly. The initial process of competence (i.e., movement out of the G_0 or quiescent phase) is stimulated by PDGF (Stiles et al., 1979). The G_{1a}, or entry phase, is stimulated by EGF and insulin (Leof et al., 1982), whereas IGF-I is the only growth factor required for progression (G_{1b}), a phase requiring protein synthesis (Leof et al., 1982; Pardee, 1989). G_{1c} phase, in which assembly of nuclei takes place, does not require growth factors.

The critical role of IGF-I as a progression factor was illustrated by the ability of an IGF-I-neutralizing antibody to inhibit IGF-I stimulation of thymidine incorporation into cells made competent by PDGF (Russell et al., 1984). IGF-I is also important in the G_1 phase of exponentially growing NIH- or BALB/c-3T3 cells and nontransformed fibroblasts (Campisi and Pardee, 1984). Baserga and co-workers demonstrated in BALB/c-3T3 cells that IGF-I and the IGF-I receptor have a special status in the control of cell proliferation. Cells constitutively expressing IGF-I and the IGF-I receptor can grow in serum-free medium without the addition of any exogenous growth factors (Pietrzkowski et al., 1992). Thus, PDGF and EGF may function simply by inducing enough IGF-I and IGF-I receptor to elicit a growth response. Indeed, PDGF and EGF increase IGF-I and IGF-I receptor biosynthesis in BALB/c-3T3 cells and W1-38 human diploid fibroblasts (Clemmons and Shaw, 1983; Clemmons, 1984). Similar conclusions have been drawn from studies in which overexpression of the IGF-I receptor in murine and human cells promoted a ligand-dependent malignant phenotype and allowed these cells to induce tumor formation when transplanted into nude mice (Kaleko et al., 1990). In addition, many tumors and transformed cell lines express high levels of IGF-I receptors, which play an important role in tumor growth as demonstrated by the inhibition of growth observed in the presence of antibody αIR3 (Gansler et al., 1989).

B. IN VIVO BIOLOGICAL ACTIONS

Recombinant DNA technology has allowed the large-scale production of pure human IGF-I (rhIGF-I), which has facilitated the study of the biological actions of IGF-I *in vivo*. The original somatomedin hypothesis suggested that the effect of GH on longitudinal growth is mediated via a circulating "somatomedin" that is derived from the liver; this sulfation factor, as originally termed, was subsequently shown to be IGF-I. Indeed, in syndromes of GH deficiency or GH resistance, reduction in body growth is correlated with low circulating IGF-I (somatomedin C) levels (Laron *et al.*, 1966; Merrimee *et al.*, 1981). In states in which GH is administered and body growth increases, circulating IGF-I levels rise.

The long-term anabolic effects of IGF-I have been tested in a number of models (Cotterill, 1992). Laron-type dwarfism (LTD) is a condition of GH resistance caused by a defect in the GH receptor. As a result, circulating IGF-I levels are very low and fail to increase in response to GH injections. On the other hand, subcutaneous injections of rhIGF-I over a 12-mo period significantly increase growth rate in LTD patients (Walker *et al.*, 1992; Wilton, 1992).

Comparisons of the effectiveness of rhIGF-I and rhGH in GH-deficient short children have not been made. In hypophysectomized or genetically GH-deficient rats, improvement in linear growth after GH replacement was more impressive than after IGF-I infusions (Guler *et al.*, 1988; Skottner *et al.*, 1987,1989). These results suggest that GH affects growth via liver (endocrine) production of IGF-I as well as via additional local (paracrine) effects, that is, production of IGF-I at the plates (Lindahl *et al.*, 1991). In addition, consistent with the "dual-effector" hypothesis of growth regulation, GH may "prime" cells to respond to IGF-I by increasing IGF-I receptor levels (Isaksson *et al.*, 1991). Additional effects of GH, not seen after IGF-I treatment, may include increased IGFBP-3 and insulin levels, which may be important in overall GH action.

IGF-1 infusion, in contrast to GH infusion, resulted in a disproportionate growth of the spleen, kidney, and thymus, exceeding bone growth. Short-term anabolic effects have been demonstrated in both animals and humans. Infusions of IGF-I into fasted rats can reduce tissue protein degradation (Jacob, 1989). IGF-I infusions also attenuated the tubular necrosis that follows bilateral renal artery occlusion in rats (Miller *et al.*, 1992). In calorically restricted humans, infusion of rhIGF-I reversed diet-induced catabolism (Clemmons *et al.*, 1992; Turkalj *et al.*, 1992). Thus, rhIGF-I may be useful in short-term reversal of catabolic states in which GH resistance is a common feature.

Infusions of rhIGF-I into both animals and humans have demonstrated a significant effect on metabolic parameters (Zenobi et al., 1992). Short-term infusions caused a significant fall in blood glucose levels as a result of increased peripheral glucose uptake, primarily into muscle. Hepatic glucose output was decreased by IGF-I infusion, especially in the fasted state, because of liver glycogen depletion and suppression of glycogenolysis by IGF-I. When compared with the effects of insulin, IGF-I was equally potent in increasing peripheral glucose uptake but less potent in decreasing hepatic glucose output and as an antilipolytic agent. These results have been interpreted to suggest that the effect of IGF-I on muscle (glucose uptake) is via its own receptor and that its effects on liver (glucose output) and adipocytes (reduced lipolysis) are via the insulin receptor (Guler et al., 1987; Jacob et al., 1989; Douglas et al., 1991; Boulware et al., 1992; Kerr et al., 1993).

C. DEVELOPMENT

1. *Preimplantation Embryo*

Normal growth and development of the fetus depend on many factors, including genetic constitution, nutrition, and hormonal factors. The GH system and the pituitary–thyroid axis are important postnatally. In contrast, the IGFs seem to be extremely important in controlling fetal development.

Preimplantation embryos (2-cell stage) express insulin receptors; exogenous insulin, acting at physiological concentrations via the insulin receptor, stimulates protein synthesis and blastocyst development at this early blastocyst stage (Spaventi et al., 1990; Harvey and Kaye, 1991,1992). Since no yolk sac or placental barrier exists, the endogenous source of insulin may be maternal (Travers et al., 1992). Both insulin and IGF-I stimulate blastocyst cell number by increasing inner cell mass. At this early cleavage stage, the IGF effect is probably mediated by the IGF-II/M-6-P receptor, since no IGF-I receptor mRNA is detectable.

Since IGF-II is expressed at the 2-cell stage and insulin and IGF-I appear later, locally produced IGF-II seems to be an important factor at this early stage. Thus, the IGF-II/M-6-P receptor may be involved in IGF signal transduction at this stage. Its effect may be direct, via postreceptor mechanisms, or indirect, by controlling (via internalization and degradation) local IGF-II levels (Senior et al., 1990).

2. Placenta

At the time of implantation, IGF-II mRNA is found in trophoblastic tissues that are invading the maternal decidua in the labyrinthine zone. The presence of this mRNA in this region, especially at the midgestational period (E14) in the rat, suggests a role in the exchange of nutrients and waste products between the maternal and fetal circulation (Zhou and Bondy, 1992). The junctional zone that comprises the metabolically and mitotically active growth plate also expresses high levels of IGF-II mRNA. IGFBP-2 expression, on the other hand, is not significant during placentation. IGF-II is thus unopposed in its effect on early invasion and establishment of the placenta. This effect of IGF-II on placental growth may be opposed once the placenta is established, since IGFBP-2 is now expressed at higher levels in the junctional zone. Both the IGF-I and the IGF-II/M-6-P receptor are expressed in these regions of the placenta; the IGF-I receptor presumably transmits the IGF signal since IGF-I at low physiological concentrations enhances amino acid transport in placental tissue (Fant et al., 1986). On the other hand, the IGF-II/M-6-P receptor may enhance IGF-II clearance.

3. Postimplantation Embryo

The IGFs also play important roles in growth and development in postimplantation embryos (DePablo et al., 1990; Telford et al., 1990). During early stages of organogenesis, IGF-I and IGF-II are expressed with distinctive patterns of cellular distribution (Stylianopoulou et al., 1988; Fig. 7). IGF-I mRNA is particularly abundant in undifferentiated mesenchymal tissue in the region of the developing face that represents target zones of the trigeminal nerve, as well as in surrounding developing cartilage and muscle. IGF-II mRNA was readily detected in developing vasculature, muscle, cartilage, and especially liver, as well as in the choroid plexus and associated structures in the developing brain (Bondy et al., 1990; Florance et al., 1991; Fig. 8). The IGF-I receptor is widely expressed, whereas the IGF-II/M-6-P receptor seems to be coexpressed with IGF-II in the developing cardiovascular system, skeletal muscle, and perichondrium. In the liver, IGF-II is expressed at high levels and, in the absence of IGF-I and IGF-II/M-6-P receptors, is secreted into the circulation in an endocrine manner. Similarly, IGF-II/M-6-P receptor expression in the choroid plexus is low relative to IGF-II and IGFBP-2 expression, suggesting that locally produced IGF-II is secreted into the cerebrospinal fluid for use elsewhere in the central nervous system.

The IGFBPs also play an important role in fetal development. The

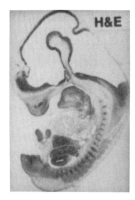  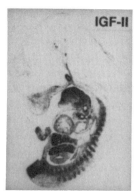

 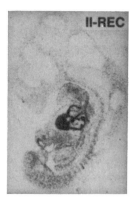

FIG. 7. *In situ* hybridization analysis of IGF-I, IGF-II,and receptor genes expression in rat embryonic tissues at day 14. I-Rec IGF-I receptor; II-Rec IGF-II/mannose-6-phosphate receptor.

apical ectoderm of the outgrowth of the limb and the facial mesodermal cells proliferate more rapidly than neighboring mesodermal cells (Streck *et al.*, 1992). In these regions, IGFBP-2 is expressed and, apparently in concert with locally produced IGFs, stimulates mesodermal cell growth. This colocalization of IGFBP-2 and IGF-II expression seems to be the rule during embryogenesis (Wood *et al.*, 1992).

IGF-II circulates at higher levels than IGF-I during most of embryogenesis and fetal development. Postnatally, IGF-II expression by most tissues falls dramatically whereas circulating IGF-I levels rise (Lund *et al.*, 1986). The IGF-I and IGF-II/M-6-P receptors are expressed in many tissues at high levels prenatally, and also show dramatic decreases postnatally. The IGFBPs have their own peculiar pattern.

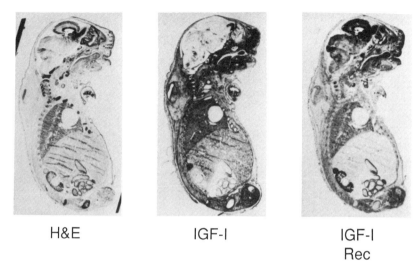

H&E IGF-I IGF-I Rec

FIG. 8. *In situ* hybridization of mRNAs for IGF-I and the IGF-I receptor genes at day 20 of rat embryonic development.

IGFBP-1 and -2 are especially abundant fetally and decrease postnatally, whereas IGFBP-3 becomes prominent during adult stages.

The importance of these factors in normal fetal growth and development is exemplified by homologous recombination–gene targeting experiments in which IGF and IGF receptor genes have been inactivated. When the IGF-II gene is inactive, the fetuses are smaller but of normal proportions with totally normal functions, suggesting a general growth function for IGF-II (DeChiara *et al.*, 1990). Spontaneous deletion mutants on chromosome 17 result in embryonic death at day 15 in the mouse. In these mutants, the IGF-II/M-6-P receptor is not expressed (Haig and Graham, 1991).

D. IN VITRO EFFECTS

In vitro effects of IGF-I include the stimulation of glucose uptake in BC3HI myocytes (Farese *et al.*, 1989). These effects may be secondary to the generation of DAG and the activation of membrane-bound PKC. These cells contain a phosphatidylinositol glycan the hydrolysis of which is enhanced by IGF-I induction of phospholipase C, leading to increased DAG and PKC (Farese *et al.*, 1988). Chronic stimulation (10 hr) of differentiated L6 myocytes by IGF-I also enhances glucose uptake, as well as glucose transporter 1 (Glut-1) gene expression (Maher *et al.*, 1989). IGF-I also stimulates glucose uptake in human fibroblasts grown in primary culture (Kato *et al.*, 1993) and in HepG2 human hepatoma cells (Verspohl *et al.*, 1988).

Other IGF-I-induced postreceptor events include stimulation of amino acid uptake in HepG2 and other cell lines (Verspohl *et al.*, 1988), and enhancement of cyclic AMP (cAMP) production in granulosa and Leydig cells (Adashi *et al.*, 1986). In FRTL-5 cells, IGF-I synergizes with thyroid stimulating hormone (TSH) to increase DAG and PKC, which in turn stimulates cAMP via G protein activation (Brenner-Gati *et al.*, 1988,1989). In primary thyroid cell cultures, IGF-I-induced cell proliferation is associated with accumulation of inositol phosphates (IP, IP_2, and IP_3) and increased cytosolic free calcium (Takasu *et al.*, 1989). Similarly, DNA synthesis and cell proliferation were shown by patch clamp studies to be related to an IGF-I-induced calcium influx in BALB/c-3T3 cells made competent by PDGF or EGF (Kojima *et al.*, 1988).

With respect to nuclear events, IGF-I induces the specific expression of a number of RNA polymerase II-transcribed genes. In FRTL-5 cells, IGF-I increases the levels of TTF-2, a thyroid transcription factor that enhances thyroglobulin gene expression (Santisteban *et al.*, 1987, 1992). Expression of the δ-1 crystallin (Alemany *et al.*, 1990) and cytochrome P-450 cholesterol side-chain cleavage enzyme (Urban *et al.*, 1990) genes is increased by IGF-I. In fibroblasts, α1 collagen chain mRNA is increased by IGF-I via a transcriptional mechanism (Goldstein *et al.*, 1989), as is oxytocin mRNA in granulosa cells (Holtorf *et al.*, 1989) and tropoelastin mRNA in pulmonary artery smooth muscle cells (Bodesh *et al.*, 1989). In the case of actin, IGF-I stimulates transcriptional activity via a serum response element-like sequence (cArG box) (Buchou *et al.*, 1991). Zumstein and Stiles (1987) calculated that, in quiescent BALB/c-3T3 cells, IGF-I regulates the expression of about 30 genes (i.e., 0.15% of the total). A minority is regulated at the level of transcription (i.e., their induction is prevented by actinomycin) and the majority at the level of mRNA stability. Some of these genes encode cytosolic proteins and others encode nuclear proteins. Of the nuclear proteins, IGF-I induces c-*myc* (Banskota *et al.*, 1989), c-*fos* (Damante *et al.*, 1988; Merriman *et al.*, 1990), and c-*jun* (Oemar *et al.*, 1991). For c-*jun*, the effect involves both transcription and mRNA stability (Chiou and Chang, 1992). IGF-I may also inhibit gene expression. This effect has been demonstrated for the GH gene in pituitary cells (Yamashita *et al.*, 1987) and for *myf*-5, a gene expressed early in the differentiation of L6 myoblasts (Mangiacapra *et al.*, 1992).

Nuclear events that follow IGF-I stimulation result in DNA synthesis and cell division. A commonly used measure of this effect is thymidine incorporation. Although most of these events are the culmination of activation of the IGF-I receptor at the cell membrane, leading to a cascade of cytosolic events, some of these events may occur more directly at the nuclear level. Phosphoinositidase C (PIC) has two isoforms, PICγ being the cytoplasmic isoform and PICβ the nuclear isoform (Divechia et al., 1991). IGF-I stimulation of Swiss 3T3 cells results in rapid activation of PICβ, following activation of nuclear lamina-related G proteins, and ultimately causes release of nuclear IP_2 and IP. This phenomenon results in a rise of nuclear DAG and leads to mitosis (Cocco et al., 1989; Martelli et al., 1992).

IGF-I also may stimulate mitosis by inducing *ras* proteins. The c-Ha-*ras* proto-oncogene in BALB/c-3T3 murine fibroblasts is expressed maximally during G_1. The IGF-I effect on DNA synthesis in G_1 is blocked by injecting an anti-*ras* antibody (Lu and Campisi, 1992). In addition to mitosis, IGF-I also induces meiosis. Thus, prophase-arrested *Xenopus* oocytes can be induced to mature by IGF-I. This effect is mediated by endogenous *ras* protein and tyrosine phosphorylation of a MAP kinase prior to germinal vesicle breakdown (Campa et al., 1992).

VI. Organ/System-Involvement of IGFs

In this section we describe three well-characterized model systems in which the IGFs play important physiological or pathological roles: the reproductive system, the nervous system, and cancer.

A. Reproductive System

1. *Introduction*

For many years, researchers have observed that the delayed puberty associated with GH deficiency may be overcome by GH replacement therapy (Christman and Halme, 1992), suggesting a role for the GH–IGF-I endocrine axis in gonadal differentiation. In clinical studies, rhGH administration to women undergoing ovulation induction decreased the requirement for menopausal gonadotropins. These studies have suggested the importance of GH, IGF-I, or both in ovarian function (Christman and Halme, 1992). Other studies in humans and animals have demonstrated the local synthesis of virtually all the components of the IGF system (peptides, binding proteins, and receptors) by specific tissues in the male and female reproductive system. This result suggests that, in addition to their endocrine role, IGFs may affect reproductive function via autocrine and paracrine actions.

2. Ovarian Physiology

IGF-I and IGF-II are expressed in ovarian tissue. In humans, IGF-I is expressed by theca cells, whereas IGF-II is expressed in granulosa cells (Hernandez et al., 1992a). In murine and porcine ovaries, on the other hand, IGF-I expression occurs in granulosa cells whereas IGF-II expression is restricted to the theca-interstitial cell (Hsu and Hammond, 1987; Hernandez et al., 1989). During postnatal development in the rat, the expression of ovarian IGF-I and IGF-II is differentially regulated. IGF-I mRNA demonstrates a peak at about day 20, which closely parallels the peak of serum FSH levels, whereas IGF-II mRNA, which is high prenatally, falls dramatically to very low adult levels (Levy et al., 1992). IGF-I mRNA is most abundant in granulosa cells lining the antrum of the follicle and surrounding the oocyte. Expression is markedly decreased in luteinized granulosa cells of the corpora lutea (Zhou et al., 1991). The production of IGF-I by porcine and murine granulosa cells is stimulated by FSH (Adashi et al., 1986; Hatey et al., 1992). Further, treatment of porcine granulosa cells in culture with gonadotropins (FSH, LH) or cAMP agonists [dibutyryl(db)-cAMP or forskolin] increases IGF-I mRNA levels. This increase is abolished by actinomycin D and α-amanitin, but not by cycloheximide, suggesting that the effect is at the level of transcription (Hatey et al., 1992).

The IGF-I and IGF-II/M-6-P receptors are widely expressed in ovarian tissue. During follicular growth, mitral and antral granulosa cells express the IGF-I receptor, as do granulosa cells in atretic follicles and in the corpora lutea, where expression is especially high. Oocytes also express the IGF-I receptor. Hypophysectomy results in reduced IGF-I receptor expression, which is corrected by replacement treatment using gonadotropins (Zhou et al., 1991). In vitro studies have shown that FSH, via increased intracellular cAMP production, increases the concentration of cell-surface IGF-I receptors (Adashi et al., 1986). Theca-interstitial cells also express IGF-I receptors, which may regulate androgen biosynthesis (Cara and Rosenfeld, 1988).

Of the six known IGFBPs, the expression of IGFBPs 1–5 has been demonstrated in human, rat, and porcine ovarian tissue (Samaras et al., 1992). In human granulosa cells, the mRNAs encoding IGFBPs 1–4 and their corresponding proteins have been detected. IGFBP-1 and -2 mRNA levels are stimulated further by human chorionic gonadotropin (hCG) (Giudice et al., 1991). IGFBP-1 production by granulosa cells is increased during luteinization (Angervo *et al.,* 1991). In addition, insulin and IGFs (Grimes and Hammond, 1992), as well as EGF (an inhibitor of FSH action in the ovary), stimulate IGFBP-1 secretion by human granulosa cells (Angervo *et al.,* 1992; Dor *et al.,* 1992).

In the rat ovary, IGFBP-1 is not expressed, whereas IGFBP-2 is theca-interstitial cell specific. IGFBP-3 is localized specifically to corpora lutea, and IGFBP-4 is localized to granulosa cells of atretic follicles. In contrast, IGFBP-5 is expressed in granulosa cells of atretic follicles, corpora lutea, and secondary interstitial cells (Nakatani *et al.,* 1991; Erickson *et al.,* 1992a,b). In addition to the distinct cellular localization of IGFBPs, estrous cycling causes an alteration in expression of the IGFBPs. Hypophysectomy results in decreased IGFBP-2 mRNA; estrogen replacement reverses this effect (Ricciarelli *et al.,* 1991).

Studies on the effects of the IGFs on ovarian function have mainly utilized granulosa and luteal cells in primary culture. IGF-I elicits a mitogenic response in human, porcine, bovine, and rat granulosa cells as measured by increased thymidine incorporation. In rat granulosa cells, this response is enhanced by low doses of FSH or db-cAMP (Bley *et al.,* 1992). IGFBP-1 inhibits IGF-I-induced thymidine incorporation in human granulosa-luteal cells (Angervo *et al.,* 1991). On the other hand, IGFs elicit differentiation-related responses in granulosa cells (Monniaux and Pisselet, 1992). IGF-I induces cytochrome P-450 aromatase gene expression and enzymatic activity in luteinized human granulosa cells, resulting in increased progesterone synthesis (Talavera and Menon, 1991). This effect is augmented by FSH and hCG (Christman *et al.,* 1991). IGF-I also increases androgen (Hernandez *et al.,* 1988) and estradiol synthesis by granulosa cells; FSH also augments this response (Erickson *et al.,* 1991). To a certain degree, the augmentation by FSH is secondary to the ability of IGF-I to increase FSH and luteinizing hormone (LH) receptors on ovarian tissues (Adashi *et al.,* 1985). These effects are inhibited in the presence of IGFBPs (Hernandez *et al.,* 1992b; Ricciarelli *et al.,* 1992). Oxytocin expression by granulosa cells is stimulated by FSH, an effect that is enhanced by IGF-I (McArdle *et al.,* 1991). The role of oxytocin in ovarian physiology is, however, as yet unclear.

3. *Uterus*

Uterine tissue expresses IGF-I (and IGF-II), the IGF-I receptor, and multiple IGFBPs. The expression of these genes is hormonally regulated. In the rodent, uterine IGF-I mRNA levels increase dramatically following estrogen treatment, an effect that is opposed by simultaneous GH administration (Murphy *et al.,* 1987). Myometrium and stroma are especially rich in IGF-I and mRNA (Ghahary *et al.,* 1990) with little expression by the luminal epithelium. In the estrous cycle, uterine IGF-I mRNA levels are highest during late diestrus, coincident with rising estradiol levels.

IGF-II gene expression in uterine tissue is much less impressive than that of IGF-I, although it may also increase following estrogen administration. IGF-I receptors are expressed by myometrial and endometrial tissues in all species studied (Geisert *et al.,* 1991; Hofig *et al.,* 1991; Kapur *et al.,* 1992). Levels of IGF-I receptors are especially high in proestrus. To date, IGFBP-1 and -3 have been demonstrated in uterine tissue. Estrogen increases IGFBP-1 mRNA levels and, during diestrus, IGFBP-1 levels are highest. IGF-I enhances the estrogen effect on uterine proliferation. Presumably the increased expression of IGF-I and IGF-I receptors, in conjunction with decreased levels of potentially inhibitory IGFBP-1 following estrogen administration, coordinates some of the effects of estrogen on the uterus. These effects may also be affected synergistically by enhanced expression of early response genes, including c-*myc,* c-*jun,* and c-*fos,* all of which are independently increased by IGF-I and estrogen.

During the peri-implantation period in the mouse, estrogen enhances IGF-I expression in epithelial cells, whereas progesterone has similar effects in the stroma (Kapur *et al.,* 1992). IGF-II and IGFBP-2 expression by the bovine endometrium increases during early pregnancy (Geisert *et al.,* 1991). Thus, the IGF system is fully operative in the uterus during normal menstrual/estrous cycles as well as during the peri-implantation period and early pregnancy.

4. *Testes*

The Sertoli cell is primarily involved in spermatogenesis, but may also regulate the function of the testosterone-synthesizing Leydig cells. Sertoli and Leydig cells synthesize IGF-I. In addition, Leydig cells express the IGF-I receptor and IGFBPs (Lin *et al.,* 1992; Skalli *et al.,* 1992). IGF-I expression by Leydig cells is down-regulated by interleukin 1, which also inhibits many of the effects of gonadotropin on Leydig cells (Lin *et al.,* 1992). Expression of the IGF-I receptor is upregulated by hCG or db-cAMP, and may help explain the synergistic effects of hCG and IGF-I on Leydig cell steroidogenesis (Nagpal *et al.,* 1991). IGFBP-3 gene expression is decreased by db-cAMP, whereas IGF-I has no effect on expression but may prevent degradation of this binding protein (Smith *et al.,* 1992). In addition to its role in stimulating testosterone production by Leydig cells, IGF-I induces a mitogenic response in spermatogonial cells that proliferate during spermatogenesis (Bernier et al., 1986; Soder et al., 1992). IGF-I also increases the levels of c-*fos*, *jun*-B, and c-*myc* mRNAs and potentiates the effect of hCG on c-*fos* and *jun*-B (Hall et al., 1991). Thus, the IGF system is involved in the regulation of the male reproductive system.

B. NERVOUS TISSUE

1. *Ligands*

The source of brain insulin is still controversial. Circulating plasma insulin reaches the brain via the circumventricular organs; under certain circumstances, insulin may be produced locally (LeRoith et al., 1993b). In contrast, IGF-I expression in the central nervous system is very abundant, especially during early development, Early in embryogenesis, IGF-I mRNA is localized in peripheral target zones (mesenchymal tissues) of trigeminal and sympathetic nerves during the period of their innervation. IGF-I is also expressed at the time of dendritic maturation and synaptic formation in cerebellar relay centers and in the developing olfactory, auditory, visual, and somatosensory systems (Bondy et al., 1990,1992b; Bondy, 1991; Lee et al., 1992). Postnatally, IGF-I is expressed by Purkinje cells of the cerebellum and by the neuronal cells of the olfactory bulb and hippocampus that are still undergoing intensive neurogenesis (Bartlett et al., 1991).

In the cerebellum and retina, a coordinate transient expression of IGF-I in Purkinje and retinal ganglion cells appears to occur with IGFBP-2 expression by Bergmann and Miller cells. These latter cells are specialized astroglial cells in close proximity to the cerebellar Purkinje and retinal ganglion cells (Aguado et al., 1992). This coordinate temporal expression suggests that IGFBP-2 expression may prevent IGF-I from stimulating proliferation in these neuroglia cells, thereby allowing them to continue their specialized function(s) (Bondy et al., 1992b).

Immunocytochemical staining has shown that IGF-I may be found in postnatal neuronal cells and is transported by axons and dendrites (Garcia-Segura et al., 1991), whereas adult glial cells are devoid of IGF-I peptide. The presence of IGF-I peptide in the choroid plexus in the absence of IGF-I mRNA suggests that the peptide here is most likely derived from the circulation.

IGF-II, in contrast to IGF-I, is expressed primarily by tissues at the vascular interface with the brain, that is, choroid plexus and organum vasculosum (Bondy et al., 1990,1992b; McKelvie et al., 1992). A variant

IGF-I peptide lacking the first three amino acids has been detected in fetal and adult brain and represents the major form of the peptide (Carlsson-Skwirut et al., 1986; Sara et al., 1986). Since brain expresses full-length mRNAs for IGF-I, this truncated form presumably arises from posttranslational modification (Sandberg-Nordqvist et al., 1992). Since the N terminus of the IGF-I peptide is partially responsible for its interaction with IGFBPs, this truncated form [des(1-3)IGF-I] binds the receptor with higher affinity and is more biologically potent than intact IGF-I in terms of neurotropic and proliferative effects (Giacobini et al., 1990).

The predominant IGFBP expressed by the developing and adult brain is IGFBP-2. This protein is also synthesized by the choroid plexus and leptomeninges, and is secreted into the cerebrospinal fluid (Tseng et al., 1989; Roghani et al., 1991) IGFBP-2 is also expressed by primary cultures of astrocytes (Klempt et al., 1992).

2. Receptors

Phylogenetically, the earliest organism expressing the insulin receptor in nervous tissue is *Drosophila* (Petruzzelli et al., 1986). Brains of lower vertebrates (e.g., cyclostomata) express specific IGF-I receptors but not the IGF-II receptor (Drakenberg et al., 1993). In the developing mammalian brain, insulin, IGF-I, and IGF-II/M-6-P receptors are found at the highest levels during prenatal development, with a marked decrease in expression during the postnatal stages (Brennan, 1988; Sklar et al., 1989; Werner et al., 1989a,1992b).

From early in ontogenesis, a homogeneous pattern of expression of the IGF-I receptor is observed throughout the central nervous system. Superimposed on this pattern is high expression in specific regions. Thus, during mid-embryogenesis, high levels are evident in the ventral floorplate of the hindbrain, where the IGF-I receptor is probably involved in the process of axonal targeting by the neuroendocrine cells.

Postnatally, IGF-I receptor mRNA is expressed by glomerular and mitral cells of the olfactory bulb, granule cells of the dentate gyrus and cerebellum, pyramidal cells of the piriform cortex, and the choroid plexus (Marks et al., 1991a). Insulin receptor mRNA is coexpressed in many of these regions (Marks et al., 1991b). Particularly high levels of IGF-I receptor mRNA are found in projection neurons of sensory and cerebellar relay centers, paralleling local IGF-I expression and suggesting a role for the IGF-I/IGF-I receptor system in maintenance of these synaptic circuitries (Bondy et al., 1992a). The IGF-II/M-6-P receptor is expressed by neurons in the hippocampal pyramidal and granule cell layers (Couce et al., 1992), but its specific function is not known.

Intriguingly, brain receptors differ from peripheral nonneural receptors in structure. Insulin and IGF-I receptors migrate faster on SDS–PAGE than do their peripheral counterparts, suggesting a 5- to 10-kDa smaller size. The difference in size is attributable to a difference in glycosylation. Brain receptors are devoid of terminal sialic acid residues or the composition of the sialic acid residues differs from that of those on peripheral receptors (Hendricks et al., 1984; LeRoith et al., 1988). These specific differences are phylogenetically conserved, since they are observed in receptors from brains of amphibians, reptiles, and birds (LeRoith et al., 1988). The lower molecular weight receptors are also observed in neural-derived cell lines (Ota et al., 1988). The significance of these structural differences remains unknown.

3. Function

Insulin and the IGFs play a number of roles in the nervous system. These molecules are involved in growth, development, and metabolic and neurotropic functions. Insulin enhances ornithine decarboxylase activity in fetal cell cultures, potentiates neurite outgrowth, activates serine phosphorylation of ribosomal protein S6 in fetal neurons, and enhances glucose transport in glial cells (Yang and Fellows, 1980; Recio-Pinto et al., 1984; Heidenreich and Toledo, 1989; Werner et al., 1989b).

The effects of the IGFs on neuronal tissue are also variable (Table II). During embryonic and early postnatal development, IGF-I is important for growth and development. The mechanisms by which IGF-I stimulates nervous tissue growth have been modeled using cultured astrocytes and may include activation of intermediates such as PKC (Tranque et al., 1992). Once neuronal-derived cells undergo differentiation and maturation, IGF-I-induced proliferation ceases (Påhlman et al., 1991). Differentiation is also a major function of IGF-I in neuronal cells, in which it results in neurite outgrowth (Ishii and Recio-Pinto, 1987), synapse formation, and myelin synthesis (Mozell and McMorris, 1991). Thus, when regions of the brain are artificially demyelinated, the remyelination process is associated with increased IGF-I expression by astrocytes (Komoly et al., 1992).

IGFs are also involved in regeneration following sciatic nerve crush (Sjoberg and Kanje, 1989; Near et al., 1992). Following hypoxic–ischemic brain injury, IGF-I expression by astrocytes is enhanced in the injured region (Garcia-Estrada et al., 1992). IGFBP-3 expression by

TABLE II
POTENTIAL BIOLOGICAL ACTIONS OF INSULIN AND IGFS
IN THE NERVOUS SYSTEM

Metabolic actions
   Increase glucose uptake in glial cells
   Increase *glut*-1 mRNA in neuronal and glial cells
   Activate pyruvate dehydrogenase
   Stimulate protein, fatty acid, sulfolipid, and cholesterol synthesis
   Central action to lower plasma glucose and free fatty acid levels
Neuromodulatory actions
   Alter neuronal firing rates
   Promote electrical couplings
   Inhibit norepinephrine reuptake
   Stimulate serotonin biosynthesis and uptake
   Maintain Na^+/K^+ pump in synaptosomes
Growth/differentiation actions
   Stimulate synaptogenesis
   Promote neurite outgrowth and survival, and nerve regeneration
   Mitogenesis in oligodendrocytes, astrocytes, neuronal cells and sympathetic neuroblasts
   Promote glial cell development
Neuroendocrine actions
   Inhibition of pituitary GH secretion

invading macrophages is also increased (Gluckman *et al.*, 1992). Intracerebroventricular injection of rhIGF-I reduces the neuronal loss that normally follows this type of injury (Gluckman and Ambler, 1992). Further, IGFs can protect cultured neurons against hypoglycemic damage (Cheng and Mattson, 1992).

Thus, IGFs are powerful mitogenic and neurotropic agents that stimulate growth and differentiation in normally developing organisms as well as compensatory growth secondary to pathological insults (Torres-Aleman *et al.*, 1992; Wozniak *et al.*, 1993). Their effectiveness as neutropic agents is enhanced by other growth factors such as basic fibroblast growth factor (FGF) (Drago *et al.*, 1991; Pons and Torres-Aleman, 1992) and EGF (Han *et al.*, 1992), and is inhibited by IGF-I-specific antisense oligonucleotides (Trojan *et al.*, 1993).

C. CANCER

IGF-I has been recognized for many years to be essential for progression of cells through the cell cycle. Once quiescent cells have been made competent, IGF-I alone is sufficient for completion of the rest of the cycle. Indeed, insulin at $1\mu g/ml$, a concentration sufficient to activate the IGF-I receptor, has long been used in serum-free cell culture conditions. The insulin and IGF-I receptors have intrinsic tyrosine kinase activity that mediates their biological actions, including stimulation of cell growth. Thus, the IGFs are important growth regulators in normal cells and, by extrapolation, may be involved in the malignant growth of certain cells (Osborne et al., 1990). As previously mentioned, overexpression of the normal IGF-I receptor in NIH-3T3 cells leads to ligand-dependent neoplastic transformation and to tumor formation in nude mice (Kaleko et al., 1990).

1. Ligands

IGF-I and IGF-II are expressed by a variety of tumor-derived cell lines. Immunoreactive IGF-I is detectable in extracts of small cell lung cancer (SCLC) and non-SCLC cells and their conditioned media (Jaques et al., 1992). The size of IGF-I in these extracts is 16 kDa, probably representing an incompletely processed form of the peptide (Macaulay, 1992). IGF-I is also expressed by some ovarian carcinoma cell lines (Yee et al., 1991), central nervous system tumors (Antoniades et al., 1992; Glick et al., 1992), and thyroid capillary cancer (Onoda et al., 1992).

Breast cancer cell lines express IGF-II but not IGF-I; IGF-I is expressed by stromal tissues in breast cancer specimens (Yee et al., 1989). Thus, IGF-I may act in a paracrine mode to stimulate breast cancer growth. MCF-7 and T47D breast cancer cell lines express IGF-II. Estrogen treatment of these cells results in a 4- to 5-fold increase in IGF-II mRNA (Osborne et al., 1989).

IGF-II is also synthesized by 30–40% of colon carcinoma cell lines. HT29 cells express immunoreactive IGF-II peptides of 17 kDa and 7.5 kDa in almost equal proportions, suggesting that the processing of IGF-II by the tumor is incomplete (Lambert et al., 1992).

A correlation between an abnormality in the IGF-II locus on chromosome 11 and certain childhood tumors associated with IGF-II overexpression has been demonstrated. These malignancies include Wilms' tumor, rhabdomyosarcoma, neuroblastoma and adrenocortical carcinoma. The overexpression of IGFs in most adult tumors studied is not the result of structural alterations such as deletions or gene amplification (Schneid et al., 1992). Demethylation of the IGF-II gene has been noted more often in those tumors overexpressing and secreting IGF-II, including pleural fibromas, hepatocarcinomas, and leiomyosarcomas (Schneid et al., 1992).

The autocrine effects of IGFs on certain tumors have also been demonstrated by the neutralization of their actions using specific antisense oligonucleotides to the IGFs (Moats-Staats *et al.*, 1993) or suramin, which interferes with the binding of the ligands to their receptors (Minniti *et al.*, 1992).

2. Receptors

The IGF-I receptor is expressed by virtually every tissue and almost every cell line known. In the case of cancer cell lines, the IGF-I receptor mediates the autocrine growth effect of IGF-I and IGF-II. In cell culture systems, antibody αIR3 blocks the proliferative effects of endogenously produced (and secreted) IGFs or those added exogenously to breast cancer cells (Arteaga and Osborne, 1989) or Wilms' tumors (Gansler *et al.*, 1989), as well as blocking the motility response of melanoma cells (Stracke *et al.*, 1989).

In certain tumors such as Wilms' tumor (Gansler *et al.*, 1988; Werner *et al.*, 1993) and in neoplastic endometrial tissues (Talavera *et al.*, 1990), expression of the IGF-I receptor is significantly increased above that in the adjacent normal tissue. This result has supported the concept that the IGFs and the IGF-I receptor are important in neoplastic growth. On the other hand, correlations between IGF-I receptor expression and the prognosis of breast cancer have suggested that increased receptor levels are associated with a better clinical prognosis (Bonneterre *et al.*, 1990).

Given the strong correlation of IGF-I receptor expression with neoplastic growth, we have begun to examine how IGF-I receptor gene transcription is regulated in tumors. Characterization of the promoter region of the mammalian IGF-I receptor (Werner *et al.*, 1992a) has revealed the presence of at least 10 potential binding sites for transcription factors of the early growth response (EGR) family, including the WT1 tumor suppressor gene product. Normally, WT1 protein binds to these elements and suppresses IGF-I receptor transcription. In Wilms' tumor, however, the WT1 protein is often mutated and, as a result, fails to suppress IGF-I receptor transcription. Consequently, the levels of IGF-I receptor mRNA are significantly increased in Wilms' tumor (Werner *et al.*, 1993). IGF-II expression is pathologically upregulated by a similar mechanism in these tumors. Thus, tumor growth in this model may be supported by increased IGF-II binding to constitutively up-regulated IGF-I receptor.

3. *IGF Binding Proteins*

IGFBPs are secreted by most cancer cell lines studied. IGFBP-1, -2, and -3 are expressed by SCLC (Reeve *et al.*, 1992) and non-SCLC cell lines (Jaques *et al.*, 1992) as well as by ovarian cancer cell lines (Yee *et al.,* 1991). Endometrial carcinoma cells express IGFBP-1, -2, -3, and -5 (Gong *et al.,* 1992). Among the human breast carcinoma cell lines, which are probably the most extensively studied cancer cells, a specific combination of IGFBPs is expressed constitutively by each cell line. Thus, MCF-7, an estrogen receptor-positive (ER$^+$) cell line, expresses IGFBP-2 whereas MDA-MD-231 and H$_S$578T, estrogen receptor-negative (ER$^-$) cell lines, secrete IGFBP-1 and -3. Indeed, in biopsy specimens from breast cancer patients, *in situ* hybridization reveals significantly greater expression of IGFBP-3 in those patients with ER$^-$ tumors (Shao *et al.,* 1992).

Expression of IGFBPs may be regulated by tumor-promoting phorbol esters via a PKC pathway. However, the nature of the regulation is cell-line specific. Phorbol esters enhance the expression of IGFBP-3 in HEC-50 cells and inhibit expression of other IGFBPs in HEC-1B cells (Gong *et al.,* 1992). Retinoic acid (RA), a powerful inhibitor of cellular proliferation, has been shown to inhibit the IGF-I-induced proliferation of a number of breast cancer cell lines (Fontana *et al.,* 1991). This effect is associated with marked alteration in the accumulation of IGFBPs in the conditioned media (Adamo *et al.,* 1992). In MCF-7 cells, RA induces a significant increase in IGFBP-2 and -3 accumulation, concomitant with the inhibition of cell proliferation (LeRoith *et al.,* 1993a) Interestingly, when recombinant IGFBP-2 and -3 are added to these cultures, these molecules individually enhance the proliferative effect of IGF-I, suggesting that they promote MCF-7 cell growth. Their increase after RA treatment may represent a compensatory increase to overcome the RA-induced inhibition of growth (LeRoith *et al.,* 1993a). Thus, as do the IGFs and their receptors, the IGFBPs play an important, still poorly defined, role in cancer development and growth.

4. *Tumor Hypoglycemia*

The association between non-islet cell tumors and fasting hypoglycemia has been known for some time. These large mesenchyme-derived tumors include hemangiopericytomas, mesotheliomas, and fibrosarcomas. A possible pathogenic mechanism for the non-islet cell tumor-induced hypoglycemia syndrome has been established (see Fig. 9). The tumors demonstrate high levels of IGF-II mRNA and protein (Lowe *et al.,* 1989b). The IGF-II protein is released in two forms: a 7.5-kDa and a 15-kDa form. The latter form probably results from incomplete processing of the precursor IGF-II molecule by the tumor, as determined by specific antibodies against the precursor molecule (Eastman *et al.,* 1992). The larger uncleaved IGF-II molecule may not be completely neutralized in the serum by IGFBP-3, or may be more

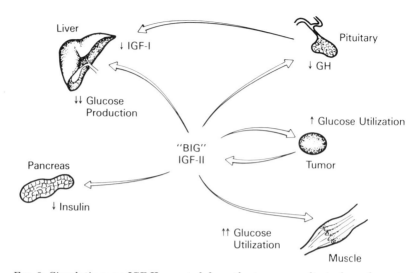

FIG. 9. Circulating pro-IGF-II secreted from the tumor results in hypoglycemia by suppressing hepatic glucose output and enhancing glucose utilization by muscle and tumor tissue. "Free" IGF-II also negatively regulates pituitary GH secretion and, secondarily, liver IGF-I production and pancreatic insulin secretion.

efficiently sequestered by the increased level of IGFBP-2 found in these patients (Zapf *et al.*, 1992). This "big" IGF-II would, therefore, be available to interact with insulin receptors on various tissues, resulting in hypoglycemia (Baxter and Daughaday, 1991). Successful removal of the tumors relieves the hypoglycemia and is associated with a change in the circulating forms of IGF-II; the fully processed 7.5-kDa form now predominates. This form is totally neutralized by the IGFBP-3 complex and, thereby, is prevented from interacting with insulin receptors (Baxter and Daughaday, 1991).

VII. Conclusions

The preceding discussion demonstrates that the IGF system is innately involved in many, if not most, aspects of cellular and tissue regulation. The numerous and complex interactions between the multiple ligands, receptors, and binding proteins described here provide ample opportunity for regulation of IGF activity (ultimately channeled through activation of the IGF-I receptor) by different factors and stimuli operating at various levels of control. These controls may include regulation of the biosynthesis of the components of the IGF system at transcriptional, posttranscriptional, translational, and posttranslational levels as well as regulation of the levels and activities of these components by degradation or modulation of the various second messenger pathways thought to be involved in IGF signal transduction.

Among the many outstanding issues still to be addressed in the control of IGF action are the exact molecular mechanisms responsible for the regulatory processes described in this chapter and the potential interactions of the IGF system with other signaling systems, including the extracellular matrix and other hormones and growth factors. The wealth of information obtained to date suggests that this area of investigation will be fruitful for years to come.

REFERENCES

Abbott, A. M., Bueno, R., Pedrini, M. T., Murray, J. M., and Smith, R. J. (1992). Insulin-like growth factor I receptor gene structure. *J. Biol. Chem.* **267,** 10759–10763.

Adamo, M. L., Lowe, W. L., Jr., LeRoith, D., and Roberts, C. T., Jr. (1989). Insulin-like growth factor I messenger ribonucleic acids with alternative 5'-untranslated regions are differentially expressed during development of the rat. *Endocrinology* **124,** 2737–2744.

Adamo, M. L., Ben-Hur, H., LeRoith, D., and Roberts, C. T., Jr. (1991a). Transcription initiation in the two leader exons of the rat IGF-I gene occurs from disperse versus localized sites. *Biochem. Biophys. Res. Commun.* **176,** 887–893.

Adamo, M. L., Ben-Hur, H., Roberts, C. T., Jr., and LeRoith, D. (1991b). Regulation of start site usage in the two leader exons of the rat insulin-like growth factor I gene by development, fasting and diabetes. *Mol. Endocrinol.* **5,** 1677–1686.

Adamo, M., Shao, Z-M., Lanau, F., Chen, J-C., Clemmons, D. R., Roberts, C. T., Jr., LeRoith, D., and Fontana, R. A. (1992). Insulin-like growth factor-I (IGF-I) and retinoic acid modulation of IGF-binding proteins (IGFBPs): IGFBP-2, -3, and -4 gene expression and protein secretion in a breast cancer cell line. *Endocrinology* **131,** 1858–1866.

Adashi, E. Y., Resnick, C. E., Brodie, A. M. H., Svoboda, M. E., and Van Wyk, J. J. (1985). Somatomedin-C enhances induction of luteinizing hormone receptors by follicle-stimulating hormone in cultured rat granulosa cells. *Endocrinology* **116,** 2369–2375.

Adashi, E. Y., Resnick, C. E., Svoboda, M. E., and Van Wyk, J. J. (1986). Follicle-stimulating hormone enhances somatomedin-C binding to cultured rat granulosa cells: Evidence for cAMP-dependence. *J. Biol. Chem.* **261,** 2923–2926.

Aguado, F., Sanchez-Franco, F., Cacidedo, L., Fernandez, T., Rodrigo, J., and Martinez-Murillo, R. (1992). Subcellular localization of insulin-like growth factor I (IGF-I) in Purkinje cells of the adult rat: An immunocytochemical study. *Neurosci. Lett.* **135,** 171–174.

Alemany, J., Borras, T., and De Pablo, F. (1990). Transcriptional stimulation of the δ_1-crystallin gene by insulin-like growth factor I and insulin requires DNA cis-elements in chicken. *Proc. Natl. Acad. Sci. USA* **87,** 3353–3357.

Angervo, M., Koistinen, R., Suikkari, A-M., and Seppala, M. (1991). Insulin-like growth factor binding protein-1 inhibits the DNA amplification induced by insulin-like growth factor I in human granulosa-luteal cells. *Hum. Reprod.* **6,** 770–773.

Angervo, M., Koistinen, R., and Seppala, M. (1992). Epidermal growth factor stimulates production of insulin-like growth factor-binding protein-1 in human granulosa-luteal cells. *J. Endocrinol.* **134,** 127–131.

Antoniades, H. N., Galanopoulos, T., Neville-Golden, J., and Maxwell, M. (1992). Expression of insulin-like growth factors I and II and the receptor mRNAs in primary human astrocytomas and meningiomas; *in vivo* studies using *in situ* hybridization and immunocytochemistry. *Int. J. Cancer* **50,** 215–222.

Arteaga, C. L., and Osborne, C. K. (1989). Growth inhibition of human breast cancer cells in vitro with an antibody against the type I somatomedin receptor. *Cancer Res.* **49,** 6237–6241.

Banskota, N. K., Taub, R., Zellner, K., Olsen, P., and King, G. L. (1989). Characterization of induction of protooncogene c-*myc* and cellular growth in human vascular smooth muscle cells by insulin and IGF-I. *Diabetes* **38,** 123–129.

Bar, R. S., Bocs, M., Clemmons, D. R., Busby, W. H., Sandra, A., Dake, B. L., and Booth, B. A. (1990). Insulin differentially alters transcapillary movement of intravascular IGFBP-1 and IGFBP-2 and endothelial cell IGF binding protein in the rat heart. *Endocrinology* **127,** 497–499.

Bartlett, W. P., Li, X-S., Williams, M., and Benkovic, S. (1991). Localization of insulin-like growth factor-1 mRNA in murine central nervous system during postnatal development. *Dev. Biol.* **147,** 239–250.

Baxter, R. C., and Daughaday, W. H. (1991). Impaired formation of the ternary insulin-like growth factor-binding protein complex in patients with hypoglycemia due to nonislet cell tumors. *J. Clin. Endocrinol. Metab.* **73,** 696–702.

Baxter, R. C., and Martin, J. I. (1989). Binding proteins for insulin-like growth factors: Structure, regulation and function. *Prog. Growth Fact. Res.* **1,** 49–69.

Beguinot, F., Kahn, C. R., Moses, A. C., White, M. F., and Smith, R. J. (1989). Differentiation-dependent phosphorylation of a 175,000 molecular weight protein in response to insulin and insulin-like growth factor I in L6 skeletal muscle cells. *Endocrinology* **125,** 1599–1603.

Bell, G. I., Stempien, M. M. Fong, N. M., and Rall, L. B. (1986). Sequences of liver cDNAs encoding two different mouse insulin-like growth factor I precursors. *Nucleic Acids Res.* **14,** 7973–7982.

Bernier, M., Chatelain, P. G., Mather, J. P., and Saez, J. M. (1986). Regulation of gonadotropin responsiveness, and cell multiplication by somatomedin-C and insulin in cultured pig Leydig cells. *J. Cell Physiol.* **129,** 257–263.

Bichell, D. P., Kikuchi, K., and Rotwein, P. (1992). Growth hormone rapidly activates insulin-like growth factor I gene transcription in vivo. *Mol. Endocrinol.* **6,** 1899–1908.

Bley, M. A., Simon, J. C., Estevez, A. G., De Asua, L. J., and Baranao, J. L. (1992). Effect of follicle-stimulating hormone on insulin-like growth factor-I-stimulated rat granulosa cell deoxyribonucleic acid synthesis. *Endocrinology* **131,** 1223–1229.

Blum, W. F., Jennem, E. W., Reppin, F., Kietzmann, K., Ranke, M. B., and Bierich, J. R. (1989). Insulin-like growth factor I (IGF-I) binding protein complex is a better mitogen than free IGF-I. *Endocrinology* **125,** 766–772.

Blundell, T. L., Bedarkar, S., and Humbel, R. E. (1983). Tertiary structures, receptor binding, and antigenicity of insulin-like growth factors. *Fed. Proc.* **42,** 2592–2597.

Bodesch, D. B., Lee, P. D. K., Parks, W. C., and Stenmark, K. R. (1989). Insulin-like

growth factor I stimulates elastin synthesis by bovine pulmonary arterial smooth muscle cells. *Biochem. Biophys. Res. Commun.* **160**, 382–387.
Bondy, C. A. (1991). Transient IGF-I gene expression during the maturation of functionally related central projection neurons. *J. Neurosci.* **11**, 3442–3455.
Bondy, C. A., Werner, H., Roberts, C. T., Jr., and LeRoith, D. (1990). Cellular pattern of insulin-like growth factor-I (IGF-I) and type I IGF receptor gene expression in early organogenesis: Comparison with IGF-II gene expression. *Mol. Endocrinol.* **4**, 1386–1398.
Bondy, C. A., Bach, M. A., and Lee, W-H. (1992a). Mapping of brain insulin and insulin-like growth factor receptor gene expression by *in situ* hybridization. *Neuroprotocols* **1**, 240–249.
Bondy, C., Werner, H., Roberts, C. T., Jr., and LeRoith, D. (1992b). Cellular pattern of type-I insulin-like growth factor receptor gene expression during maturation of the rat brain: Comparison with insulin-like growth factors I and II. *Neuroscience* **46**, 909–923.
Bonneterre, J., Peyrat, J. P., Beuscart, R., and Demaille, A. (1990). Prognostic significance of insulin-like growth factor I receptors in human breast cancer. *Cancer Res.* **50**, 6931–6935.
Boulware, S. D., Tamborlanem, W. V., Matthews, L. S., and Sherwin, R. S. (1992). Diverse effects of insulin-like growth factor I on glucose, lipid, and amino acid metabolism. *Am. J. Physiol.* **262**, E130–E133.
Brennan, W. A., Jr. (1988). Developmental aspects of the rat brain insulin receptor. Loss of sialic acid and fluctuation in number characterize fetal development. *Endocrinology* **122**, 2364–2370.
Brenner-Gati, L., Berg, K. A., and Gershengorn, M. C. (1988). Thyroid-stimulating hormone and insulin-like growth factor-I synergize to elevate 1,2-diacylglycerol in rat thyroid cells: Stimulation of DNA synthesis via interaction between lipid and adenylyl cyclase signal transduction systems. *J. Clin. Invest.* **82**, 1144–1148.
Brenner-Gati, L., Berg, K. A., and Gershengorn, M. C. (1989). Insulin-like growth factor-I potentiates thyrotropin stimulation of adenylyl cyclase in FRTL-5 cells. *Endocrinology* **125**, 1315–1320.
Buchou, T., Gaben, A-M., Phan-Dihn-Tuy, F., and Mester, J. (1991). Insulin/insulin-like growth factor I induce actin transcription in mouse fibroblasts expressing constitutively *myc* gene. *Mol. Cell. Endocrinol.* **75**, 181–187.
Cahill, A. L., and Perlman, R. L. (1991). Activation of a microtubule-associated protein-2 kinase by insulin-like growth factor-I in bovine chromaffin cells. *J. Neurochem.* **57**, 1832–1839.
Camacho, Hubner, C., Busby, W. H., Jr., McCusker, R. H., Wright, G., and Clemmons, D. R. (1992). Identification of the forms of insulin-like growth factor-binding proteins produced by human fibroblasts and the mechanisms that regulate their secretion. *J. Biol. Chem.* **267**, 11949–11956.
Campa, M. J., Glickman, J. F., Yamamoto, K., and Chang, K-J. (1992). The antibiotic azatyrosine suppresses progesterone or [Val^{12}]p21 Ha-*ras*/insulin-like growth factor I-induced germinal vesicle breakdown and tyrosine phosphorylation of *Xenopus* mitogen-activated protein kinase in oocytes. *Proc. Natl. Acad. Sci. USA* **89**, 7654–7658.
Campisi, J., and Pardee, A. B. (1984). Post-transcriptional control of the onset of DNA synthesis by an insulin-like growth factor. *Mol. Cell. Biol.* **4**, 1807–1810.
Cara, J. F., and Rosenfeld, R. (1988). Insulin-like growth factor I and insulin potentiate

luteinizing hormone-induced androgen synthesis by rat ovarian thecal-interstitial cells. *Endocrinology* **123**, 733–739.

Carlsson-Skwirut, C., Jornvall, H., Holmgren, A., Anderson, C., Bergman, T., Lundquist, G., Sjogren, B., and Sara, V. R. (1986). Isolation and characterization of variant IGF-I as well as IGF-II from adult human brain. *FEBS Lett.* **201**, 46–50.

Cheng, B., and Mattson, M. P. (1992). IGF-I and IGF-II protect cultured hippocampal and septal neurons against calcium-mediated hypoglycemic damage. *J. Neurosci.* **12**, 1558–1566.

Chiou, S. T., and Chang, W. C. (1992). Insulin-like growth factor 1 stimulates transcription of the c-*jun* proto-oncogene in BALB/c3T3 cells. *Biochem. Biophys. Res. Commun.* **183**, 524–531.

Christman, G. M., and Halme, J. K. (1992). Growth hormone: Revisited. *Fertil. Steril.* **57**, 12–13.

Christman, G. M., Randolph, J. F., Jr., Peegel, H., and Menon, K. M. J., (1991). Differential responsiveness of luteinized human granulosa cells to gonadotropins and insulin-like growth factor I for induction of aromatase activity. *Am. Fertil. Steril.* **55**, 1099–1106.

Clemmons, D. R. (1984). Multiple hormones stimulate the production of somatomedin by cultured human fibroblasts. *J. Clin. Endocrinol. Metab.* **58**, 850–856.

Clemmons, D. R., and Shaw, D. S. (1983). Variables controlling somatomedin production by cultured human fibroblasts. *J. Cell Physiol.* **115**, 137–142.

Clemmons, D. R., Smith-Banks, A., and Underwood, L. E. (1992). Reversal of diet-induced catabolism by infusion of recombinant insulin-like growth factor-I in humans. *J. Clin. Endocrinol. Metab.* **75**, 234–238.

Cocco, L. Martelli, A. M., Gilmour, R. S., Ognibene, A., Manzoli, F. A., and Irvine, R. F. (1989). Changes in nuclear inositol phospholipids induced in intact cells by insulin-like growth factor I. *Biochem. Biophys. Res. Commun.* **159**, 720–725.

Condorelli, G., Formisano, P., Villone, G., Smith, R. J., and Beguinot, F. (1989). Insulin and insulin-like growth factor I (IGF-I) stimulate phosphorylation of a 175,000 cytoskeleton-associated protein in intact FRTL5 cells. *J. Biol. Chem.* **264**, 12633–12639.

Cooke, D. W., Bandert, L. A., Roberts, C. T., Jr., LeRoith, D., and Casella, S. J. (1991). Analysis of the human type I insulin-like growth factor receptor promoter region. *Biochem. Biophys. Res. Commun.* **117**, 1113–1120.

Cotterill, A. M. (1992). The therapeutic potential of recombinant human insulin-like growth factor-I. *Clin. Endocrinol.* **37**, 11–16.

Couce, M. E., Weatherington, A. J., and McGinty, J. F. (1992). Expression of insulin-like growth factor-II (IGF-II) and IGF-II/mannose-6-phosphate receptor in the rat hippocampus: An *in situ* hybridization and immunocytochemical study. *Endocrinology* **131**, 1636–1642.

Czech, M. P. (1989). Signal transmission by the insulin-like growth factors. *Cell* **59**, 235–238.

Damante, G., Cox, F., and Rapoport, B. (1988). IGF-I increases c-*fos* expression in FRTL5 rat thyroid cells by activating the c-*fos* promoter. *Biochem. Biophys. Res. Commun.* **151**, 1194–1199.

Daughaday, W. H., and Rotwein, P. (1989). Insulin-like growth factors I and II. Peptide, messenger ribonucleic acid and gene structures, serum, and tissue concentrations. *Endocrine Rev.* **10**, 68–91.

DeChiara, T. M., Efstratiadis, A., and Robertson, E. J. (1990). A growth-deficiency phenotype in heterozygous mice carrying an insulin-like growth factor II gene disrupted by targeting. *Nature (London)* **345,** 78–80.

De Mellow, J. S. M., and Baxter, R. C. (1988). Growth hormone dependent insulin-like growth factor binding protein both inhibits and potentiates IGF-I stimulated DNA synthesis in skin fibroblasts. *Biochem. Biophys. Res. Commun.* **156,** 199–206.

De Pablo, F., Scott, L. A., and Roth, J. (1990). Insulin and insulin-like growth factor I in early development: Peptides, receptors and biological events. *Endocrine Rev.* **4,** 558–577.

de Pagter-Holthuizen, P., Jansen, M., van Schaik, F. M. A., van der Kammen, R. A., Oosterjk, C., Van der Brande, J. L., and Sussenbach, J. S. (1987). The human insulin-like growth factor II contains two development-specific promoters. *FEBS Lett.* **214,** 259–264.

de Pagter-Holthuizen, P., Jansen, M., van der Kammen, R. A., van Schaik, F. M. A., and Sussenbach, J. S. (1988). Differential expression of the human insulin-like growth factor II gene. *Biochem. Biophys. Acta* **950,** 282–295.

Divechia, N. Banfic, H., and Irvine, R. F. (1991). The polyphosphoinositide cycle exists in the nuclei of Swiss 3T3 cells under the control of a receptor (for IGF-I) in the plasma membrane, and stimulation of the cycle increases nuclear diacylglycerol and apparently induces translocation of protein kinase C to the nucleus. *EMBO J.* **10,** 3207–3214.

Dor, J., Costritsci, N. Pariente, C., Rabinovici, J., Mashiach, S. Lunenfeld, B., Kaneti, H., Seppala, M., Koistinen, R., and Karasik, A. (1992). Insulin-like growth factor-I and follicle-stimulating hormone suppress insulin-like growth factor binding protein-1 secretion by human granulosa-luteal cells. *J. Clin. Endocrinol. Metab.* **75,** 969–975.

Douglas, R. G., Breier, B. H., Gallaher, B. W., Koea, J. B., Shaw, J. H. F., and Gluckman, P. D. (1991). The circulating molecular weight forms of infused recombinant insulin-like growth factor-I and effects on glucose and fat metabolism in lambs. *Diabetologia* **34,** 790–795.

Drago, M., Murphy, M., Carroll, S. M., Harvey, R. P., and Bartlett, P. F. (1991). Fibroblast growth factor-mediated proliferation of central nervous system precursors depends on endogenous production of insulin-like growth factor I. *Proc. Natl. Acad. Sci. USA* **88,** 2199–2203.

Drakenberg, K., Sara, V. R., Falkner, S., Gammeltoft, S., Maake, C., and Reinecke, M. (1993). Identification of IGF-1 receptors in primitive vertebrates. *Regul. Peptides* **43,** 73–81.

Drop, S. I. S., Schuller, A. G. P., Lindenbergh-Kortleve, D. J., Groffen, C., Brinkman, A., and Zwarthoff, E. C. (1992). Structural aspects of the IGFBP family. *Growth Regul.* **3,** 80–87.

Dull, T. J., Gray, A., Hayflick, J. S., and Ullrich, A. (1984). Insulin-like growth factor II precursor gene organization in relation to the insulin gene family. *Nature (London)* **310,** 777–781.

Eastman, R. C., Carson, R. E., Orloff, D. G., Cochran, C. S., Perdue, J. F., Rechler, M. M., Lanau, F., Roberts, C. T., Jr., Shapiro, H., Roth, J., and LeRoith, D. (1992). Glucose utilization in a patient with hepatoma and hypoglycemia. *J. Clin. Invest.* **89,** 1958–1963.

Elgin, R. G., Busby, W. H., Jr., and Clemmons, D. R. (1987). An insulin-like growth factor (IGF) binding protein enhances the biologic response to IGF-I. *Proc. Natl. Acad. Sci. USA* **84,** 3254–3258.

Erickson, G. F., Garzo, V. G., and Magoffin, D. A. (1991). Progesterone production by

human granulosa cells cultured in serum free medium: Effects of gonadotrophins and insulin-like growth factor I (IGF-I). *Hum. Reprod.* **6,** 1074–1081.
Erickson, G. F., Nakatani, A., Ling, N., and Shimasaki, S. (1992a). Cyclic changes in insulin-like growth factor binding protein-4 messenger ribonucleic acid in the rat ovary. *Endocrinology* **130,** 625–636.
Erickson, G. F., Nakatani, A., Ling, N., and Shimasaki, S. (1992b). Localization of insulin-like growth factor-binding protein-5 messenger ribonucleic acid in rat ovaries during the estrous cycle. *Endocrinology* **130,** 1867–1878.
Fant, M., Munro, H., and Moses, A. C. (1986). An autocrine/paracrine role for insulin-like growth factors in the regulation of human placental growth. *J. Clin. Endocrinol. Metab.* **63,** 499–506.
Farese, R. V., Nair, G. P., Standaert, M. L., and Cooper, D. R. (1988). Epidermal growth factor and insulin-like growth factor-I stimulate the hydrolysis of the insulin-sensitive phosphatidylinositol-glycan in BC3H-1 myocytes. *Biochem. Biophys. Res. Commun.* **156,** 1346–1352.
Farese, R. V., Nair, G. P., Sierra, C. G., Standaert, M. L., Pollet, R. J., and Cooper, D. R. (1989). Insulin-like effects of epidermal growth factor and insulin-like growth factor-I on [^{3}H]2-deoxyglucose uptake, diacylglycerol generation and protein kinase C activation in BC3H-1 myocytes. *Biochem. J.* **261,** 927–934.
Florance, R. S. K., Senior, P. V., Byrne, S., and Beck, F. (1991). The expression of IGF-II in the early post-implantation rat conceptus. *J. Anat.* **175,** 169–179.
Fontana, F. A., Burrows-Mezu, A., Clemmons, D. R., and LeRoith, D. (1991). Retinoid modulation of insulin-like growth binding proteins and inhibition of breast carcinoma proliferation. *Endocrinology* **128,** 1115–1122.
Foyt, H. L., LeRoith, D., and Roberts, C. T., Jr. (1991). Differential association of insulin-like growth factor I mRNA variants with polysomes in vivo. *J. Biol. Chem.* **266,** 7300–7305.
Foyt, H. L., Lanau, F., Woloschak, M., LeRoith, D., and Roberts, C. T., Jr. (1992). Effect of growth hormone on levels of differentially processed insulin-like growth factor I mRNAs in total and polysomal mRNA populations. *Mol. Endocrinol.* **6,** 1881–1888.
Frost, R. A., and Tseng, L. (1991). Insulin-like growth factor binding protein 1 (IGFBP-1) is phosphorylated on serine by cultured human endothelial stromal cells and protein kinase A *in vitro. J. Biol. Chem.* **266,** 18082–18086.
Gansler, T., Allen, K. D., Burant, C. F., Inabnett, T., Scott, A., Buse, M. G., Sens, D. A., and Garvin, A. J. (1988). Detection of type 1 insulin-like growth factor (IGF) receptors in Wilms' tumors. *Am. J. Pathol.* **130,** 431–435.
Gansler, T., Furlanetto, R., Gramling, T. S., Robinson, K. A., Blocker, N., Buse, M. G., Sens, D. A., and Garvin, A. J. (1989). Antibody to type I IGF-I receptor inhibits growth of Wilms' tumor in culture and athymic mice. *Am. J. Pathol.* **135,** 961–966.
Garcia-Estrada, J., Garcia-Segura, L. M., and Torres-Aleman, I. (1992). Expression of insulin-like growth factor I by astrocytes in response to injury. *Brain Res.* **592,** 343–347.
Garcia-Segura, L. M., Perez, J., Pons, S., Rejas, M. T., and Torres-Aleman, I. (1991). Localization of insulin-like growth factor I (IGF-I)-like immunoreactivity in the developing and adult rat brain. *Brain Res.* **560,** 167–174.
Geisert, R. D., Lee, C-Y., Simmen, F. A., Zavy, M. T., Fliss, A. E., Bazer, F. W., and Simmen, R. C. M. (1991). Expression of messenger RNAs encoding insulin-like growth factor-I, -II, and insulin-like growth factor binding protein-2 in bovine endometrium during the estrous cycle and early pregnancy. *Biol. Reprod.* **45,** 975–983.
Ghahary, A., Chakrabarti, S., and Murphy, T. J. (1990). Localization of the sites of

synthesis and action of insulin-like growth factor-I in the rat uterus. *Mol. Endocrinol.* **4,** 191–195.
Giacobini, M. M. J., Olson, L., Hoffer, B. J., and Sara, V. R. (1990). Truncated IGF-I exerts trophic effects on fetal brain tissue grafts. *Exp. Neurol.* **108,** 33–37.
Giudice, L. C., Milkj, A. A., Milkowski, D. A., and Danasouri, I. E. (1991). Human granulosa contain messenger ribonucleic acids encoding insulin-like growth factor-binding proteins (IGFBPs) and secrete IGFBPs in culture. *Fertil. Steril.* **56,** 475–480.
Glick, R. P., Unterman, T. G., Van Der Woude, M., and Zollner-Blaydes, L. (1992). Insulin and insulin-like growth factors in central nervous system tumors. *J. Neurosurg.* **77,** 445–450.
Gluckman, P. D., and Ambler, G. R. (1992). Therapeutic use of insulin-like growth factor I: Lessons from *in vivo* animal studies. *Acta Paediatr. Suppl.* **383,** 134–136.
Gluckman, P., Klempt, N., Guan, J., Mallard, C., Sirimanne, E., Dragunow, M., Klempt, M., Singh, K., Williams, C., and Nikolics, K. (1992). A role for IGF-I in the rescue of CNS neurons following hypoxic-ischemic injury. *Biochem. Biophys. Res. Commun.* **182,** 593–599.
Goldstein, R. H., Poliks, C. F., Pilch, P. F., Smith, B. D., and Fine, A. (1989). Stimulation of collagen formation by insulin and insulin-like growth factor I in cultures of human lung fibroblasts. *Endocrinology* **124,** 964–970.
Gong, Y., Ballejo, G., Alkhalaf, B., Molnar, P., Murphy, L. C., and Murphy, L. J., (1992). Phorbol esters differentially regulate the expression of insulin-like growth factor-binding proteins in endometrial carcinoma cells. *Endocrinology* **131,** 2747–2754.
Grimes, R. W., and Hammond, J. M. (1992). Insulin and insulin-like growth factors (IGFs) stimulate production of IGF-binding proteins by ovarian granulosa cells. *Endocrinology* **131,** 553–558.
Guler, H-P., Zapf, J., and Froesch, E. R. (1987). Short-term metabolic effects of recombinant human insulin-like growth factor I in healthy adults. *N. Engl. J. Med.* **317,** 137–140.
Guler, H-P., Zapf, J., Scheiwiller, E., and Froesch, E. R. (1988). Recombinant human insulin-like growth factor I stimulates growth and has distinct effects on organ size in hypophysectomized rats. *Proc. Natl. Acad. Sci. USA* **85,** 4889–4893.
Haig, D., and Graham, C. (1991). Genomic imprinting and the strange case of the insulin-like growth factor II receptor. *Cell* **64,** 1045–1046.
Hall, L. J., Kajimoto, Y. Bichell, D., Kim, S-W., James, P. L., Counts, D., Nixon, L. J., Tobin, G., and Rotwein, P. (1992). Functional analysis of the rat insulin-like growth factor I gene and identification of an IGF-I gene promoter. *DNA Cell Biol.* **11,** 301–313.
Hall, S. H., Berthelon, M-C., Avallet, O., and Saez, J. M. (1991). Regulation of c-*fos*, c-*jun*, *jun*-B, and c-*myc* messenger ribonucleic acids by gonadotropin and growth factors in cultured pig Leydig cell. *Endocrinology* **129,** 1243–1249.
Han, V. K. M., Smith, A., Myint, W., Nygard, K., and Bradshaw, S. (1992). Mitogenic activity of epidermal growth factor on newborn rat astroglia: Interaction with insulin-like growth factors. *Endocrinology* **131,** 1134–1142.
Hari, J., Pierce, S. B., Morgan, D. O., Sara, V., Smith, M. C., and Roth, R. A. (1987). The receptor for insulin-like growth factor II mediates an insulin-like response. *EMBO J.* **6,** 3367–3371.
Harvey, M. B., and Kaye, P. L. (1991). Mouse blastocysts respond metabolically to short-term stimulation by insulin and IGF-I through the insulin receptor. *Mol. Reprod. Dev.* **29,** 253–258.
Harvey, M. B., and Kaye, P. L. (1992). Mediation of the actions of insulin and insulin-like

growth factor-1 on preimplantation mouse embryos *in vitro*. *Mol. Reprod. Dev.* **33**, 270–275.

Hatey, F., Langlois, I., Mulsant, P., Bonnet, A., Benne, F., and Gasser, F. (1992). Gonadotropins induce accumulation of insulin-like growth factor I mRNA in pig granulosa cells *in vitro*. *Mol. Cell. Endocrinol.* **86**, 205–211.

Heidenreich, K. A., and Toledo, S. P. (1989). Insulin receptors mediate growth effects in cultured fetal neurons. II. Activation of a protein kinase that phosphorylates ribosomal protein S6. *Endocrinology* **125**, 1458–1463.

Hendricks, S. A., Agardh, C.-D., Taylor, S. I., and Roth, J. (1984). Unique features of the insulin receptor in rat brain. *J. Neurochem.* **43**, 1302–1309.

Heppler, J. E., Van Wyk, J. J., and Lund, P. K. (1990). Different half-lives of insulin-like growth factor mRNAs that differ in length of 3'-untranslated sequence. *Endocrinology* **127**, 1550–1552.

Hernandez, E. R., Resnick, C. E., Svoboda, M. E., Van Wyk, J. J., Payne, D. W., and Adashi, E. Y. (1988). Somatomedin-C/insulin-like growth factor-I (Sm-C/IGF-I) as an enhancer of androgen biosynthesis by cultured rat ovarian cells. *Endocrinology* **122**, 1603–1613.

Hernandez, E. R., Roberts, C. T. Jr, LeRoith, D., and Adashi, E. Y. (1989). Rat ovarian insulin-like growth factor I (IGF-I) gene expression is granulosa cell-selective: 5 prime UT mRNA variant representation and hormonal regulation. *Endocrinology* **125**, 572–574.

Hernandez, E. R., Hurwitz, A., Vera, A., Pellicer, A., Adashi, E. Y., LeRoith, D., and Roberts, C. T., Jr. (1992a). Expression of the genes encoding the insulin-like growth factors and their receptors in the human ovary. *J. Clin. Endocrinol. Metab.* **74**, 419–425.

Hernandez, E. R., Rosenfeld, R. G., Carlsson-Skwirut, C., and Francis, G. L. (1992b). Granulosa cell-derived insulin-like growth factor (IGF) binding proteins are inhibitory to IGF-I hormonal action. *J. Clin. Invest.* **90**, 1593–1599.

Hofig, A., Michel, F. J., Simmen, F. A., and Simmen, R. C. M. (1991). Constitutive expression of uterine receptors for insulin-like growth factor-I during the periimplantation period in the pig. *Biol. Reprod.* **45**, 533–539.

Holthuizen, P., van der Lee, F. M., Ikejiri, K., Yamamoto, M., and Sussenbach, J. S. (1990). Identification and initial characterization of a fourth leader exon and promoter of the human IGF-II gene. *Biochim. Biophys. Acta* **1087**, 341–343.

Holtorf, A.-P., Furuya, K., Ivell, R., and McArdle, C. A. (1989). Oxytocin production and oxytocin messenger ribonucleic acid levels in bovine granulosa cells are regulated by insulin-like growth factor-I: Dependence on the developmental status of the ovarian follicle. *Endocrinology* **125**, 2612–2618.

Hossenlopp, P., Segovia, B., Lassarre, C., Roghani, M., Bredon, M., and Binoux, M. (1990). Evidence of enzymatic degradation of insulin-like growth factor-binding proteins in the 150K complex during pregnancy. *J. Clin. Endocrinol. Metab.* **71**, 797–803.

Hoyt, E. C., Van Wyk, J. J., and Lund, P. K. (1988). Tissue and development specific regulation of a complex family of rat insulin-like growth factor I messenger ribonucleic acids. *Mol. Endocrinol.* **2**, 1077–1086.

Hsu, C. J., and Hammond, J. M. (1987). Concomitant effects of growth hormone on secretion of insulin-like growth factor I and progesterone by cultured porcine granulosa cells. *Endocrinology* **121**, 1343–1348.

Ikari, N., Yoshino, H., Moses, A. C., and Flier, J. S. (1988). Evidence that receptor aggregation may play a role in transmembrane signaling through the insulin-like growth factor-I receptor. *Mol. Endocrinol.* **2**, 831–837.

Irminger, J. C., Rosen, K. M., Humbel, R. E., and Vila-Komaroff, L. (1987). Tissue-specific expression of insulin-like growth factor II mRNAs with distinct 5' untranslated regions. *Biochemistry* **84,** 6330–6334.

Isaksson, O. G. P., Lindahl, A., Isgaard, J., Nilsson, A., Törnell, J., and Carlsson, B. (1991). Dual regulation of cartilage growth. *In* "Modern Concepts of Insulin-Like Growth Factors" (E. M. Spencer, ed.), pp. 121–130. Elsevier, New York.

Ishii, D. N., and Recio-Pinto, E. (1987). Role of insulin, insulin-like growth factors and nerve growth factor in neurite formation. *In* "Insulin, Insulin-Like Growth Factors and Their Receptors in the CNS" (M. K. Raizada, M. I. Phillips, D. LeRoith, eds.), pp. 315–348. Plenum Press, New York.

Jacob, R., Barrett, E., Plewe, G., Fagin, K. D., and Sherwin, R. S. (1989). Acute effects of insulin-like growth factor I on glucose and amino acid metabolism in the awake fasted rat. *J. Clin. Invest.* **83,** 1717–1723.

Jacobs, S., and Cuatrecasas, P. (1986). Phosphorylation of receptors for insulin and insulin-like growth factor I. *J. Biol. Chem.* **261,** 934–939.

Jansen, E., Steenbergh, P. H., LeRoith, D., Roberts, C. T., Jr., and Sussenbach, J. S. (1991). Identification of multiple transcription start sites in the human IGF-I gene. *Mol. Cell. Endocrinol.* **78,** 115–125.

Jaques, G., Kiefer, P., Schoneberger, H. J., Wegmann, B., Kaiser, U., Brandscheid, D., and Havemann, K. (1992). Differential expression of insulin-like growth factor binding proteins in human non-small cell lung cancer cell lines. *Eur. J. Cancer* **28A,** 1899–1904.

Jones, J. I., D'Ercole, A. J., Camacho-Hubner, C., and Clemmons, D. R. (1991). Phosphorylation of insulin-like growth factor binding protein I in cell culture and in vivo: Effects on affinity for IGF-I. *Proc. Natl. Acad. Sci. USA* **88,** 7481–7485.

Kadowaki, T., Koyasu, S., Nishida, E., Tobe, K., Izumi, T., Takaku, F., Sakai, H., Yahara, I., and Kasuga, M. (1987). Tyrosine phosphorylation of common and specific sets of cellular proteins rapidly induced by insulin, insulin-like growth factor-I, and epidermal growth factor in an intact cell. *J. Biol. Chem.* **262,** 7342–7350.

Kaleko, M., Rutter, W. J., and Miller, A. D. (1990). Overexpression of the human insulin-like growth factor I receptor promotes ligand-dependent neoplastic transformation. *Mol. Cell. Biol.* **10,** 464–470.

Kapur, S., Tamada, H., Dey, S. K., and Andrews, G. K. (1992). Expression of insulin-like growth factor-I (IGF-I) and its receptor in the peri-implantation mouse uterus, and cell-specific regulation of IGF-I gene expression by estradiol and progesterone. *Biol. Reprod.* **46,** 208–219.

Kato, H., Faria, T. N., Stannard, B., Roberts, C. T., Jr., and LeRoith, D. (1993). Role of tyrosine kinase activity in signal transduction by the insulin-like growth factor-I (IGF)-I receptor: Characterization of kinase-deficient IGF-I receptors and the action of an IGF-I-mimetic antibody (αIR-3). *J. Biol. Chem.* **268,** 2655–2661.

Kenner, K. A., and Heidenreich, K. A. (1991). Insulin and Insulin-like growth factors stimulate *in vivo* receptor autophosphorylation and tyrosine phosphorylation of a 70K substrate in cultured fetal chick neurons. *Endocrinology* **129,** 301–311.

Kerr, D., Tamborlane, W. V., Rife, F., and Sherwin, R. S. (1993). Effect of insulin-like growth factor-1 on the responses to and recognition of hypoglycemia in humans. *J. Clin. Invest.* **91,** 141–147.

Kiess, W., Thomas, C. L., Greenstein, L. A., Lee, L., Sklar, M. M., Rechler, M. M., Sahagian, G. G., and Nissley, S. P. (1989). Insulin-like growth factor II (IGF-II) inhibits both the cellular uptake of β-galactosidase and the binding of β-galactosidase to purified IGF-II/mannose 6-phosphate receptor. *J. Biol. Chem.* **264,** 4710–4714.

Kiess, W., Thomas, C. L., Sklar, M. M., and Nissley, S. P. (1990). Beta-galactosidase decreases the binding affinity of the insulin-like growth factor II/mannose 6-phosphate receptor for the insulin-like growth factor II. *Eur. J. Biochem.* **190**, 71–77.

Kim, S-W., Lajara, R., and Rotwein, P. (1991). Structure and function of a human insulin-like growth factor-I gene promoter. *Mol. Endocrinol.* **5**, 1964–1972.

Klarlund, J. K., and Czech, M. P. (1988). Insulin-like growth factor I and insulin rapidly increase casein kinase II activity in BALB/c 3T3 fibroblasts. *J. Biol. Chem.* **263**, 15872–15875.

Klempt, N. D., Klempt, M., Gunn, A. J., Singh, K., and Gluckman, P. D. (1992). Expression of insulin-like growth factor-binding protein 2 (IGF-BP 2) following transient hypoxia–ischemia in the infant rat brain. *Mol. Brain Res.* **15**, 55–61.

Kikuchi, K., Bichell, D. P., and Rotwein, P. (1992). Chromatin changes accompany the developmental activation of insulin-like growth factor I gene transcription. *J. Biol. Chem.* **267**, 21505–21511.

Kojima, I., Matsunaga, H., Kurokawa, K., Ogata, E., and Nishimoto, I. (1988). Calcium influx: An intracellular message of the mitogenic action of insulin-like growth factor-I. *J. Biol. Chem.* **263**, 16561–16567.

Komoly, S., Hudson, L. D., Webster, H. D., and Bondy, C. A. (1992). Insulin-like growth factor I gene expression is induced in astrocytes during experimental demyelination. *Proc. Natl. Acad. Sci. USA* **89**, 1894–1898.

Lambert, S., Carlisi, A., Collette, J., Franchimont, P., and Gol-Winkler, R. (1992). Insulin-like growth factor II in two human colon-carcinoma cell lines: Gene structure and expression, and protein secretion. *Int. J. Cancer* **52**, 404–408.

Laron, Z., Pertzelan, A., and Mannheimer, S. (1966). Genetic pituitary dwarfism with high serum concentration of growth hormone: A new inborn error of metabolism? *Isr. J. Med. Sci.* **62**, 152–155.

LaTour, D., Mohan, S., Linkhart, T. A., Baylink, D. J., and Strong, D. D. (1990). Inhibitory insulin-like growth factor-binding protein: Cloning, complete sequence, and physiological regulation. *Mol. Endocrinol.* **4**, 1806–1814.

Lee, W-H., Javedan, S., and Bondy, C. A. (1992). Coordinate expression of insulin-like growth factor system components by neurons and neuroglia during retinal and cerebellar development. *J. Neurosci.* **12**, 4737–4744.

Leof, E. B., Wharton, W., Van Wyk, J. J., and Pledger, W. J. (1982). Epidermal growth factor (EGF) and somatomedin C regulated G1 progression in competent BALBc/3T3 cells. *Exp. Cell Res.* **141**, 107–115.

LeRoith, D., Lowe, W., L., Jr., Shemer, J., Raizada, M., and Ota, A. (1988). Development of brain insulin receptors. *Int. J. Biochem.* **5**, 225–230.

LeRoith, D., Adamo, M. L., Shemer, J., Lanau, F., Shen-Orr, Z., Yaron, A., Roberts, C. T., Jr., Clemmons, D. R., Sheikh, M. S. Shao, Z. M., Chen, J-C., and Fontana, J. (1993a). Retinoic acid inhibits growth of breast cancer cell lines: The role of insulin-like growth factor binding proteins. *Growth Regul.* **3**, 78–80.

LeRoith, D., Roberts, C. T., Jr., Werner, H., Bondy, C., Raizada, M., and Adamo, M. L. (1993b). Insulin-like growth factors in the brain. *In* "Neurotrophic Factors." (S. E. Loughlin and J. H. Fallon, eds.), pp. 391–414. Academic Press, San Diego.

Levy, M. J., Hernandez, E. R., Adashi, E. Y., Stillman, R. J., Roberts, C. T., Jr., and LeRoith, D. (1992). Expression of the insulin-like growth factor (IGF)-I and -II and the IGF-I and -II receptor genes during postnatal development of the rat ovary. *Endocrinology* **131**, 1202–1206.

Lin, T., Wang, D., Nagpal, M. L., Chang, W., and Calkins, J. H. (1992). Down-regulation of Leydig cell insulin-like growth factor-I gene expression by interleukin-1. *Endocrinology* **130**, 1217–1224.

Lindahl, A., Isgaard, J., and Isaksson, O. G. P. (1991). Growth and differentiation. *J. Clin. Endocrinol. Metab.* **5,** 671–687.
Lobel, P., Dalmas, N. M., and Kornfeld, S. (1988). Cloning and sequence analysis of the cation-independent mannose 6-phosphate receptor. *J. Biol. Chem.* **263,** 2563–2570.
Lowe, W. L., Roberts, C. T., Jr., Lasky, S. R., and LeRoith, D. (1987). Differential expression of alternative 5' untranslated regions in mRNAs encoding rat insulin-like growth factor I. *Proc. Natl. Acad. Sci. USA* **84,** 8946–8950.
Lowe, W. L., Lasky, S. R., LeRoith, D., and Roberts, C. T., Jr. (1988). Distribution and regulation of rat insulin-like growth factor I messenger ribonucleic acids encoding alternative carboxy terminal E-peptides: Evidence for differential processing and regulation in liver. *Mol. Endocrinol.* **2,** 528–535.
Lowe, W. L., Adamo, M., Werner, H., Roberts, C. T., Jr., and LeRoith, D. (1989a). Regulation by fasting of rat insulin-like growth factor I and its receptor: Effects on gene expression and kidney. *J. Clin. Invest.* **84,** 619–626.
Lowe, W. L., Jr., Roberts, C. T., Jr., LeRoith, D., Rojeski, M. T., Merimee, T. J., Fui, S. T., Keen, H., Mersey, J., Gluzman, S., Spratt, D., Eastman, R., Arnold, D., and Roth, J. (1989b). Insulin-like growth factor II in non-islet tumors associated with hypoglycemia increased levels of messenger ribonucleic acid. *J. Clin. Endocrinol. Metab.* **69,** 1153–1159.
Lu, K., and Campisi, J. (1992). Ras proteins are essential and selective for the action of insulin-like growth factor 1 late in the G_1 phase of the cell cycle in BALB/c murine fibroblasts. *Proc. Natl. Acad. Sci. USA* **89,** 3889–3893.
Lund, P. K., Moats-Staats, B. M., Hynes, M. A., Simmons, J. G., Jansen, M., D'Ercole, A. J., and Van Wyk, J. J. (1986). Somatomedin C/insulin-like growth factor I and insulin-like growth factor II mRNAs in rat fetal and adult tissues. *J. Biol. Chem.* **261,** 14539–14544.
Lund, P. K., Hoyt, E. C., and Van Wyk, J. J. (1989). The size heterogeneity of rat insulin-like growth factor I mRNAs is due primarily to differences in the length of 3'-untranslated sequence. *Mol. Endocrinol.* **3,** 2054–2061.
Luo, J. M., Reid, R. E., and Murphy, L. J. (1990). Dexamethasone increases hepatic insulin-like growth factor binding protein-1 (IGFBP-1) mRNA and serum IGFBP-1 concentrations in the rat. *Endocrinology* **127,** 1456–1462.
Macaulay, V. M. (1992). Insulin-like growth factors and cancer. *Br. J. Cancer* **65,** 311–320.
MacDonald, R. G., Pfeffer, S. R., Coussens, L., Tepper, M. A., Brocklebank, C. M., Mole, J. E., Andersen, J. K., Chen, E., Czech, M. P., and Ullrich, A. (1988). A single receptor binds both insulin-like growth factor II and mannose-6-phosphate. *Science* **239,** 3334–3357.
Maher, F., Clark, S., and Harrison, L. C. (1989). Chronic stimulation of glucose transporter gene expression in L6 myocytes mediated via the insulin-like growth factor-I receptor. *Mol. Endocrinol.* **3,** 2128–2135.
Mangiacapra, F. J., Roof, S. L., Ewton, D. Z., and Florini, J. R. (1992). Paradoxical decrease in *myf*-5 messenger RNA levels during induction of myogenic differentiation by insulin-like growth factors. *Mol. Endocrinol.* **6,** 2038–2044.
Marks, J. L., Porte, D., Jr., and Baskin, D. G. (1991a). Localization of type I insulin-like growth factor receptor messenger RNA in the adult rat brain by *in situ* hybridization. *Mol. Endocrinol.* **5,** 1158–1168.
Marks, J. L., Porte, D., Jr., Stahl, W. L., and Baskin, D. G. (1991b). Localization of insulin receptor mRNA in rat brain by *in situ* hybridization. *Endocrinology* **27,** 3234–3236.
Martelli, A. M., Gilmour, R. S., Bertagnoio, V., Nerl, L., M., Manzoli, L., and Cocco, L.

(1992). Nuclear localization and signalling activity of phosphoinositidase C_β in Swiss 3T3 cells. *Nature (London)* **358**, 242–244.

Mathews, L. S., Norstedt, G., and Palmiter, R. D. (1986). Regulation of insulin-like growth factor I gene expression by growth hormone. *Proc. Natl. Acad. Sci. USA* **83**, 9343–9347.

Matsmaga, H., Nishimoto, L., Kojima, L., Yamashita, N., Knrokawa, K., and Ogata, E. (1988). Activation of a calcium-permeable cation channel by insulin-like growth factor II in BALB/c 3T3 cells. *Am. J. Physiol.* **255**, C442–C446.

Matsuguchi, T., Takahashi, K., Ikejiri, K., Ueno, T., Endo, H., and Yamamoto, M. (1990). Functional analysis of multiple promoters of the rat insulin-like growth factor II gene. *Biochim. Biophys. Acta* **1048**, 165–170.

McArdle, C. A., Kohl, C., Rieger, K., Groner, I., and Wehrenberg, U. (1991). Effects of gonadotropins, insulin and insulin-like growth factor I on ovarian oxytocin and progesterone production. *Mol. Cell. Endocrinol.* **78**, 211–220.

McKelvie, P. A., Rosen, K. M., Kinney, H. C., and Villa-Komaroff, L. (1992). Insulin-like growth factor II expression in the developing human brain. *J. Neuropathol. Exp. Neurol.* **51**, 464–471.

Meinsma, D., Holthuizen, P. E., Van den Brande, J. L., and Sussenbach, J. S. (1991). Specific endonucleolytic cleavage of IGF-II mRNAs. *Biochem. Biophys. Res. Commun.* **179**, 1509–1516.

Merimee, T. J., Zapf, J., and Froesch, E. R. (1981). Dwarfism in the pygmy. An isolated deficiency of insulin-like growth factor I. *N. Engl. J. Med.* **305**, 965–968.

Merriman, H. L., La Tour, D., Linkhart, T. A., Mohan, S., Baylink, D. J., and Strong, D. D. (1990). Insulin-like growth factor-I and insulin-like growth factor-II induce c-*fos* in mouse osteoblastic cells. *Calcif. Tissue Int.* **46**, 258–262.

Miller, S. B., Martin. D. R., Kissane, J., and Hammerman, M. R. (1992). Insulin-like growth factor I accelerates recovery from ischemic acute tubular necrosis in the rat. *Proc. Natl. Acad. Sci. USA* **89**, 11876–11880.

Minniti, C. P., Maggi, M., and Helman, L. J. (1992). Suramin inhibits the growth of human rhabdomyosarcoma by interrupting the insulin-like growth factor II autocrine growth loop. *Cancer Res.* **52**, 1830–1835.

Moats-Staats, B., Retsh-Bogart, G. Z., Price, W. A., Jarvis, H. W., D'Ercole, A. J., and Stiles, A. D. (1993). Insulin-like growth factor-I (IGF-I) antisense oligonucleotide mediated inhibition of DNA synthesis by WI-38 cells: Evidence for autocrine actions of IGF-I. *Mol. Endocrinol.* **7**, 171–180.

Mohan, S., Bautista, C. M., Wergedal, J., and Baylink, D. J. (1989). Isolation of inhibitory insulin-like growth factor (IGF) binding protein from bone cell conditioned medium: A potential local regulator of IGF action. *Proc. Natl. Acad. Sci. USA* **86**, 8338–8342.

Monniaux, D., and Pisselet, C. (1992). Control of proliferation and differentiation of bovine granulosa cells by insulin-like growth factor-I and follicle-stimulating hormone *in vitro*. *Biol. Reprod.* **46**, 109–119.

Mooney, R. A., Freund, G. G., Way, B. A., and Bordwell, K. L. (1992). Expression of a transmembrane phosphotyrosine phosphatase inhibits cellular response to platelet-derived growth factor and insulin-like growth factor-I. *J. Biol. Chem.* **287**, 23443–23448.

Morgan, D. O., Edman, J. C., Standring, D. N., Fried, V. A., Smith, M. C., Roth, R. A., and Rutter, W. J. (1987). Insulin-like growth factor II receptor as a multifunctional binding protein. *Nature (London)* **329**, 301–307.

Moxham, C. P., Duronio, V., and Jacobs, S. (1989). Insulin-like growth factor I receptor-subunit heterogeneity. Evidence for hybrid tetramers composed of insulin-like growth factor I and insulin-receptor heterodimers. *J. Biol. Chem.* **264**, 13238–13244.

Mozell, R. L., and McMorris, F. A. (1991). Insulin-like growth factor I stimulates oligodendrocyte development and myelination in rat brain aggregate cultures. *J. Neurosci. Res.* **30,** 382–390.
Murphy, I. J., Murphy, L. C., and Freisen, F. I. G. (1987). Estrogen induces insulin-like growth factor-I expression in the rat uterus. *Mol. Endocrinol.* **1,** 445–450.
Murphy, I. J., Seneviratne, C., Moreira, P., and Reid, R. E. (1991). Enhanced expression of insulin-like growth factor-binding protein-1 in the fasted rat, the effects of insulin and growth hormone administration. *Endocrinology* **128,** 689–696.
Nagpal, M. L., Wang, D., Calkins, J. H., Chang, W., and Lin, T. (1991). Human chorionic gondotropin up-regulates insulin-like growth factor-I receptor gene expression of Leydig cells. *Endocrinology* **129,** 2820–2826.
Nakatani, A., Shimasaki, S., Erickson, G. F., and Ling, N. (1991). Tissue-specific expression of four insulin-like growth factor-binding proteins (1, 2, 3, and 4) in the rat ovary. *Endocrinology* **129,** 1521–1529.
Near, S. L., Whalen, L. R., Miller, J. A., and Ishii, D. N. (1992). Insulin-like growth factor II stimulates motor nerve regeneration. *Proc. Natl. Acad. Sci. USA* **89,** 11716–11720.
Nielsen, F. C., Gammeltoft, S., and Christiansen, J. (1990). Translational discrimination of mRNAs coding for human insulin-like growth factor II. *J. Biol. Chem.* **265,** 13431–13434.
Nishimoto, I., Ogata, E., and Kojima, I. (1987). Pertussis toxin inhibits the action of insulin-like growth factor I. *Biochem. Biophys. Res. Comun.* **148,** 403–411.
Nishimoto, I., Murayama, Y., Katada, T., Ui, M., and Ogata, E. (1989). Possible direct linkage of insulin-like growth factor II receptor with guanine nucleotide-binding proteins. *J. Biol. Chem.* **264,** 14029–14038.
Oemar, B. S., Law, N. M., and Rosenzweig, S. A. (1991). Insulin-like growth factor-1 induces tyrosyl phosphorylation of nuclear proteins. *J. Biol. Chem.* **266,** 24241–24244.
Okamoto, T., Katada, T., Murayama, Y., Ui, M., Ogata, E., and Nishimoto, I. (1990). A simple structure encodes G protein-activating function of the IGF-II/mannose 6-phosphate receptor. *Cell* **62,** 709–717.
Onoda, N., Ohmura, E., Tsushima, T., Ohba, Y., Emoto, N., Isozaki, O., Sato, Y., Shizume, K., and Demura, H. (1992). Autocrine role of insulin-like growth factor (IGF)-I in a human thyroid cancer cell line. *Eur. J. Cancer* **28A,** 1904–1909.
Ooi, G. T., Orlowski, C. C., Brown, A. I., Becker, R. E., Unterman, T. G., and Rechler, M. M. (1990). Different tissue distribution and hormonal regulation of mRNAs encoding rat insulin-like growth factor binding proteins rIGFBP-1 and rIGFBP-2. *Mol. Endocrinol.* **4,** 321–328.
Osborne, C. K., Coronado, E. B., Kitten, L. J., Arteaga, C. L., Fuqua, S. A. W., Ramasharma, K., Marshall, M., and Li, C. H. (1989). Insulin-like growth factor-II (IGF-II): A potential autocrine/paracrine growth factor for human breast cancer acting via the IGF-I receptor. *Mol. Endocrinol.* **3,** 1701–1709.
Osborne, C. K., Clemmons, D. R., and Arteaga, C. L. (1990). Regulation of breast cancer growth by insulin-like growth factors. *J. Steroid Biochem. Mol. Biol.* **37,** 805–809.
Ota, A., Wilson, G. L., and LeRoith, D. (1988). Insulin-like growth factor I receptors on mouse neuroblastoma cells. Two β subunits are derived from differences in glycosylation. *Eur. J. Biochem.* **174,** 521–530.
Påhlman, S., Meyerson, G., Lindgren, E., Schalling, M., and Johansson, I. (1991). Insulin-like growth factor I shifts from promoting cell division to potentiating maturation during neuronal differentiation. *Proc. Natl. Acad. Sci. USA* **88,** 9994–9998.

Pardee, A. B. (1989). G_1 events and regulation of cell proliferation. *Science* **246**, 603–608.
Peraldi, P., Hauguel-De Mouzon, S., Alengrin, F., and Van Obberghen, E. (1992). Dephosphorylation of human insulin-like growth factor I (IGF-I) receptors by membrane-associated tyrosine phosphatases. *Biochem. J. (U.K.)* **285**, 71–78.
Petruzzelli, L., Herrera, R., Arenas-Garcia, R., and Rosen, O. M., (1986). Isolation of a *Drosophila* genomic sequence homologous to the kinase domain of the human insulin receptor and detection of the phosphorylated *Drosophila* receptor with an antipeptide antibody. *Proc. Natl. Acad. Sci. USA* **83**, 4710–4714.
Pietrzkowski, Z., Lammers, R., Carpenter, G., Soderquist, A. M., Limardo, M., Phillips, P. D., Ullrich, A., and Baserga, R. (1992). Constitutive expression of insulin-like growth factor 1 and insulin-like growth factor 1 receptor abrogates all requirements for exogenous growth factors. *Cell Growth Diff.* **3**, 199–205.
Pillay, T. S., Whittaker, J., Lammers, R., Ullrich, A., and Siddle, K. (1991). Multisite serine phosphorylation of the insulin and IGF-I receptors in transfected cells. *FEBS Lett.* **288**, 206–211.
Pons, S., and Torres-Aleman, I. (1992). Basic fibroblast growth factor modulates insulin-like growth factor-I, its receptors, and its binding proteins in hypothalamic cell cultures. *Endocrinology* **131**, 2271–2278.
Powell, D. R., Suwanichkul, A., Cubbage, M. I., DePaolis, L. A., Snuggs, M. B., and Lee, P. D. K. (1991). Insulin inhibits transcription of the human gene for insulin-like growth factor binding protein-1. *J. Biol. Chem.* **266**, 18868–18876.
Rechler, M. M. (1991). Insulin-like growth factor II: gene structure and expression into messenger RNA and protein. *In* "Insulin-Like Growth Factors: Molecular and Cellular Aspects" (D. LeRoith, ed.), pp. 87–110. CRC Press, Boca Raton, Florida.
Rechler, M. M., and Nissley, S. P. (1990). Insulin-like growth factors. *In* "Peptide Growth Factors and Their Receptors" (M. H., Sporn and A. B. Roberts, eds.), pp. 263–282. Springer-Verlag, Berlin.
Recio-Pinto, E., Lang, F. F., and Ishii, D. N. (1984). Insulin and insulin-like growth factor II permit nerve growth factor binding and the neurite formation response in cultured human neuroblastoma cells. *Proc. Natl. Acad. Sci. USA* **81**, 2562–2566.
Reeve, J. G., Brinkman, A., Hughes, S., Mitchell, J., Schwander, J., and Bleehen, N. M. (1992). Expression of insulinlike growth factor (IGF) and IGF-binding protein genes in human lung tumor cell lines. *J. Natl. Cancer Inst.* **84**, 628–634.
Ricciarelli, E., Hernandez, E. R., Hurwitz, A., Kokia, E., Rosenfeld, R. G., Schwander, J., and Adashi, E. Y. (1991). The ovarian expression of the antigonadotropic insulin-like growth factor binding protein-2 is theca-interstitial cell selective: Evidence for hormonal regulation. *Endocrinology* **129**, 2266–2268.
Ricciarelli, E., Hernandez, E. R., Tedeschi, C., Botero, L. F., Kokia, E., Rohan, R. M., Rosenfeld, R. G., Albiston, A. L., Herington, A. C., and Adashi, E. Y. (1992). Rat ovarian insulin-like growth factor binding protein-3: A growth hormone-dependent theca-interstitial cell-derived antigonadotropin. *Endocrinology* **130**, 3092–3094.
Roberts, C. T., Jr., Lasky, S. R., Lowe, W. L., Jr., and LeRoith, D. (1987a). Rat IGF-I cDNAs contain multiple 5′-untranslated regions. *Biochem. Biophys. Res. Commun.* **146**, 1154–1159.
Roberts, C. T., Jr., Lasky, S. R., Lowe, W. L., Seaman, W. T., and LeRoith, D. (1987b). Molecular cloning of rat insulin-like growth factor I complementary deoxyribonucleic acids, differential messenger ribonucleic acid processing and regulation by growth hormone in extrahepatic tissues. *Mol. Endocrinol.* **1**, 243–248.
Rogers, S. A., and Hammerman, M. R. (1988). Insulin-like growth factor II stimulates

production of inositol triphosphate in proximal tubular basolateral membranes from canine kidney. *Proc. Natl. Acad. Sci. USA* **85,** 4037–4041.
Roghani, M., Lassarre, C., Zapf, J., Povoa, G., and Binoux, M. (1991). Two insulin-like growth factor (IGF)-binding proteins are responsible for the selective affinity for IGF-II of cerebrospinal fluid binding proteins. *J. Clin. Endocrinol. Metab.* **73,** 658–666.
Ross, M., Francis, G. L., Szabo, I., Wallace, J. C., and Ballard, F. J. (1989). Insulin-like growth factor (IGF)-binding proteins inhibit the biological activities of IGF-I and IGF-2 but not des-(1–3)-IGF-I. *Biochem. J.* **258,** 267–271.
Rothenberg, P. L., Lane, W. S., Karasik, A., Backer, J., White, M., and Kahn, C. R. (1991). Purification and partial sequence analysis of pp185, the major cellular substrate of the insulin receptor tyrosine kinase. *J. Biol. Chem.* **266,** 8302–8311.
Rotwein, P. S. (1986). Two insulin-like growth factor I messenger RNAs are expressed in human liver. *Proc. Natl. Acad. Sci. USA* **83,** 77–81.
Rotwein, P. S., Pollock, K. M., Didier, D. K., and Krivi, G. G. (1986). Organization and sequence of the human insulin-like growth factor I gene. *J. Biol. Chem.* **261,** 4828–4832.
Rotwein, P., Foltz, R. J., and Gordon, J. L. (1987). Biosynthesis of the human insulin-like growth factor I (IGF-I). The primary translation product of the IGF-I mRNA contains an unusual 48-amino acid signal peptide. *J. Biol. Chem.* **262,** 11807–11812.
Ruoslathi, E., and Pierschlaber, M. D. (1987). New perspectives in cell adhesion: RGD and integrins. *Science* **238,** 491–497.
Russell, W. E., Van Wyk, J. J., and Pledger, W. J. (1984). Inhibition of the mitogenic effects of plasma by a monocloncal antibody to somatomedin-C. *Proc. Natl. Acad. Sci. USA* **81,** 2389–2392.
Samaras, S. E., Hagen, D. R., Shimasaki, S., Ling, N., and Hammond, J. M. (1992). Expression of insulin-like growth factor-binding protein-2 and -3 messenger ribonucleic acid in the porcine ovary, localization and physiological changes. *Endocrinology* **130,** 2739–2744.
Sandberg-Nordqvist, A-C., Ståhlbom, P-A., Lake, M., and Sara, V. R. (1992). Characterization of two cDNAs encoding insulin-like growth factor 1 (IGF-1) in the human fetal brain. *Mol. Brain Res.* **12,** 275–277.
Santisteban, P., Kohn, L. D., and Di Lauro, R. (1987). Thyroglobulin gene expression is regulated by insulin and insulin-like growth factor-I as well as thyrotropin, in FRTL-5 thyroid cells. *J. Biol. Chem.* **262,** 4048–4052.
Santisteban, P., Acebron, A., Polycarpou-Schwartz, M., and DiLauro, R. (1992). Insulin and insulin-like growth factor 1 regulate a thyroid-specific nuclear protein that binds to the thyroglobulin promoter. *Mol. Endocrinol.* **6,** 1310–1317.
Sara, V. R., Carlsson-Skwirut, C., Andersson, C., Hall, E., Sjogren, B., Holmgren, A., and Jornvall, H. (1986). Characterization of somatomedins from human fetal brain: Identification of a variant form of insulin-like growth factor I. *Proc. Natl. Acad. Sci. USA* **83,** 4904–4907.
Schneid, H., Seurin, D., Noguiez, P., and Le Bouc, Y. (1992). Abnormalities of insulin-like growth factor (IGF-I and IGF-II) genes in human tumor tissue. *Growth Regul.* **2,** 45–54.
Senior, P. V., Byrne, S., Brammar, W. J., and Beck, F. (1990). Expression of the IGF-II/mannose-6-phosphate receptor mRNA and protein in the developing rat. *Development* **109,** 67–73.
Shao, Z-M., Sheikh, M. S., Ordonez, J. V., Feng, P., Kute, T., Chen, J-C., Alsner, S., Schnaper, L., LeRoith, D., Roberts, C. T., Jr., and Fontana, J. (1992). IGFBP-3 gene

expression and estrogen receptor status in human breast carcinoma. *Cancer Res.* **52**, 5100–5103.

Shemer, J., Adamo, M., Wilson, G. L., Heffez, D., Zick, Y., and LeRoith, D. (1987). Insulin and insulin-like growth factor-I stimulate a common endogenous phosphoprotein substrate (pp185) in intact neuroblastoma cells. *J. Biol. Chem.* **202**, 15476–15482.

Shemer, J., Adamo, M. L., Roberts, C. T., Jr., and LeRoith, D. (1992). Tissue specific transcription start site usage in the leader exons of the rat insulin-like growth factor-I gene: Evidence for differential regulation in the developing kidney. *Endocrinology* **131**, 2793–2799.

Shimasaki, S., Gao, L., Shimonaka, M., and Ling, N. (1991a). Isolation and molecular cloning of IGF binding protein 6. *Mol. Endocrinol.* **5**, 938–948.

Shimasaki, S., Shimonaka, S., Zhany, H. P., and Ling, N. (1991b). Identification of five different IGFBPs from adult rat serum and molecular cloning of a novel IGFBP-5 in rat and human. *J. Biol. Chem.* **266**, 10646–10653.

Shimatsu, A., and Rotwein, R. (1987a). Mosaic evolution of the insulin-like growth factors: Organization, sequence, and expression of the rat insulin-like growth factor I gene. *J. Biol. Chem.* **262**, 7894–7900.

Shimatsu, A., and Rotwein, P. (1987b). Sequence of two rat insulin-like growth factor I mRNAs differing within the 5' untranslated region. *Nucleic Acids Res.* **15**, 7196–7204.

Sjoberg, J., and Kanje, M. (1989). Insulin-like growth factor (IGF-I) as a stimulator of regeneration in the freeze-injured rat sciatic nerve. *Brain Res.* **485**, 102–108.

Skalli, M., Avallet, O., Vigier, M., and Saez, J. M. (1992). Opposite vectorial secretion of insulin-like growth factor-I and its binding proteins by pig sertoli cells cultured in the bicameral chamber system. *Endocrinology* **131**, 985–987.

Sklar, M. M., Kiess, W., Thomas, C. L., and Nissley, S. P. (1989). Developmental expression of the tissue insulin-like growth factor II/mannose-6-phosphate receptor in the rat: Measurement by quantitative immunoblotting. *J. Biol. Chem.* **264**, 16733–16738.

Skottner, A., Clark, R. G., Robinson, I. C. A. F., and Fryklund, L. (1987). Recombinant human insulin-like growth factor: Testing the somatomedin hypothesis in hypophysectomized rats. *J. Endocrinol.* **112**, 123–132.

Skottner, A., Clark, R. G., Fryklund, L., and Robinson, I. C. A. F. (1989). Growth responses in mutant dwarf rat to human growth hormone and recombinant human insulin-like growth factor I. *Endocrinology* **124**, 2519–2526.

Smith, E. P., Cheung, P. T., Ferguson, A., and Chernausek, S. D. (1992). Mechanisms of sertoli cell insulin-like growth factor (IGF)-binding protein-3 regulation by IGF-I and adenosine 3',5'-monophosphate. *Endocrinology* **131**, 2733–2741.

Soares, M. B., Ishii, D. N., and Efstratiadis, A. (1985). Developmental and tissue-specific expression of a family of transcripts related to rat insulin-like growth factor II mRNA. *Nucleic Acids Res.* **13**, 1119–1134.

Soder, O., Bang, P., Wahab, A., and Parvinen, M. (1992). Insulin-like growth factors selectively stimulate spermatogonial, but not meiotic, deoxyribonucleic acid synthesis during rat spermatogenesis. *Endocrinology* **131**, 2344–2350.

Spaargaren, M., Defize, L. H. K., Boonstra, J., and de Laat, S. W. (1991). Antibody-induced dimerization activates the epidermal growth factor receptor tyrosine kinase. *J. Biol. Chem.* **266**, 1733–1739.

Spaventi, R., Antica, M., and Pavelic, K. (199). Insulin and insulin-like growth factor I (IGF-I) in early mouse embryogenesis. *Development* **108**, 491–495.

Steenbergh, P. H., Koonen-Reemst, A. M. C. B., Clentjens, C. B. J. M., and Sussenbach, J. S. (1991). Complete nucleotide sequence of the high molecular weight human IGF-I mRNA. *Biochem. Biophys. Res. Commun.* **175**, 507–514.
Stiles, C. D., Capone, G. T., Scher, C. D., Antoniades, N. H., Van Wyk, J. J., and Pledger, W. J. (1979). Dual control of cell growth by somatomedins and platelet derived growth factor. *Proc. Natl. Acad. Sci USA* **76**, 1279–1283.
Stracke, M. L., Engel, J. D., Wilson, L. W., Rechler, M. M., Liotta, L. A., and Schiffmann, E. (1989). The type I insulin-like growth factor receptor is a motility receptor in human melanoma cells. *J. Biol. Chem.* **264**, 21544–21549.
Streck, R. D., Wood, T. L., Hsu, M-S., and Pintar, J. E. (1992). Insulin-like growth factor I and II and insulin-like growth factor binding protein-2 RNAs are expressed in adjacent tissues within rat embryonic and fetal limbs. *Dev. Biol.* **151**, 586–596.
Stylianopoulou, F., Efstratiadis, A., Herbert, J., and Pintar, J. (1988). Pattern of the insulin-like growth factor II gene expression during rat embryogenesis. *Development* **103**, 497–506.
Sun, X-J., Rothenberg, P., Kahn, C. R., Backer, J. M., Araki, E., Wilden, P. A., Cahill, D. A., Goldstein, B. J., and White, M. F. (1991). Structure of the insulin receptor substrate IRS-1 defines a unique signal transduction protein. *Nature (London)* **352**, 73–77.
Surmacz, E., Kaczmarek, L., Ronning, O., and Baserga, R. (1987). Activation of ribosomal DNA promoter in cells exposed to insulin-like growth factor-1. *Mol. Cell. Biol.* **7**, 657–663.
Sussenbach, J. S. (1989). The gene structure of the insulin-like growth factor family. *Prog. Growth Factor Res.* **1**, 33–48.
Szabo, L., Mottershead, D. G., Ballard, F. J., and Wallace, J. C. (1988). The bovine insulin-like growth factor (IGF) binding protein purified from conditioned medium requires the N-terminal tripeptide in IGF-I for binding. *Biochem. Biophys. Res. Commun.* **151**, 207–214.
Takasu, N., Takasu, M., Komiya, I., Nagasawa, Y., Asawa, T., Shimizu, Y., and Yamada, T. (1989). Insulin-like growth factor I stimulates inositol phosphate accumulation, a rise in cytoplasmic free calcium, and proliferation in cultured porcine thyroid cells. *J. Biol. Chem.* **264**, 18485–18488.
Talavera, F., and Menon, K. M. J. (1991). Studies on rat luteal cell response to insulin-like growth factor I (IGF-I): Identification of a specific cell membrane receptor for IGF-I in the luteinized rat ovary. *Endocrinology* **129**, 1340–1346.
Talavera, F., Reynolds, R. K., Roberts, J. A., and Menon, K. M. J. (1990). Insulin-like growth factor I receptors in normal and neoplastic human endometrium. *Cancer Res.* **50**, 3019–3024.
Taylor, B. A., and Grieco, D. (1991). Localization of the gene encoding insulin-like growth factor I on mouse Chromosome 10. *Cell* **56**, 57–58.
Telford, N. A., Hogan, A., Franz, C. R., and Schultz, G. A. (1990). Expression of genes of insulin and insulin-like growth factors and receptors in early post-implantation mouse embryos and embryonal carcinoma cells. *Mol. Reprod. Dev.* **27**, 81–92.
Torres-Aleman, I., Pons, S., and Santos-Benito, F. F. (1992). Survival of Purkinje cells in cerebellar cultures is increased by insulin-like growth factor I. *Eur. J. Neurosci.* **4**, 864–869.
Tranque, P. A., Calle, R., Naftolin, F., and Robbins, R. (1992). Involvement of protein kinase-C in the mitogenic effect of insulin-like growth factor-I on rat astrocytes. *Endocrinology* **131**, 1948–1954.

Travers, J. P., Exell, L., Huany, B., Town, E., Lammiman, M. J., Pratten, M. K., and Beck, F. (1992). Insulin and insulinlike growth factors in embryonic development: Effects of a biologically inert insulin (Guinea pig) on rat embryonic growth and development *in vitro. Diabetes* **41,** 318-324.

Treadway, J. L., Morrison, B. D., Goldfine, I. D., and Pessin, J. E. (1989). Assembly of insulin/insulin-like growth factor I hybrid receptors in vitro. *J. Biol. Chem.* **264,** 21450-21453.

Tricoli, J. V., Rall, L. B., Scott, J., Bell, G. I., and Shows, T. B. (1984). Localization of insulin-like growth factor genes to human chromosomes 11 and 12. *Nature (London)* **310,** 784-786.

Trojan, J., Johnson, T. R., Rudin, S. D., Ilan, J., Tykocinski, M. L., and Ilan, J. (1993). Treatment and prevention of rat glioblastoma by immunogenic C6 cells expressing antisense insulin-like growth factor I RNA. *Science* **259,** 94-97.

Tseng, L. Y-H., Brown, A. L., Yang, Y. W-H., Romanus, J. A., Orlowski, C. C., Taylor, T., and Rechler, M. M. (1989). The fetal rat binding protein for insulin-like growth factors is expressed in the choroid plexus and cerebrospinal fluid of adult rats. *Mol. Endocrinol.* **3,** 1559-1568.

Turkalj, I., Keller, U., Ninnis, R., Vosmeer, S., and Stauffacher, W. (1992). Effect of increasing doses of recombinant human insulin-like growth factor-I on glucose, lipid, and leucine metabolism in man. *J. Clin. Endocrinol. Metab.* **75,** 1186-1191.

Ueno, T., Takahashi, K., Matsuguchi, T., Ikejiri, K., Endo, H., and Yamamoto, M. (1989). Multiple polyadenylation sites in a large 3'-most exon of the rat insulin-like growth factor II gene. *Biochim. Biophys. Acta* **1009,** 27-34.

Ullrich, A., Gray, A., Tam, A. W., Yang-Feng, T., Tsubokawa, M., Collins, C., Henzel, W., Le Bon, T., Kathuria, S., Chen, E., Jacobs, S., Francke, U., Ramachandran, J., and Fujita-Yamaguchi, T. (1986). Insulin-like growth factor I receptor primary structure: Comparison with insulin receptor suggests structural determinants that define functional specificity. *EMBO J.* **5,** 2503-2512.

Urban, R. J., Garmey, J. C., Shupnik, M. A., and Veldhuis, J. D. (1990). Insulin-like growth factor type I increases concentrations of messenger ribonucleic acid encoding cytochrome P450 cholesterol side-chain cleavage enzyme in primary cultures of porcine granulosa cells. *Endocrinology* **127,** 2481-2488.

Van den Brande, J. L. (1990). Somatomedins on the move. *Hormone Res.* **32,** 58-68.

Verspohl, E. J., Maddux, B. A., and Goldfine, I. D. (1988). Insulin and insulin-like growth factor I regulate the same biological functions in HEP-G2 cells via their own specific receptors. *J. Clin. Endocrinol. Metab.* **67,** 169-176.

Walker, J. L., Van Wyk, J. J., and Underwood, L. E. (1992). Stimulation of statural growth by recombinant insulin-like growth factor I in a child with growth hormone insensitivity syndrome (Laron type). *J. Pediatr.* **121,** 641-646.

Werner, H., Woloschak, M., Adamo, M., Shen-Orr, Z., Roberts, C. T., Jr., and LeRoith, D. (1989a). Developmental regulation of the rat insulin-like growth factor I receptor gene. *Proc. Natl. Acad. Sci. USA* **86,** 7451-7455.

Werner, H., Raizada, M. K., Mudd, L. M., Foyt, H. L., Simpson, I. A., Roberts, C. T., Jr., and LeRoith, D. (1989b). Regulation of rat brain/HepG2 glucose transport gene expression by insulin and insulin-like growth factor-I in primary cultures of neuronal and glial cells. *Endocrinology* **125,** 314-320.

Werner, H., Shen-Orr, Z., Stannard, B., Burguera, B., Roberts, C. T., Jr., and LeRoith, D. (1990a). Experimental diabetes increases insulin-like growth factor I and II receptor concentration and gene expression in kidney. *Diabetes* **39,** 1490-1497.

Werner, H., Stannard, B., Bach, M. A., LeRoith, D., and Roberts, C. T., Jr. (1990b). Cloning and characterization of the proximal promoter region of the rat insulin-like growth factor I (IGF-I) receptor gene. *Biochem. Biophys. Res. Commun.* **169**, 1021–1027.

Werner, H., Bach, M. A., Stannard, B., Roberts, C. T., Jr., and LeRoith, D. (1992a). Structural and functional analysis of the insulin-like growth factor I receptor gene promoter. *Mol. Endocrinol.* **6**, 1545–1558.

Werner, H., Roberts, C. T., Jr., Raizada, M. K., Bondy, C. A., Adamo, M., and LeRoith, D. (1992b). Developmental regulation of the insulin and insulin-like growth factor receptors in the central nervous system. In "Receptors in the Developing Nervous System" (I. S. Zagon and P. J. McLaughlin, eds.), pp. 109–127. Chapman & Hall, London.

Werner, H., Re, G. G., Drummond, I. A., Sukhatme, V. P., Rauscher, F. J., III, Sens, D. A., Garvin, A. J., LeRoith, D., and Roberts, C. T., Jr. (1993). Increased expression of the insulin-like growth factor I receptor gene, IGF1R, in Wilms' tumor is correlated with modulation of IGF1R promoter activity by the WT1 Wilms' tumor gene product. *Proc. Natl. Acad. Sci. USA* **90**, 5828–5832.

White, M. F., Maron, R., and Kahn, C. R. (1985). Insulin rapidly stimulates tyrosine phosphorylation of a M_r 185,000 protein in intact cells. *Nature (London)* **318**, 183–185.

Wilton, P. (1992). Treatment with recombinant human insulin-like growth factor I of children with growth hormone receptor deficiency (Laron syndrome). *Acta Paediatr. Suppl.* **383**, 137–141.

Wood, T. L., Streck, R. D., and Pintar, J. E. (1992). Expression of the IGFBP-2 gene in post-implantation rat embryos. *Development* **114**, 59–66.

Wozniak, M., Rydzewski, B., Baker, S. P., and Raizada, M. K. (1993). The cellular and physiological actions of insulin in the central nervous system. *Neurochem. Int.* **22**, 1–10.

Yamamoto, K., Lapetina, E. G., and Moxham, C. P. (1992). Insulin like growth factor-I induces limited association of phosphatidylinositol 3-kinase to its receptor. *Endocrinology* **130**, 1490–1498.

Yamashita, S., Ong, J., and Melmed, S. (1987). Regulation of human growth hormone gene expression by insulin-like growth factor I in transfected cells. *J. Biol. Chem.* **262**, 13254–13257.

Yang, J. W., and Fellows, R. E. (1980). Characterization of insulin stimulation of the incorporation of radioactive precursors into macromolecules in cultured rat brain cells. *Endocrinology* **107**, 1717–1722.

Yee, D. Paik, S., Libovic, G. S., Marcus, R. R., Favoni, R. E., Cullen, K. J., Lippman, M. E., and Rosen, N. (1989). Analysis of insulin-like growth factor I gene expression in malignancy: Evidence for a paracrine role in human breast cancer. *Mol. Endocrinol.* **3**, 509–517.

Yee, D., Morales, F. R., Hamilton, T. C., and Von Hoff, D. D. (1991). Expression of insulin-like growth factor I, its binding proteins, and its receptor in ovarian cancer. *Cancer Res.* **51**, 5107–5112.

Zapf, J., Futo, E., Peter, M., and Froesch, E. R. (1992). Can "big" insulin-like growth factor II in serum of tumor patients account for the development of extrapancreatic tumor hypoglycemia? *J. Clin. Invest.* **90**, 2574–2584.

Zenobi, P. D., Graf, S., Ursprung, H., and Froesch, E. R. (1992). Effects of insulin-like growth factor-I on glucose tolerance, insulin levels, and insulin secretion. *J. Clin. Invest.* **89**, 1908–1913.

Zhou, J., and Bondy, C. (1992). Insulin-like growth factor-II and its binding proteins in placental development. *Endocrinology* **131**, 1230–1240.
Zhou, J., Chin, E., and Bondy, C. (1991). Cellular pattern of insulin-like growth factor-I (IGF-I) and IGF-I receptor gene expression in the developing and mature ovarian follicle. *Endocrinology* **129**, 3281–3288.
Zumstein, P., and Stiles, C. D. (1987). Molecular cloning of gene sequences that are regulated by insulin-like growth factor I. *J. Biol. Chem.* **262**, 11252–11260.

Heterologous Expression of G Protein-Linked Receptors in Pituitary and Fibroblast Cell Lines

PAUL R. ALBERT

Department of Pharmacology and Therapeutics
McGill University
Montreal, Quebec, Canada H3G 1Y6

I. Introduction
II. Heterologous Expression Systems
   A. Advantages/Disadvantages
   B. Vectors for Eukaryotic Expression
III. GH Pituitary Cells as Models for Signal Transduction
   A. General Properties
   B. GH Cells as Models of Hormone Action—TRH
   C. $G_{i/o}$-Coupled Receptors in GH Cells
   D. Heterologous Expression of $G_{i/o}$-Coupled Receptors in GH Cells
   E. G Protein Specificity: Antisense Knockouts
IV. Fibroblast Cells as Models for Signal Transduction
   A. Cell-Specific Signaling of $G_{i/o}$-Coupled Receptors
   B. Cell-Specific Signaling of G_s-Coupled Receptors
   C. The Cellular Milieu
   D. Pathway-Selective Modulation
V. Conclusion and Future Prospects
References

I. Introduction

Receptors that couple with heterotrimeric G proteins have a characteristic heptahelical structure, and constitute a class of receptors for hundreds of endogenous amines, peptides, hormones, and odorants (Birnbaumer *et al.*, 1990; Collins *et al.*, 1991; Ostrowski *et al.*, 1992). For many of these receptors, particularly receptors for small molecules such as amines, the transmembrane domains dictate ligand binding specificity, in addition to providing the scaffolding for the three-dimensional structure of the receptor. The intracellular peptide loops, particularly those regions apposed to the plasma membrane, provide the contact points for specific noncovalent interactions that form the receptor–G protein ternary complex. On binding of the agonist to the receptor, an exchange of GTP for GDP at the G protein is catalyzed. Binding of GTP to the G protein α subunit induces dissociation of the ternary complex, releasing an activated α subunit and βγ dimer that interact with various effector molecules to increase or decrease their activity (Gilman, 1987; Conklin and Bourne, 1993). Interaction with the effector enhances the GTPase activity of the G protein to generate the inactive GDP-bound α subunit, which dissociates from the effector and complexes with receptor and βγ dimer to restart the cycle. Thus, the three essential components of receptor-induced signal initiation in this system are the recognition protein (receptor), transducer (G protein), and output amplifier (effector molecules) (Ross, 1989). The specificity of the system appears to be absolute. For example, in all tissues and species examined, the β-adrenergic receptor is coupled to stimulation of adenylyl cyclase by G_s, a cholera toxin-sensitive G protein. However, new findings have diversified this simplistic view of G protein-coupled signal transduction (Birnbaumer et al., 1990; Birnbaumer, 1992).

The application of molecular biology techniques such as low-stringency hybridization and polymerase chain reaction (PCR)-based cloning to the isolation of receptors (Ostrowski et al., 1992), G proteins (Simon et al., 1991), and effectors (Rhee et al., 1988; Tang and Gilman, 1992) has helped reveal the diversity and complexity of signal transduction systems. Although the number of subtypes has increased, common structures (e.g., homology of specific domains) and functions (e.g., coupling to specific G protein families) have allowed classification of receptors and G proteins into discrete families. For example, the family of adrenergic receptors includes both α- and β-adrenergic receptors; G proteins are groups into G_s, G_i, and G_q families; and effectors such as phospholipase C (PLC) are divided into α, β, γ, and δ subgroups. Each of these subgroups has multiple subtypes, for example, β1, β2, and β3 adrenergic receptors, G_i1-3 and G_oA/B, and PLCγ1/γ2. Differences in structure suggest levels of specialization that may play crucial roles in vivo, but have not yet been fully appreciated. At a molecular level, these differences may encode specificity of ligand and protein–protein interactions. These interactions can be assessed in selected subtypes using heterologous expression techniques.

This chapter explores the approaches that have been taken to analyze the specificity and regulation of receptors, G proteins, and effectors. The basic issue is how receptors send signals into cells to direct and coordinate the appropriate cellular responses. Starting at the receptor and progressing into the cell, many advances based on the availability of cloned receptors, G protein subunits, and effectors are resulting in a more coherent and complete view of cellular signaling. Toward this end, the most common technique currently used is the heterologous expression of individual receptor subtypes in receptor-negative cell lines. Several reviews on the diversity of G proteins and the receptors to which they couple have been published. The emphasis of this chapter is on four basic concepts in receptor signaling that have been elucidated through the use of heterologous expression systems: (1) multiple signaling of individual receptors, that is, mediation of multiple signal transduction pathways by a single receptor; (2) G-protein specificities involving differential coupling of receptor subtypes to selected G proteins and G-protein-specific activation of discrete effector pathways; (3) the cell specificity of receptor signaling, emphasizing the contribution of the host cell phenotype (such as endocrine or nonendocrine) to the signaling phenotype of individual receptors; and (4) pathway-selective modulation of receptor signaling, or the diverse effects of modulators such as protein kinases on different signals mediated by a single receptor subtype. Finally, a section on future prospects in receptor signaling and receptor-mediated biological activities concludes the review.

II. Heterologous Expression Systems

A. Advantages and Disadvantages

Although, this chapter focuses on receptors, analogous considerations apply to transfection of any cloned cDNA. Heterologous expression of gene products in clonal cell lines has proven to be a powerful method for the characterization of the activities of these molecules. Cloned DNA encoding specific gene products can be introduced into cells by several methods including calcium phosphate co-precipitation, cationic lipophilic carrier reagents (e.g., Lipofectin, GIBCO; Transfectam, Promega), or electroporation. Numerous investigators have utilized this approach to express cloned, mutated, or chimeric receptors, channels, G proteins, and so on in a variety of cell lines. The procedures we have used for calcium phosphate transfection, isolation, and screening of eukaryotic transfectants are detailed elsewhere (Albert, 1992).

The transfection procedure has allowed studies at a molecular level that could never before have been completed. With respect to receptor–G protein studies, some of the new analyses include:

1. *Receptor specificity.* The study of single receptor subtypes in isolation has led to the appreciation of multiple signals generated by individual receptor subtypes.

2. *Ligand specificity.* The pharmacology and activity (e.g., full or partial agonism, or antagonism) of various compounds on individual receptor subtypes have been characterized.
3. *Cell specificity.* The dependence of receptor pharmacology or responsiveness on the cellular environment has led to the concept of cell-specific signaling.
4. *Biochemical specificity.* The modulation of receptors or their responses by biochemical processes (such as phosphorylation or isoprenylation) has been studied.
5. *Knock-out analysis.* The expression of antisense RNA, or of dominant negative mutants, to produce specific knockouts of individual components of the receptor signaling pathway has allowed for a molecular dissection of the pathways of receptor-induced responses.
6. *Mutational analysis.* Using site-directed mutagenesis and chimeric receptor constructs, structure–function analysis of receptors and G proteins has been carried out delineating sites important for ligand binding and receptor–G protein interactions.

The utility of such approaches is that they allow a hitherto unattainable molecular characterization of receptor–G protein interactions, structure–function relationships, and receptor-mediated signaling pathways. Such molecular dissection should facilitate the understanding of normal and abnormal receptor interactions and the development of new, more selective receptor–G protein-modifying pharmacological agents, as discussed in the following sections.

B. Vectors for Eukaryotic Expression

Heterologous expression of proteins in eukaryotic cells requires isolation of the full coding sequence of the gene of interest (preferably as a cDNA or intronless gene) to obtain the correct translation product. In contrast, for antisense expression, a partial clone or even a strategically placed antisense oligonucleotide may be sufficient to block endogenous protein synthesis (see Section IIIE). For heterologous expression, the cDNA is subcloned in the sense orientation downstream from a viral promoter that is effective in the host cell selected for transfection. Viral promoters such as the Rous sarcoma virus (RSV), Simian virus (SV-40), or cytomegalovirus (CMV) long terminal repeat (LTR) promoters provide a high level of constitutive gene expression in a wide variety of cells. One notable exception is the SV-40 promoter, which competes with PIT-1 nuclear transactivation factors and has been found to be a weak promoter in cells that express PIT-1, such as GH3 or P3 cells (Coleman et al., 1991). Similarly, the herpes simplex virus thymidine kinase (HSV-TK) promoter is found to be inactive in these cells. The promoter with the greatest activity is the CMV promoter, which is found to be active in most cells tested (except NIH-3T3 cells). The RSV promoter is less active, and the mouse metallothionein (mMT) promoter has even less transcriptional activity (Coleman et al., 1991). However, high-expressing stable transfectant clones can be isolated from weak promoters, depending on the upstream DNA elements at the site of genomic integration.

The mMT promoter is an example of an inducible promoter since, on pretreatment with Zn^{2+}, the transcription rate is augmented up to 10- or 20-fold. Another example of an inducible promoter is the glucocorticoid response element (GRE), which confers glucocorticoid-sensitive gene induction. Inducible promoters require the presence of the appropriate inducers (e.g., glucocorticoids and their receptors) for the transcriptional response, so the cell lines chosen must possess the appropriate inducer proteins. The property of inducibility is useful, particularly if the protein to be expressed is constitutively active and could irreversibly alter the host cell (e.g., constitutively active G protein mutants that induce transformation; Kroll et al., 1992). Addition of the inducer allows controlled expression of such proteins for analysis of their signaling pathways.

Finally, a host cell line for transfection must be selected. Several considerations—including the absence or presence of the cDNA to be transfected or its targets, growth rate of the cell, efficiency of transfection, biochemical properties, cell lineage, and inducibility—should be considered. For transient transfection, when high efficiency is crucial to obtaining detectable protein levels, SV-40-transformed cell lines (e.g., COS-1 or COS-7 green monkey kidney cells) are particular useful. These cells express high levels of SV-40 *trans*-acting DNA binding proteins that promote high level expression of cDNA molecules cloned into plasmids containing an upstream SV-40 promoter. For receptor expression, levels of 10–20 pmol/mg protein have been reported in these cells, approximately 10 times higher than in other cell systems that have been stably transfected (e.g., Fargin et al., 1988, 1989; Albert et al., 1990b). However, at such high receptor numbers, the relevance of signal transduction studies to normal cells is unclear. In our studies, we have utilized GH pituitary cells and various fibroblast-derived cell lines, such as Ltk⁻ cells, as hosts for the expression of various receptors to study their signaling pathways in different cellular environments. As discussed in the following sections, the GH cells provide a model of the neuroendocrine epithelial-derived cell environment, whereas the L cells provide a model of the mesenchymal cell environment. We have utilized stable transfection to obtain new transfected cell lines in which to perform detailed biochemical characterizations at reasonable levels of receptor expression.

III. GH Pituitary Cells as Models for Signal Transduction

A. General Properties

1. Hormone Secretion

Initially, GH cells were isolated by Tashjian and colleagues (1968) from pituitary tumors of rats that had been X-irradiated. The cells were cloned, and passaged back and forth in culture and as tumors in animals; the less adherent cells were cloned and examined. The cell lines were initially named on the basis of the synthesis and secretion of growth hormone (GH), but subsequently have been found to secrete prolactin (PRL), and several associated substances such as secretogranin II and sulfated glycoproteins. Since PRL and GH are co-produced and co-secreted by the same individual cells, GH cell tumors are suggested to have arisen from transformation of somatomammotrophs, dual- secreting pituitary cells that predominate in the developing pituitary gland (Frawley and Boockfor, 1991). Additional selection of the GH cells has resulted in the GC and $GH_1 2C_1$ cell lines, which secrete only GH, and the GH3-derived cell lines GH3, GH3B6, and GH4C1 (or GH4) cells, which co-secrete GH and PRL. Although initially subcloned from GH3 cells and selected for high PRL/low GH production, GH4C1 cells now appear to secrete levels similar to those of the parental GH3 cells. However, the GH4C1 cells exhibit a more rounded morphology than GH3 cells. Also, some electrophysiological differences exist between GH3 and GH4C1 cells (see subsequent discussion). GH cells differ from normal pituitary cells in several ways. Unlike normal pituitary cells, GH cells are transformed and cause tumors in injected animals (Tashjian et al., 1968). GH cells store little hormone and have a high basal rate of secretion, which may be the result of a lack of negative regulation since, unlike normal lactotrophs, GH cells lack inhibitory dopamine-D2 receptors (Tashjian, 1979). In certain strains of GH3 cells, but not in GH4C1 cells, dopamine-D2 receptor expression can be induced by treatment with epidermal growth factor (EGF) (Missale et al., 1991a,b). Finally, unlike normal lactotrophs, GH cells lack the cyclooxygenase enzyme and do not metabolize arachidonic acid to prostaglandins (Osborne and Tashjian, 1981).

TABLE I
Characteristics of G Protein-Coupled Receptors in GH Cells

Receptor and ligand	Subtype	K_d (nM)	cAMP	PI	gCa	gK	PRL/GH release	PRL mRNA
G_q-coupled								
TRH	rTRH[a]	0.3[b]	↔[c]	↑[d]	↑[e]	↑K_{Ca}[f], ↓K_V[g]	↑[h]	↑[i]
Bombesin		1.2[j]		↑[k]			↑[l]	
$G_{i/o}$-coupled								
Somatostatin	SSTR 1,2[m]	0.6[n]	↓[o]	↔[p]	↓[q]	↑[r]	↓[s]	↓[t]
Muscarine	M4[u]		↓[v]		↓[w]	↑[x]	↓[v]	
Adenosine	A1		↓[y]		↓[z]		↓[y]	
G_S-coupled								
VIP		2.2[aa]	↑[bb]	↔[cc]	↑[dd]		↑[ee]	↑[ff]
PGE$_2$							↑[gg]	↑[gg]

The characteristics and actions of various ligand receptors in GH3 or GH4C1 cells with respect to specific aspects of GH cell function are indicated by designation or by the direction of arrows. Abbreviations are as in the text; PGE$_2$, prostaglandin E$_2$.

References: [a]Zhao et al., 1992; Sellar et al., 1993; [b]Hinkle and Tashjian, 1973; [c]Drust et al., 1982; Dorflinger and Schonbrunn, 1983a; [d]Martin, 1983; Rebecchi et al., 1983; Schlegel et al., 1984; Aub et al., 1986; Martin et al., 1986; Tashjian et al., 1987; [e]Albert and Tashjian, 1984a,b; Tan and Tashjian, 1984a,b; Armstrong and Eckert, 1987; Gollasch et al., 1991; but see Simasko, 1991b; [f]Ozawa, 1981; Dubinski and Oxford, 1985; Ritchie, 1987; Mollard et al., 1988b; [g]Barros et al., 1985; Dubinski and Oxford, 1985; [h]Albert and Tashjian, 1984a,b; Aizawa and Hinkle, 1985; [i]Murdoch et al., 1983, 1985; [j]Westendorf and Schonbrunn, 1983; [k]Drust and Martin, 1984; [l]Westendorf and Schonbrunn, 1982; [m]Schonbrunn and Loose-Mitchell, 1993; [n]Schonbrunn and Tashjian, 1978; [o]Dorflinger and Schonbrunn, 1983a; Yajima et al., 1986; [q]Mollard et al., 1988a; Kleuss et al., 1991; [r]Yatani et al., 1987; Koch and Schonbrunn, 1988; Mollard et al., 1988b; [s]Dorflinger and Schonbrunn, 1983b; Koch and Schonbrunn, 1988; [t]Schonbrunn and Tashjian, 1978; [u]Pinkas-Kramarski et al., 1990; [v]Wojcikiewicz et al., 1984; Dorflinger and Schonbrunn, 1985; [w]Schlegel et al., 1987; Kleuss et al., 1991; [x]Yatani et al., 1987; Mollard et al., 1988a; White et al., 1991, 1993; [y]Dorflinger and Schonbrunn, 1985; [z]Mollard et al., 1991; [aa]Bjøro et al., 1987; [bb]Gourdji et al., 1979; Dorflinger and Schonbrunn, 1983a; Guild and Drummond, 1984; [cc]Sutton and Martin, 1984; [dd]Albert and Tashjian, 1985; [ee]Gourdji et al., 1979; Dorflinger and Schonbrunn, 1983a; [ff]Carrillo et al., 1985; [gg]Gautvik and Kriz, 1976.

In GH3 and GH4C1 cells, PRL and GH as well as secretogranin I and II are co-secreted in a regulated fashion (Hinkle et al., 1992). The secretory rate is increased by releasing hormones such as thyrotropin releasing hormone (TRH) or vasoactive intestinal peptide (VIP) and is decreased by inhibitory hormones such as somatostatin (Table I). Several investigators have used the cells as vehicles in which to study the regulated secretion of various peptide hormones, and found cosecretion of the transfected gene products with endogenous PRL and GH (e.g., Albert and Liston, 1993). In addition, several of these hormones also regulate the synthesis and gene transcription of PRL and GH (Tashjian, 1979; Murdoch *et al.*, 1985), utilizing PIT-1 promoter elements found on these genes (Day and Maurer, 1990; Iverson *et al.*, 1990; Yan *et al.*, 1991).

2. Receptors

Because of the variety of receptors expressed, GH cells have also provided models in which to study receptor-induced signal transduction mechanisms. A compilation of the G protein-coupled receptors and responses elicited by them in GH3 or GH4C1 cells is presented in Table I. Members of the G_q-coupled (TRH, bombesin), $G_{i/o}$-coupled (somatostatin, muscarinic, adenosine-A1), and G_s-coupled (VIP, PGE_2) receptor families are expressed in GH cells, as are the G proteins to which these receptors couple (Paulssen *et al.*, 1991). The cGMP-mediated opening of potassium channels in response to atrial natriuretic peptide has been described (White *et al.*, 1993). In addition, the GH cells respond to a variety of growth factors including EGF, transforming growth factors α and β (TGFα,β), insulin, and platelet-derived growth factor (PDGF), and to steroid hormones such as estrogen, glucocorticoids, vitamin D, and thyroid hormone (Tashjian, 1979; Schonbrunn *et al.*, 1980; Ramsdell, 1991). In particular, the TRH and somatostatin receptors have provided insight into the mechanisms of stimulus–secretion coupling. For example, TRH is a prototypical G_q-coupled receptor (Aragay *et al.*, 1992) that induces immediate phosphatidyl inositol (PI) turnover to generate inositol trisphosphate (IP_3) and diacylglycerol (DAG). These products, IP3 and DAG act as bifurcating second messengers to induce a spike in increase of cytosolic free calcium concentration ($[Ca^{2+}]_i$) and activation of protein kinase C (PKC), respectively (Berridge and Irvine, 1984,1989), both of which synergize to increase secretion (Albert and Tashjian, 1984b,1985; Martin and Kowalchyk, 1984a,b). Somatostatin receptors couple to $G_{i/o}$ proteins to induce opposite "inhibitory" actions on 3′,5′-cyclic AMP (cAMP) levels, membrane potential, $[Ca^{2+}]_i$, and hormone secretion (Koch *et al.*, 1985). On the other hand, VIP receptors couple to G_s to increase cAMP by activating adenylyl cyclase, and induce increases in $[Ca^{2+}]_i$ and hormone secretion (Koch *et al.*, 1985).

3. Ion Channels and Transporters

One of the key properties of GH cells that distinguishes them from other cell lines is the abundance of voltage- and ion-gated channels.

EXPRESSION OF G PROTEIN-LINKED RECEPTORS   67

Because of the rounded morphology (particularly GH4C1 cells), and small size (14–18 μm diameter) of GH cells these cells provide the ideal environment for whole-cell patch clamp recording, which has been used to characterize many of the ion channels present on the cells. Evidence from ion flux experiments supports the electrophysiological evidence of potassium and calcium channels and the actions of receptor activation on them (Tan and Tashjian, 1984a,b; Koch et al., 1988). Some of the electrophysiological properties of these channels are listed in Table II. These properties strongly resemble the properties of channels expressed in electrically excitable neuronal cells (Spedding and Kenny, 1992). In addition to the ion conductances listed in Table II, a tetrodotoxin-sensitive sodium conductance (Dubinsky and Oxford, 1984) and a calcium-sensitive chloride conductance (Rogawski et al., 1988) have been detected in GH cells. The presence of sodium and calcium channels on the GH cell membranes account for the spontaneous action potentials observed in these cells. The basal activity of these channels regulates resting level of calcium influx and sets the

TABLE II
CHARACTERISTICS OF ION CHANNELS IN GH CELLS

Channel type	Conductance (pS)	Ca^{2+} sensitivity (μM)	V_{act} (mV)	τ (msec)
Ca^{2+} channels				
LVA (T/N)	8–10[a,b]		−33 to	0.3–1.2[a−e]
HVA (L)	23[b]		−40[a−f]	2.5–20[a,c]
			−20[b−f]	
K+ channels				
K(Ca)				
BK (Ca)	150–300[c,g−j]	10[g]		> 1000[c,g−j]
SK (Ca)	10–15[g,k]	1[g]		> 1000[g,k]
VSK (Ca)	4[g]			
K (V)				
A_s				500–3,000[k,l]
A_f				20–80[c,k,l]

The characteristics of calcium and potassium channels identified by patch clamp in GH3 or GH4C1 cells are indicated numerically. Abbreviations: V_{act}, activation potential; τ, half-time of inactivation; LVA, low voltage activated; HVA, high voltage activated; BK (Ca), large conductance, calcium dependent; SK (Ca), small conductance, calcium dependent; VSK (Ca), very small conductance, calcium dependent; K(V) voltage-dependent K+ channels.

References: [a]Hagiwara and Ohmori, 1983; [b]Armstrong and Eckert, 1987; [c]Dubinsky and Oxford, 1984; [d]Matteson and Armstrong, 1986; [e]Cohen and McCarthy, 1987; [f]Herrington et al., 1991; [g]Lang and Ritchie, 1987; [h]Rogawski, 1989; [i]Dufy and Barker, 1982; [j]White et al., 1991; [k]Simasko, 1991a; [l]Rogawski, 1988.

level of basal $[Ca^{2+}]_i$ and resting hormone secretion rate. GH4C1 cells express less sodium current and more calcium current than GH3 cells, and also have predominantly calcium-dependent potassium currents (Dubinsky and Oxford, 1984). Several of the channels present on GH cells are regulated by activation of receptors. For example, calcium-dependent potassium channels mediate the transient hyperpolarization on TRH addition, which is caused by calcium-induced opening of these channels (Dubinsky and Oxford, 1985; Mollard et al., 1988b). Closure of these channels during the sustained phase of TRH action is implicated in the enhancement of calcium influx during this phase (Dubinsky and Oxford, 1985). Thus, the ion channels present on GH cells are analogous to neuronal channels in their electrophysiological characteristics and are modulated by extracellular signaling molecules, such as TRH, as observed in neurons and endocrine cells.

Several ion transporters have also been described in GH cells. A calcium–ATPase activity has been identified (Kaczorowski et al., 1984), as have a sodium/calcium exchanger (Barros and Kaczorowski, 1984) and a sodium/proton exchanger (Hallam and Tashjian, 1987; Tornquist and Tashjian, 1992; Mariot et al., 1993) that is activated by agents that increase PKC activity such as TRH or 12α-tetradecanoyl-4β-phorbol 13-acetate (TPA), a specific activator of PKC (Nishizuka, 1988). Also, the presence of a ouabain-sensitive sodium/potassium–ATPase has been implicated (Scammell and Dannies, 1983). Considering all their features, GH cells provide a reasonable model of neuroendocrine cells because of the numerous properties they share with cells found *in vivo*.

B. GH Cells as Models of Hormone Action—TRH

Studies of the receptors endogenously expressed in GH cells have led to new insights into the pathways of G protein-linked receptor actions, particularly with respect to stimulus–secretion coupling. The TRH receptor is an example of the family of receptors that couples via pertussis toxin (PTX)-insensitive G proteins to the immediate phospholipase C (PLC)-mediated hydrolysis of phosphatidyl inositol bisphosphate (PIP_2), leading to the formation of the dual second messengers IP_3 and DAG (Aub et al., 1986; Martin et al., 1986; Straub and Gershengorn, 1986), as illustrated in Fig. 1. The specific G proteins involved have been identified as G_q and G_{11} using specific anti-α subunit antibodies to block coupling of the receptor to PLC (Aragay et al., 1992). The concept of two temporal phases of PI turnover and signaling was first appreciated in studies of TRH actions on $[Ca^{2+}]_i$ in these cells (Tan and Tashjian, 1981; Albert and Tashjian, 1984a,b; Gershengorn

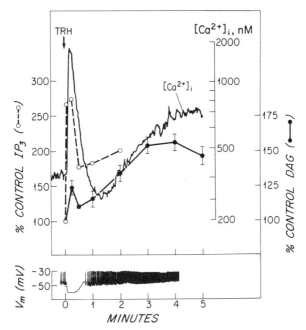

FIG. 1. Time course of the acute actions of thyrotropin releasing hormone (TRH). TRH (0.1 or 1μM) was added at time zero. (Top) Continuous fluorescence recording of Quin 2-loaded GH4C1 cells, as described by Albert and Tashjian (1984a). The level of [^{3}H]I(1,4,5)P$_3$ ○ of duplicate samples is depicted as % control level and represents data replotted with permission from Drummond et al. (1984). Copyright © American Society of Pharmacology and Experimental Therapeutics. The level of [^{3}H]DAG ● represents the mean ± SE of triplicate samples and is redrawn with permission from Macphee and Drummond (1984). Copyright © American Society of Pharmacology and Experimental Therapeutics. (Bottom) Electrophysiological recording of membrane potential (V_m) of a single GH3 cell, as recorded by Ozawa (1981) (redrawn with permission of Elsevier Science Publishers).

and Thaw, 1985; Fig. 1). The first phase involves an acute spike in [Ca^{2+}]$_i$ due to IP$_3$-induced release of calcium stores and subsequent extrusion of the calcium released, resulting in a burst of hormone secretion. The calcium increase is not sufficient to induce the burst in secretion, and synergism between the calcium spike and DAG-induced PKC activation must occur. The second phase is characterized by a prolonged 2-fold increase in PI turnover, which correlated with a sustained increase in calcium influx and sustained enhancement of [Ca^{2+}]$_i$ and hormone secretion. A more detailed model of the actions of TRH discussed in the next sections is presented in Albert and Tashjian

(1984b). An alternative view of TRH action is presented by Gershengorn (1986).

1. *First Phase of TRH Action*

After the addition of TRH, the receptor induces an immediate and dramatic "spike" of increase in $[Ca^{2+}]_i$ and a burst of hormone secretion, which characterizes the first or "spike" phase of TRH action. The TRH-induced increase in $[Ca^{2+}]_i$ appears to coincide with a peak generation of IP_3 and DAG (Fig. 1; Tashjian *et al.*, 1987), the products of TRH-induced PI turnover. These second messengers activate parallel and interacting pathways. IP_3, by activating specific intracellular receptors (Berridge and Irvin, 1989), induces the release of nonmitochondrial calcium stores (Biden *et al.*, 1986), transiently increasing $[Ca^{2+}]_i$ to augment the activity of a variety of calcium-dependent protein kinases, enzymes, and ion channels. DAG mediates activation of PKC and its translocation to the membrane (Fearon and Tashjian, 1987; Martin *et al.*, 1990). Several lines of evidence indicate that the increase in $[Ca^{2+}]_i$ and the generation of DAG induced by TRH-induced PI turnover are both required for the initial burst of PRL and GH secretion:

1. Selective blockade of the calcium spike, but not of IP_3 and DAG formation, blocks TRH-induced secretion (Albert and Tashjian, 1984b).
2. A calcium spike induced by release of cellular calcium using ionomycin (which selectively liberates calcium stores) without concomitant generation of DAG fails to induce the burst of hormone secretion (Albert and Tashjian, 1984b,1985,1986).
3. Mimicking the elevation of DAG using analogs or phorbol esters that specifically activate PKC alone fails to induce burst secretion (Albert and Tashjian, 1985; Albert *et al.*, 1987).
4. Simultaneous addition of ionomycin (to release cellular calcium) and a PKC activator reconstitutes burst secretion, whereas neither agent is effective on its own (Albert and Tashjian, 1985).

These results imply that synergy between the spike of $[Ca^{2+}]_i$ and DAG must occur to induce burst secretion. At what level does this synergy occur? Studies of the translocation of PKC to the membrane, a step thought to be essential for activation of secretion, indicate that calcium and DAG synergize at the level of PKC activation (Fearon and Tashjian, 1987). Thus ionomycin, which induces a calcium spike without altering PI turnover, does not induce PKC translocation. However, ionomycin pretreatment, which depletes TRH-induced release of calcium stores, completely blocks the rapid PKC translocation induced by TRH. The ability of TRH to induce PKC translocation requires simultaneous elevation of $[Ca^{2+}]_i$ and generation of DAG, and strongly correlates with the ability of TRH to induce the burst of hormone secretion. Thus, calcium and DAG synergize to activate and translocate PKC to active membrane sites and to generate the TRH-induced burst of secretion. However, the exact components with which calcium and PKC interact to cause secretion remain to be delineated. A clear understanding of the burst of secretion awaits the discovery of the proteins involved in vesicle fusion.

The immediate and transient membrane hyperpolarization (Fig. 1) that is observed during the spike phase of TRH action appears to result directly from the transient elevation of $[Ca^{2+}]_i$ that characterizes this phase (Mollard et al., 1988b). The elevated $[Ca^{2+}]_i$ triggers the opening of calcium-dependent potassium channels, enhancing gK_{Ca} and resulting in membrane hyperpolarization (Dubinsky and Oxford, 1985; Lang and Ritchie, 1987; Ritchie, 1987). In addition, the elevation of $[Ca^{2+}]_i$ induces a temporary calcium-dependent inactivation of calcium-channnl conductance (Levitan and Kramer, 1990; Kramer et al., 1991). As $[Ca^{2+}]_i$ decays to basal level, gK_{Ca} decreases and the membrane repolarizes. The significance of the transient hyperpolarization in GH cells is unclear, since spike phase secretion appears to result from synergy between increased $[Ca^{2+}]_i$ and DAG, as discussed earlier. However, in neurons, in which the site of receptor activation and the site of secretion are separated, the hyperpolarization would inhibit action potential firing transiently, thus inhibiting secretion from the nerve terminal.

2. Second Phase of TRH Action

The second phase of TRH-induced PRL secretion has been called the "plateau" or sustained phase because of the prolonged enhancement of $[Ca^{2+}]_i$ and hormone secretion that occurs after the transient increase in the first phase (Fig. 1). The enhancements of $[Ca^{2+}]_i$ and secretion during the second phase are blocked by reduction in $[Ca^{2+}]_e$ or by calcium channel antagonists such as dihydropyridines, suggesting the involvement of calcium influx via dihydropyridine-sensitive calcium channels. A corresponding TRH-induced enhancement of calcium-dependent action potential frequency is observed (Fig. 1). A portion of the TRH-induced calcium influx is dihydropyridine resistant (Albert and Tashjian, 1984b) and may involve receptor-operated mechanisms, perhaps mediated by the prolonged elevation of IP_3 after TRH administration (Fig. 1; Tashjian et al., 1987). These data indicate that calcium influx plays a key role in the generation of the second, sustained phase of TRH-induced calcium influx and hormone secretion (Albert and Tashjian, 1984a,b; Tan and Tashjian, 1984b; Dubinsky and Oxford, 1985).

Investigators have proposed that DAG-mediated activation of PKC during the second phase of TRH action induces the electrophysiological changes observed. TRH induces a sustained elevation of DAG (Fig. 1) that could mediate prolonged PKC activation. Activation of PKC by phorbol esters and DAG analogs mimics TRH action on ion channels (Haymes et al., 1992) and $[Ca^{2+}]_i$ (Albert and Tashjian, 1984b,1985). The precise mechanism for the enhancement of calcium-dependent action potentials by TRH is unresolved. Researchers have suggested that TRH-induced inhibition of potassium conductances (gK_v or gK_{Ca}) during the plateau phase leads to enhanced repolarization and the increased frequency of action potentials (Barros et al., 1985; Dubinsky and Oxford, 1985). PKC activation may induce this phase of increase in $[Ca^{2+}]_i$ and secretion, since PKC agonists induce a similar plateau in $[Ca^{2+}]_i$ and secretion without inducing the spike phase. However, mediation of this action of TRH by PKC has been questioned.

1. PKC actually inhibits the magnitude of calcium conductance by 30% (Simasko, 1991b; Haymes et al., 1992), which can account for the transient decrease in $[Ca^{2+}]_i$ caused by TPA and TRH (Albert and Tashjian, 1984b,1985; Drummond, 1985). However, this decrease is transient (lasting 2–3 min) and is followed by a sustained increase in dihydropyridine-sensitive calcium influx during the second phase, which could result from potassium channel inhibition.
2. During second phase, PKC translocates from the membrane fraction (the active state) back to the cytosolic fraction (the presumably inactive state), although the activity of the enzyme during this phase has not been assessed directly (Fearon and Tashjian, 1987; Martin et al., 1990). In contrast, TPA induces a sustained translocation of PKC.

Clearly, the actions of TPA and TRH during this phase are not identical. A calcium dependence exists for sustained secretion induced by TRH that is not seen for TPA (Albert and Tashjian, 1984b; Martin and Kowalchyk, 1984a). Nevertheless, PKC activation by TRH may act indirectly via potassium channel inactivation to promote calcium influx and calcium-dependent hormone secretion. In conclusion, the molecular mechanisms involved in generating the plateau phase of TRH action will require further clarification. However, clearly calcium influx is a necessary component of the ability of TRH to enhance secretion during the second phase.

C. $G_{i/o}$-COUPLED RECEPTORS IN GH CELLS

In GH cells, the most extensively studied inhibitory response is that induced by somatostatin, a well-known inhibitory regulator of GH secretion *in vivo*. Like other $G_{i/o}$-coupled receptors, somatostatin inhibits G_s stimulation of adenylyl cyclase activity by VIP, and consequently inhibits VIP-induced hormone secretion. However, the observation that somatostatin inhibits both basal and VIP-stimulated PRL secretion by cAMP-independent and cAMP-dependent mechanisms, respectively, suggests that somatostatin has actions other than the inhibition of adenylyl cyclase that could account for cAMP-independent inhibition by somatostatin of basal or TRH-induced hormone secretion (Dorflinger and Schonbrunn, 1983a,b; Koch *et al.*, 1985; Yajima *et al.*, 1986). For example:

1. somatostatin hyperpolarizes GH cells, preventing spontaneous action potentials (Schlegel *et al.*, 1985,1987; Mollard *et al.*, 1988a).
2. somatostatin-induced hyperpolarization is mediated by an increase in potassium current due to increased probability of opening of the large conductance (BK) calcium-dependent potassium channel, mediated by a dephosphorylation event (Yatani *et al.*, 1987; Koch *et al.*, 1988; White *et al.*, 1991).
3. somatostatin causes an immediate and sustained decrease in $[Ca^{2+}]_i$ as a result of decreased calcium influx (Schlegel *et al.*, 1985; Koch *et al.*, 1985,1988), mediated in part by the hyperpolarization-induced decrease in calcium influx and in part by a G_o-linked action to close calcium channels (Mollard *et al.*, 1988a; Kleuss *et al.*, 1991).
4. all somatostatin-induced responses are blocked by pretreatment with PTX (Koch *et al.*, 1985), which specifically inactivates $G_{i/o}$ proteins by catalyzing ADP-ribosylation of a conserved C-terminal cysteine residue (Birnbaumer *et al.*, 1990).

The step between G protein and channel has been portrayed as a direct interaction between the G protein α subunit and the ion channel or an associated regulatory subunit, on the basis of reconstitution assays in which purified α subunits activate the potassium channel in membrane patches (Yatani *et al.*, 1987,1988; Birnbaumer *et al.*, 1990). The finding that the serine phosphatase inhibitor okadaic acid blocks somatostatin-induced opening of potassium channels suggests that the relevant α subunits activate a regulatory protein phosphatase that dephosphorylates the potassium channel or associated protein to open the channel (White et al., 1991,1993). Researchers have not yet demonstrated that somatostatin or appropriate G proteins induce dephosphorylation of potassium channels, or whether dephosphorylation of the channel induces channel opening.

Evidence indicates that the actions of somatostatin on ion currents play a key role in somatostatin-induced inhibition of hormone secretion in GH cells.

1. The hyperpolarizing action of somatostatin depends on a potassium gradient, and is abolished by manipulations that eliminate the electrochemical gradient. These manipulations do not affect somatostatin-induced inhibition of cAMP accumulation (Koch and Schonbrunn, 1988).
2. The actions of somatostatin on potassium conductance, $[Ca^{2+}]_i$, and secretion are not abrogated in the presence of excessive elevation of intracellular cAMP, ruling out cAMP inhibition as a necessary mediator of somatostatin action (Koch et al., 1985,1988; White et al., 1991).
3. Blockade of calcium channels abolishes somatostatin action on $[Ca^{2+}]_i$ and secretion, suggesting that the somatostatin-induced closing of calcium channels mediates inhibition of $[Ca^{2+}]_i$ and secretion (Schlegel et al., 1985; Koch et al., 1988).

Whether membrane hyperpolarization or direct closure of calcium channels plays the predominant role in receptor-mediated inhibition of $[Ca^{2+}]_i$ and hormone secretion is unclear. Manipulations that alter K^+ currents, for example, altering K^+ gradients or blocking calcium channels, also affect the basal activity of calcium channels and block somatostatin action on these channels. Thus, the relative importance of enhancement of gK and direct inhibition of gCa in the inhibitory action of somatostatin on $[Ca^{2+}]_i$ and secretion remains to be clarified. Nevertheless, clearly inhibition of calcium influx is the net outcome that mediates inhibitory regulation of hormone secretion by somatostatin.

The PTX sensitivity of all somatostatin actions is consistent with mediation of multiple signaling pathways by $G_{i/o}$ proteins (Koch et al., 1985; Yajima et al., 1986). Specific roles for distinct G proteins were first identified by reconstitution of the somatostatin/carbachol-induced potassium current response by addition of a purified G protein, G_k (i.e., G_i3; Codina et al., 1988), to membrane patches in the presence of GTP (Yatani et al., 1987). The finding that somatostatin and carbachol also induce inhibition of calcium channel opening via distinct G_o protein subtypes indicates that these receptors interact with different G proteins to induce the same action. However, the conclusion that a single receptor can induce multiple pathways is obscured because several somatostatin or muscarinic receptors are coexpressed in GH cells. To avoid this complication, we have studied the actions of individual cloned receptors in GH cells, and found that different G proteins do mediate distinct effector pathways of a single receptor (see subsequent discussion).

D. HETEROLOGOUS EXPRESSION OF $G_{I/O}$-COUPLED RECEPTORS IN GH CELLS

1. *Specificity of the Transfection Method*

To study in detail the signal transduction pathways of other receptors known to inhibit adenylyl cyclase, we took advantage of the GH cell system as a well-characterized cell line possessing endogenous receptors to which response of transfected receptors could be compared and assessed. The most logical receptor to introduce into these lactotroph cells was the dopamine-D2 receptor, a receptor absent in GH cells but present in normal lactotrophs, and the primary regulator of PRL secretion and synthesis *in vivo* (Albert et al., 1990a). Another receptor that we introduced into GH cells was the $G_{i/o}$-coupled 5HT1A receptor (Albert et al., 1990b). The GH4ZR7 line was isolated after stable transfection of the GH4C1 cell line with the rat dopamine-D2S (short form) receptor under transcriptional control of the mMT promoter; the GH4ZD10 cell line was transfected with the rat 5-HT1A receptor. As shown in Fig. 2A, in stable clones transfected with the dopamine-D2S receptor, a major RNA species of 2.5–3 kb was observed, consistent with the size predicted from the plasmid construct. The level of D2 receptor RNA was up-regulated by Zn^{2+} pretreatment, which activates the metallothionein promoter of the expression plasmid, indicating the plasmid-specific expression of the transfected receptor cDNA. Dopamine-D2 receptor RNA was not detected in nontransfected GH4C1 or GH4ZD10 cells, as shown in Fig. 2A. Likewise, 5-HT1A receptor was not present in GH4C1 or GH4ZR7 RNA samples, and was up-regulated by Zn^{2+} pretreatment (Fig. 2B). This result demonstrated the specificity of transfection, and the absence of these receptors in nontransfected cells, contrary to the finding in GH3 cells of an endogenous D2-like receptor (Missale et al., 1991a,b). In all assays,

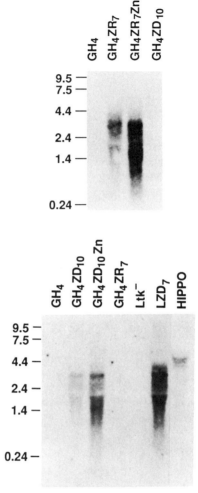

Fig. 2. Specific expression of receptors in transfected cells. 20 μg total RNA was isolated from the indicated GH4C1 transfectant cell lines and electrophoresed as described (Albert et al., 1990a,b). GH4 and Ltk⁻ cell lines are nontransfected, GH4ZR7 cell lines are transfected with the dopamine-D2S receptor; GH4ZD10 and LZD-7 cell lines are transfected with the rat 5-HT1A receptor. HIPPO represents hippocampus. Cells pretreated for 16 hr with 100 μM Zn^{2+} to induce the metallothionein promoter of the transfected cDNA construct are indicated. (**A**) Dopamine-D2 receptor RNA (2.5–3 kb) was detected with a specific cDNA probe under high stringency conditions. (**B**) 5-HT1A receptor RNA transcripts (2.5–3 kb in cell lines, 3.5–4 kb in hippocampal tissue) were detected in duplicate RNA samples using a specific genomic probe (Albert et al., 1990b) under high stringency conditions. The migration of RNA molecular weight markers is indicated.

effects produced by dopamine and serotonin were absent in nontransfected cells or in cells transfected with other neurotransmitter receptors. Thus, the transfection procedure did not induce a hypothetical increase in expression of a cryptic endogenous gene. The only receptor expressed was the one that was transfected.

2. *Multiple Signals and Receptor-Specific Efficacies*

In GH4ZR7 cells, activation of the transfected dopamine D2S receptor inhibited both basal and VIP-stimulated cAMP accumulation with the expected D2 pharmacology, and also inhibited TRH- and VIP-stimulated PRL secretion (Albert et al., 1990a). The D2 receptor also induced membrane hyperpolarization and decreased basal $[Ca^{2+}]_i$, as observed for somatostatin receptors (Vallar et al., 1990). In addition, receptor activation inhibited opening of calcium channels (Y. F. Liu et al., 1994). All these actions were blocked by PTX pretreatment, consistent with the actions anticipated for $G_{i/o}$-coupled receptors. A common set of effectors for these receptors was predicted from the ability of somatostatin, carbachol, and adenosine A1 receptors to induce this set of responses, yet the presence of multiple receptors (e.g., SSTR1 and SSTR2) complicated the interpretation. The induction of the full spectrum of inhibitory responses by a single subtype of dopamine receptor indicated that multiple signals were induced by a single receptor subtype. This phenomenon was further confirmed by the heterologous expression of other $G_{i/o}$-coupled receptors. Like the dopamine-D2 receptor, the cloned rat 5-HT1A receptor also induced the full spectrum of inhibitory responses with the appropriate pharmacology: inhibition of cAMP formation, hyperpolarization, inhibition of calcium channel opening, and decrease in $[Ca^{2+}]_i$ (Albert et al., 1990b; Liu and Albert, 1991; Fowler et al., 1992). Thus, single cloned receptor subtypes induce all the inhibitory responses induced by the endogenous receptors. These cell lines also provide excellent model systems in which to examine in detail the pharmacology of receptor agonists, partial agonists, and antagonists (Fowler et al., 1992; Hoyer and Boddeke, 1993).

The extent of inhibitory responses was greater for the D2 receptor than for somatostatin or muscarinic receptors. For example, the dopamine-D2S receptor in GH4ZR7 gave a stronger inhibition of VIP-stimulated cAMP accumulation than did somatostatin and carbachol (Fig. 3). In combination with dopamine, the inhibition by somatostatin and carbachol remained close to the level for dopamine alone. However, by increasing dopamine receptor number with Zn^{2+} treatment, the level of cAMP inhibition was significantly augmented. This result

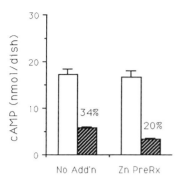
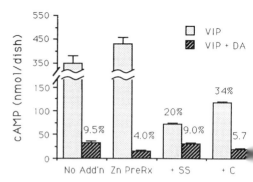

FIG. 3. Different efficacies of receptors for inhibition of cAMP accumulation in D2-expressing GH4ZR7 cells. Triplicate 35-mm dishes of GH4ZR7 cells were incubated for 30 min as described (Albert et al., 1990a). cAMP levels in the medium represent mean ± SE determined by specific radioimmunoassay. Cells were treated with dopamine (shaded bars, 2 μM) or not treated (open bars) in the absence (left) or presence (right) of 250 nM VIP. Addition of dopamine, VIP, somatostatin (SS), or carbachol (C) was during the 30-min assay. Pretreatment with 100 μM added Zn^{2+} (Zn PreRx) was for 16 hr in growth medium to augment dopamine-D2 receptor levels. The % of control for each condition is indicated.

suggested that the dopamine receptor was more effective than the other receptors in inhibiting cAMP, since the number of receptors in the uninduced cells was similar to the number of somatostatin receptors (Albert et al., 1990a). In addition, both short and long forms of the receptor for dopamine (D2S and D2L) completely inhibited cAMP levels and calcium influx induced by BAYK8644 (a calcium channel agonist). For other receptors, such as the transfected 5-HT1A receptor or endogenous somatostatin or carbachol receptors, the inhibitions of this response were partial (Albert et al., 1990b; Liu and Albert, 1991; Liu et al., 1994). These differences in the apparent efficacy of receptor-induced responses suggested that differences in receptor interactions with G proteins may determine efficacy. We used the antisense strategy described in the next section to test this possibility more directly.

E. G Protein Specificity: Antisense Knockouts

1. *Reconstitution, Antibody, and Mutational Approaches*

Several methods have been used to identify the G proteins with which various receptors interact, and to determine which pathways these G proteins mediate. The main problem that each of these approaches must overcome is the high level of conservation between the various G protein subunits (e.g., 88% amino acid identity between $\alpha_i 2$

and α_i1; 94% identity between α_i1 and α_i3; Jones and Reed, 1987), which complicates the generation of specific probes. Reconstitution of effector function with purified subunits allows for determination of the role of preactivation (e.g., using GTP analogs) of particular subunits in regulating the function of particular effectors. However, this approach is limited to electrophysiological measurements, in which small amounts of purified subunits are effective, and is complicated by contamination from related subunits. Thus, concentration-dependence relationships must be established for each subunit to draw conclusions about effector specificity from these studies. The use of cloned bacterially expressed α subunits has eliminated contamination with related subunits. However, these preparations are 10 to 20-fold less potent, presumably because of a lack of posttranslational processing in the bacterial expression systems (Yatani et al., 1988; Birnbaumer et al., 1990).

Another approach is that of G protein-selective antibodies that immunoprecipitate receptor-G protein complexes or block receptor action. Antibodies have been raised against peptides that share sequences (>50% amino acid identity) with different G protein α subunits. Although these antibodies possess subunit selectivity, conflicting results suggest inconsistencies in specificity with different antibodies in different systems. For example, in various reports describing immunoprecipitation of somatostatin receptors with anti-G-protein antibodies, researchers concluded that G_i1 and G_i3 interact with the receptor, whereas G_i2 and G_o do not (Law et al., 1991; Murray-Whelan and Schlegel, 1992). However, agonist-occupied receptor was found to interact with all $G_{i/o}$ proteins (Law and Reisine, 1992), consistent with antisense experiments indicating that G_o mediates somatostatin-induced inhibition of calcium channel opening, (see Table III). The differences in recognition of G-protein subtypes by somatostatin receptors may have arisen because certain antibodies were raised against C-terminal portions of the α subunit that interact with receptors and effectors (Conklin and Bourne, 1993). In addition, the existence of multiple somatostatin receptors with differing G-protein specificities (Law et al., 1993) complicates these antibody studies. Nevertheless, specific G-protein subunit interactions have been described using antibody approaches that are consistent other types of study (Table III).

Other approaches have utilized cell lines in which only a few subtypes of α subunit that are detectable by immunoblot analysis are presumed to be the only ones relevant (e.g., Dell'Aqua et al., 1993; Montmayeur et al., 1993). However that the absence of detection of α subunit means a lack of functional importance is not clear. Low levels

TABLE III
$G_{I/O}$ Protein Coupling Specificities

G Protein	Action/effector	Receptor
$\alpha_o A$ (α_{o1})	↓ VDC[a,b]	M4,[a] OPR,[b] $\alpha 2$AR[b]
$\alpha_o B$ (α_{o2})	↓ VDC[a]	SSTR[a,b]
α_o	↓ VDC[c–g]	OPR,[c] NPYR,[d] $\alpha 2$AR,[e] D2,[f] SSTR,[g] M4,[g] D2S,[g] D2L*[g]
	↔ AC[g–i]	SSTR,[g–i] M4,[g] D2S,[g] D2L[g]
	↔ VDK[f]	D2[f]
$\alpha_i 1$	↓ AC[h–j]	SSTR,[h–j] M4,[j] D2S,[j] D2L,[j] 5-HT1A[j]
	↔VDC,[f] VDK[f]	$\alpha 2$AR,[e] D2[f]
$\alpha_i 2$	↓ AC[g,k–o]	SSTR,[g] M4,[g] D2S,*[g] D2L,[g] OPR,[k] $\alpha 2$,[l] THR[m]
	↔ VDC,[e,f] VDK[f]	D2[f]
	↑ PLC[p]	M2[p]
$\alpha_i 3$	↑ VDK[f,q]	D2,[f] SSTR,[q] M4[q]
	↑ PLC[p]	M2[p]
	↔ VDC[f]	D2[f]
β_1	↓ VDC[r,s]	SSTR[r,s]
β_3	↓ VDC[r]	M4[r]

The results of studies on the activities of individual G protein subunits on various effectors and the receptors that triggered these actions. The direction of regulation is indicated by arrows. Asterisks indicate a partial effect. Abbreviations: VDC, voltage-dependent calcium channels; VDK, voltage-dependent potassium channels; AC, adenylyl cyclase; PLC, phospholipase C; $\alpha 2$AR, α_2-adrenergic; OPR, leu-enkephalin opiate; SSTR, somatostatin; M2, muscarinic-M2; M4, muscarinic-M4; D2, D2S, D2L, dopamine-D2 (short and long forms); THR, thrombin. The experimental approaches used were (1) antisene "knockout" in references a, g, j, m, and r; (2) reconstitution with purified G proteins in references c, d, and q; (3) complementation in reference b; (4) antibody interference in references e, f, h, i, k, l, and s; (5) constitutive mutation in references n and o.
References: [a]Kleuss et al., 1991; [b]Taussig et al., 1992; [c]Hescheler et al., 1987; [d]Ewald et al., 1989; [e]McFadzean et al., 1989; [f]Lledo et al., 1992; [g]Y. F. Liu et al., 1994; [h]Tallent and Reisine, 1992; [i]Law et al., 1993; [j]Y. F. Liu, K. H. Jacobs, M. M. Rasenick, and P. R. Albert, unpublished data; [k]Simmonds et al., 1989; [l]McKenzie and Milligan, 1990; [m]Watkins et al., 1992; [n]Wong et al., 1991; [o]Lowndes et al., 1991; [p]Dell'Acqua et al., 1993; [q]Yatani et al., 1987; [r]Kleuss et al., 1992; [s]Law and Reisine, 1992.

of certain subunits may be sufficient to mediate responses, and not all antisera are of equivalent sensitivity or selectivity in the detection of α subunits. Thus, although these analyses provide suggestive data, they do not provide conclusive evidence for specific receptor–G protein interactions.

The use of mutant α subunits followed the observations that point mutations at sites analogous to those that activate the *ras* oncogene also activate heterotrimeric G protein α subunits, and cause them to be oncogenic in a limited number of tissues and cell lines (Lyons *et al.*, 1990). In particular, the constitutively active mutant of $\alpha_i 2$ (named *gip*-2) has been shown to inhibit cAMP accumulation (Lowndes *et al.*, 1991; Wong *et al.*, 1991; Gupta *et al.*, 1992). However, since these mutants are constitutively active, they do not interact with receptors and are not useful for determining receptor specificity. Another type of mutant, at the C-terminal PTX-sensitive ADP-ribosylation site of $\alpha_o A$, has been used to complement α_o in PTX-treated cells. Since the mutated form is not inhibited by PTX, it remains active. As observed in antisense studies, $\alpha_o A$ (or $\alpha_o 1$) couples certain receptors (opioid, adrenergic-α2) but not others (somatostatin) to inhibition of calcium channels (Taussig *et al.*, 1992). The main caveat of this experiment is the possibility that the mutation has altered interaction of the G protein with receptors or effectors. However, these results are consistent with results from antisense approaches, suggesting that G protein interactions are not greatly altered by the point mutation.

2. Stable Transfection of Antisense G-Protein Subunit cDNA Constructs

An alternative approach to studying G-protein specificity is the use of antisense oligonucleotides or cDNAs to selectively hybridize to the individual mRNAs of each G-protein subunit, leading to a specific "knockout" of that subunit (Fig. 4). Because of the degeneracy of the genetic code, sequences of identical amino acids may have different nucleic acid sequence, suggesting an enhanced specificity of this technique over antibody techniques (Izant and Weintraub, 1985).

a. Antisense Oligonucleotides. The strategy of oligonucleotide injection into single cells (Kleuss *et al.*, 1991,1992,1993; Baertschi *et al.*, 1992; Lledo *et al.*, 1992,1993) is shown schematically in Fig. 4A. Oligonucleotides (15–40 bp) are generally directed 5' to the initiation codon of the target RNA sequence, where they hybridize and interfere with translational initiation. Oligonucleotides directed to unique 3' untranslated sequences are ineffective, presumably because they do not inhibit translation (Kleuss *et al.*, 1991; Lledo *et al.*, 1993). Thus, the DNA–RNA hybrids formed following injection of the oligonucleotides do not appear to be degraded by double-strand-specific RNases, since oligonucleotides that do not block translation would be effective if this mechanism were used. The disadvantage of the technique is that the specificity of the knockout has not been verified by Western blot analy-

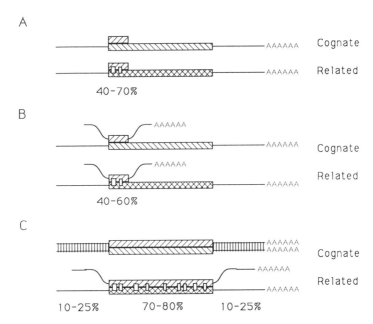

FIG. 4. Various antisense "knockout" strategies that have been used to deplete G protein α subunits are illustrated schematically. The antisense DNA oligonucleotide or RNA transcript (containing poly-A tail) is drawn above and the target RNA molecules (either cognate or related) are drawn below. The translated highly homologous regions are boxed; untranslated low homology regions are lines. Nonhybridizing portions are shown by gaps. (**A**) Injection of antisense oligonucleotides, as done by Kleuss et al. (1991), where the oligonucleotide DNA is directed against the translation initiation site. (**B**) Transfection of an antisense oligonucleotide plasmid construct, where the oligonucleotide is transcribed into RNA and utilizes a poly-A signal on the plasmid, as done by Watkins et al. (1992). (**C**) Transfection of full-length antisense cDNA plasmid contruct in which the full-length cDNA is transcribed as an antisense RNA, as done by Y. F. Liu et al., 1994.

sis, since single cells do not provide sufficient material for such analysis. With up to 70% identity to other α subunits and the requirement of 100-fold molar excess for inhibition, the oligonucleotides may cross-react. However, Lledo et al. (1993) used an elegant in vitro translation assay to assess the specificity of their oligonucleotides and found subtype-specific knockouts in this assay. In addition, Campbell et al., (1993) have reported 75% inhibition of αo immunostaining in cells injected with antisense αo oligonucleotides with no changes in αi staining. Thus, the specificity of injected oligonucleotides can be tested, and appears to be adequate to draw functional conclusions.

Malbon and colleagues (Watkins et al., 1992; Moxham et al., 1993) reported using transfection of oligonucleotides in eukaryotic expression vectors. In this case, a short RNA molecule encoding the antisense oligonucleotide, 5′ and 3′ untranslated sequence of unknown length, and a 3′ poly-A tail was the expected product, and formed an RNA–RNA hybrid at the site of oligonucleotide homology, with the predicted consequences as described for injected oligonucleotides (Fig. 4B). Relying on the hybridization of a short nucleotide sequence, the specificity this approach was limited to the specificity of the oligonucleotide, which had 40–60% identity with other α subunit sequences. However, direct measurement by immunoblot suggested that the latter method could specifically deplete individual α subunit proteins. Indirect pharmacological analyses of cells injected oligonucleotides supported the specificity of this approach.

b. *Antisense cDNA.* We have utilized a third antisense approach: eukaryotic expression of the full-length rat antisense cDNA to G protein α subunits (Y. F. Liu et al., 1994). The advantage of this approach is that hybridization of the full-length antisense cDNA should be of higher specificity because of the major contribution of subunit-specific 5′ and 3′ untranslated regions of low sequence identity (10–25%). These 5′ and 3′ regions are expected to act as anchoring regions to lock the hybridization to the cognate sense RNA and to render unfavorable the hybridization to nonhomologous related RNAs (Fig. 4C). The hybridization of full-length antisense RNA to cognate sense RNA should generate a stable full-length double-stranded RNA that may be degraded by RNase or blocked from translational initiation. Generation of stable cell lines expressing antisense α subunit RNAs has the advantage of providing stable cell lines for the study of multiple responses of various G protein-coupled receptors. With the injection method, only single cell measurements (such as patch clamp) can be performed. A standard transfection method was used, except that stringent hygromycin B resistance was used for selection since the host cell lines (GH4ZR7, GH4D2L, and GH4ZD10) were already G418 resistant from previous transfection of receptor cDNA constructs (Y. F. Liu et al., 1994). Immunoblot analysis of cells transfected with antisense α_o construct showed that α_o protein was depleted to undetectable levels (<5% of control levels). The levels of α_i and β subunits were not altered by this treatment, indicating the specificity of knockout. More convincingly, knockout of α_i2 did not alter the level of closely related α_i3 protein (Y. F. Liu et al., 1994).

Having obtained cell lines depleted by >95% of G_o protein, inhibition of BAYK8466-induced calcium channel opening was examined. This response was abolished for somatostatin, muscarinic, and dopamine receptors in GH4ZR7 (D2S-expressing) cells (Y. F. Liu et al., 1994). In the same cells, receptor-mediated inhibition of cAMP synthesis was unaltered. Thus, we observed a complete dissociation: α_o mediated complete inhibition of calcium channel activation, but was not coupled to basal or stimulated states of adenylyl cyclase in GH cells. The exception was the D2L receptor, which retained partial coupling to calcium channel inhibition in α_o-depleted cells, suggesting that other subunits may be recruited by this receptor. Interestingly, this system provides a clear example of a difference in the coupling of the two forms of D2 receptor, which has been difficult to show (but see Hayes et al., 1992). Using GH cells transfected with antisense $\alpha_i 2$ constructs, an opposite pattern of coupling was observed (Y. F. Liu et al., 1994), that is, in antisense $\alpha_i 2$ cells, somatostatin, muscarinic, and dopamine receptors displayed impaired coupling to inhibition of adenylyl cyclase, but retained complete inhibition of calcium channels. The exception was the dopamine D2S receptor, for which the impairment was only 30%, demonstrating a second difference in coupling between the two D2 receptors. Using heterologous expression of D2 receptors in $G_i 2$-deficient JEG-2 cells, a similar conclusion was reached regarding the requirement of $G_i 2$ for D2L but not D2S signaling to inhibit adenylyl cyclase activity (Montmayeur et al., 1993).

In antisense $\alpha_i 2$ GH cells, somatostatin unexpectedly enhanced VIP-induced stimulation of cAMP levels. Thus, the regulation of basal and stimulated states of adenylyl cyclase can be dissociated and may involve different G proteins. The stimulation produced by somatostatin could be due to $\beta\gamma$-mediated potentiation of α_s-activated adenylyl cyclase II or IV (Tang and Gilman, 1991, 1992). These cell lines should provide useful tools for further analysis of the biochemical actions of inhibitory hormones, such as inhibition of hormone secretion, gene transcription, or cell proliferation, which cannot be readily measured in single cells (but see LaMorte et al., 1993).

Table III summarizes the actions of different $G_{i/o}$ proteins using the approaches discussed earlier. Some conclusions can be drawn from the general agreement of results utilizing different techniques. G_o negatively regulates the activity of calcium channels, as shown by reconstitution, antibody, and antisense experiments; $G_i 1$, -2, and -3 do not, as seen in antisense experiments. $G_i 3$ positively regulates potassium channels and, in certain tissues, other G_i proteins participate. G_i proteins are all involved in the inhibition of adenylyl cyclase, but G_o is not. Perhaps most interestingly, in GH cells, somatostatin receptors couple to the specific combination of $\alpha_o 2/\beta 1/\gamma 3$ to inhibit calcium channels, whereas muscarinic receptors couple to $\alpha_o 1/\beta 3/\gamma 4$ to regulate these channels (Kleuss et al., 1993).

In addition to the actions tabulated, evidence exists of other interactions of this G protein family. G_o interacts with the neuronal cytoskeletal protein GAP-43 (Strittmatter et al., 1990, 1991) and with amyloid protein (Nishimoto et al., 1993). $G_i 3$ interacts with ras (Yatani et al., 1990a,b), a small G protein involved in growth signaling. However, the biological significance of these interactions is unclear at present.

IV. Fibroblast Cells as Models for Signal Transduction

Several fibroblast- or mesenchyme-derived tumor cell lines have been used as hosts for transfection of receptors, often because of their convenience. These cells were selected for high expression in transient and stable transfections from a variety of viral promoters, high efficiency of transfection, rapid growth, and ease of maintenance. We have used two such cell lines in our studies: transformed mouse Ltk⁻ cells and nontransformed mouse BALB/c-3T3 cells. In contrast to GH cells, these cells do not produce hormones and appear to lack a regulated secretory pathway. L cells also lack calcium channels (Perez-Reyes et al., 1990), although calcium channels have been observed in BALB/c-3T3 cells. These cells possess a limited number of G protein-coupled receptors, including thrombin (coupled to $G_{i/o}$ and to G_q), P2-purinergic (coupled to G_q), and prostaglandin E_1 (PGE_1) receptors (coupled to G_s), that can be used to study signal transduction in these cells (Liu and Albert, 1991; Abdel-Baset et al., 1992; Guderman et al., 1992). The most striking finding is that thrombin induces inhibition of cAMP accumulation, PI turnover, and intracellular calcium mobilization pathways (Pouysségur and Seuwen, 1992). In GH pituitary cells, stimulation of the PI pathway by a $G_{i/o}$-coupled receptor has never reported, but an indirect inhibition of this pathway due to decrease in calcium has been shown (Vallar et al., 1990). To study further the differences in signaling between these cells, cloned dopamine-D2 and 5-HT1A receptors were transfected into these cells.

A. Cell Specific Signaling of $G_{I/O}$-Coupled Receptors

1. Multiple Signals in Fibroblast Cells

Since the ability of thrombin to stimulate PI turnover may have been restricted to a subtype of thrombin receptor, we examined the actions of transfected $G_{i/o}$-coupled dopamine-D2 (Vallar et al., 1990; Liu et al., 1992a) and 5-HT1A receptors (Liu and Albert, 1991; Abdel-Baset et al., 1992) separately in these cells. Activation of either receptor inhibited forskolin-stimulated cAMP levels without altering the basal cAMP level. As observed in GH cells, the dopamine-D2 receptor gave a strong, nearly complete reduction in cAMP, whereas the reduction induced by the 5-HT1A receptor was about 50%. In addition, the D2 receptor induced membrane hyperpolarization in the L cells, as in the GH cells. However, unlike in GH cells, both receptors induced an immediate increase in $[Ca^{2+}]_i$ that decayed within 1–2 min. This calcium increase was associated with an increase in PI turnover, as measured by increased [^{3}H]inositol phosphate (IP_3 and IP_4) formation. All these actions were abolished by PTX, suggesting mediation by $G_{i/o}$ proteins. In addition, the EC_{50}'s for the receptor-mediated PI response and inhibition of cAMP accumulation were similar, suggesting that both responses were induced in parallel. We transfected several $G_{i/o}$-coupled receptor subtypes into L or BALB/c-3T3 cells, including human 5-HT1B and 5-HT1D receptor cDNAs (courtesy B. K. O'Dowd, Clarke Institute, Toronto, Canada), and found that these receptors also induced a PI-linked calcium increase (Fig. 5A). In transfected BALB/c-3T3 cell lines expressing different numbers of 5-HT1B receptors, we found a correlation between the magnitude of the receptor-mediated increase in $[Ca^{2+}]_i$ and inhibition of forskolin-stimulated cAMP level (Fig. 5B), supporting the notion that similar numbers of receptor regulate PI turnover and cAMP inhibition in parallel. However, since inhibition of cAMP is absent in resting cells, we proposed that the PI pathway mediated receptor-induced signaling in the basal state (Liu and Albert, 1991). This hypothesis contrasts with findings in HeLa cells, in which the EC_{50} for the PI response was 10-fold higher than that for cAMP inhibition and required a larger complement of receptors to be elicited (Fargin et al., 1989). The greatest difference in signaling was between fibroblast cells and pituitary GH cells, in which no stimulation of PI or $[Ca^{2+}]_i$ was ever observed, even after Zn^{2+} induction to increase the number of transfected receptors. The cell-specific nature of $G_{i/o}$-coupled receptors is illustrated in the model (Fig. 6), in which the inhibitory pathways of decrease in $[Ca^{2+}]_i$, cAMP, and secretion dominate in the pituitary cell whereas the stimulatory pathway of PTX-sensitive PI turnover regulates the resting fibroblast cell.

We propose that fibroblast cells express a "switch factor" that acts to couple $G_{i/o}$ proteins to PLC. This factor appears to be a protein present in fibroblasts, since fusion experiments with L and GH cells have shown that the PI pathway was activated by agonist addition in all

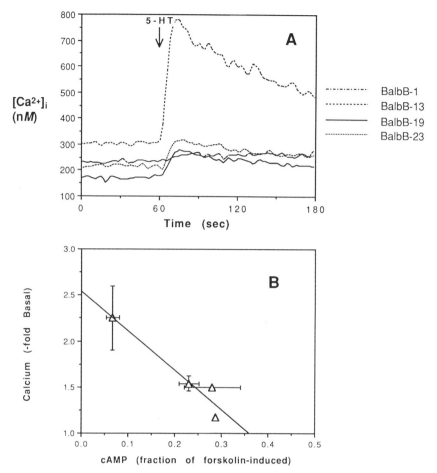

FIG. 5. Parallel between calcium increase and cAMP inhibition in different BALB/c-3T3 clones expressing the 5-HT1B receptor. (A) A series of BALB/c-3T3 cell lines transfected with the 5-HT1B receptor were isolated and tested for calcium response induced by 1 µM 5-HT addition at 60 sec using the fura-2 calcium assay, as described by Liu and Albert (1991). (B) The cell lines in A were assayed for inhibition of forskolin-induced cAMP accumulation. The ordinate plots the mean ± SE of the increase in calcium levels for each clone; the abscissa is the mean ± of normalized inhibition of forskolin-induced cAMP accumulation (level of cAMP in the presence of 1 µM 5-HT + 10 µM forskolin/cAMP level with 10 µM forskolin alone). The linear relation had $r = 0.93$, $y_{MAX} = 2.54$-fold, and $X_{int} = 0.35$.

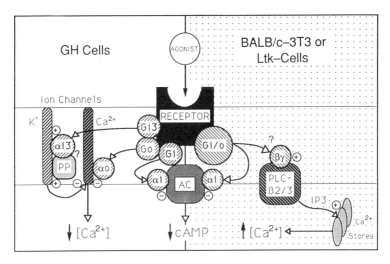

FIG. 6. Cell-specific signaling of G_i/G_o-coupled receptors. Agonist binding to receptors, such as somatostatin, muscarinic M4, dopamine-D2, 5-HT1A, and so on, initiates the GTP-dependent activation and dissociation of G_i/G_o proteins. In GH cells, $\alpha_i 3$ activates K+ channels, possibly involving intermediate activation of a serine protein phosphatase (PP), to hyperpolarize the cell and inactivate calcium channel opening; α_o directly inactivates or indirectly inactivates calcium channels; together, these actions decrease calcium influx and lower $[Ca^{2+}]_i$. Various α_is, depending on the receptor, inhibit adenylyl cyclase (AC) activity to decrease cAMP levels. In fibroblast BALB/c-3T3 or Ltk$^-$ cells, receptor activation inhibits adenylyl cyclase activity via α_i and enhances phosphoinositol (PI) turnover, perhaps via $\beta\gamma$-dependent activation of phospholipase C (PLC)-β_2 or $\beta 3$, as proposed in the text. Activation of PLC produces inositol triphosphate (IP$_3$) which releases intracellular calcium stores to increase $[Ca^{2+}]_i$. DAG, which activates protein kinase C (PKC; not shown) with feedback, inhibits coupling to PLC. See references in text. Adapted with permission from Liu and Albert (1991). Copyright © The American Society for Biochemistry & Molecular Biology.

selected fusion cell lines expressing the transfected dopamine-D2 or 5-HT1A receptor (P. R. Albert and J. Yantsulis, unpublished results). If the factor was an inhibitory protein from GH cells, no PI response in fusion cell lines would be predicted. This factor does not appear to be a new PTX-sensitive G protein α subunit, since transient transfection of 5-HT1A-expressing L cells with antisense RNA for $\alpha_i 2$, but not for the other subtypes, partially blocks the PI pathway (Y. F. Liu and P. R. Albert, unpublished results), suggesting that $\alpha_i 2$ mediates the PI response in L cells. Since $\alpha_i 2$ is functional in GH cells (i.e., for inhibition of cAMP levels) but does not mediate PI turnover, some protein that interacts with $\alpha_i 2$ in L cells must mediate the PI response.

As illustrated in the model (Fig. 6), evidence describing the activation of PLC-β2 and PLC-β3 by βγ subunits derived from PTX-sensitive G proteins suggests a pathway involving these molecules (Birnbaumer, 1992; Camps et al., 1992; Katz et al., 1992). The current hypothesis is that the "switch factor" is a specific βγ combination or PLC-β subtype found in L cells but lacking in GH cells. For example, if L cells express PLC-β2 or -β3, PTX-sensitive activation of PI turnover βγ dimers would be observed (Lee et al., 1992; Park et al., 1993). The expression of low levels of the switch factor in certain cell types (e.g., HeLa cells) might explain why high levels of receptor are required to activate the PI pathway. Lack of expression of these PLC isozymes should cause complete loss of the PI pathway, as seen in GH cells. Additional work is necessary to identify the switch factors that mediate the PI pathway in fibroblast cells.

2. *Cell-Specific Actions on Gene Transcription and Cell Proliferation*

a. *Gene Transcription.* The cell-specific signaling properties of the dopamine-D2 and 5-HT1A receptors would be expected to produce differences in cellular activity. When expressed in GH cells, the dopamine-D2 receptor mediates inhibition of endogenous PRL gene transcription (Elsholtz et al., 1991). This action requires the presence of PIT-1 DNA binding elements located in the proximal promoter of the PRL gene. Inhibition is abolished if these elements are removed. Receptor activation also decreases the activity of the PIT-1 promoter, suggesting that coordinate inhibition of PIT-1 expression and action is responsible for dopamine-induced inhibition of PRL gene transcription. Interestingly, the inhibitory action of dopamine appears to be mediated by inhibition of cAMP accumulation and by membrane hyperpolarization.

A clear contrast occurs in L cells transfected with dopamine-D2 receptors (Lew and Elsholtz, 1993). In the absence of PIT-1 (L cells lack endogenous PIT-1), the PRL promoter reporter constructs are not transcribed efficiently. Transfection and expression of PIT-1 allows efficient PRL gene transcription. In L cells expressing the D2 receptor and transfected with PIT-1, dopamine stimulates PRL gene transcription, presumably by a dopamine-induced stimulation of PI turnover. Thus, the switch in signal transduction from inhibition to stimulation translates into a stimulation of gene transcription in responsive genes. However, as yet, the actions of D2 receptor activation on endogenous gene expression in L cells and the role of the PI-linked pathway in such actions have not been examined.

b. *Cell Proliferation.* In GH cells transfected with the dopamine-D2 receptor (GH4ZR7 cells), dopamine has been shown to enhance crude tyrosine phosphatase activity (Florio et al., 1992). This pathway is blocked by pretreatment with PTX and represents another signal transduction pathway of the D2 receptor. Since tyrosine phosphorylation/dephosphorylation is associated with growth factor action (Liebow et al., 1989; Crews and Erikson, 1993; Davis, 1993; Solomon, 1993), researchers have proposed that this pathway may be associated with regulation of cell growth. Interestingly, in D2-expressing GH cells, dopamine decreases DNA synthesis and reduces cell proliferation, actions that correlate with the activation of tyrosine phosphatase (Florio et al., 1992). In contrast, in GH cells transfected with the G_s-coupled dopamine-D1 receptor (Civelli et al., 1993), dopamine does not activate tyrosine phosphatase and does not alter DNA synthesis. However, that the D2-regulated "inhibitory" growth response is solely mediated by $G_{i/o}$-coupled receptors is not clear, since receptors that stimulate secretion (e.g., EGF or TRH) also inhibit DNA synthesis in GH4C1 cells (Schonbrunn et al., 1980). Actions of these receptors on tyrosine phosphatase activity have not been reported.

To examine the actions of $G_{i/o}$-coupled receptors in fibroblast cells, we used nontransformed BALB/c-3T3 cells transfected with the 5-HT1A receptor (Abdel-Baset et al., 1992). Unlike transformed L cells, BALB/c-3T3 cells can be grown to a density-dependent growth arrest in low serum-containing medium. Thus, the actions of various agents on DNA synthesis can be measured. Both thrombin and 5-HT increase DNA synthesis in the transfected cells, but only thrombin is effective in nontransfected cells. These actions, as well as the basal rate of DNA synthesis, are greatly reduced by PTX treatment. Exposure of the 5-HT1A transfected cells to 1 μM 5-HT for 10–14 days leads to the formation of foci of morphologically transformed cells that are not observed in nontransfected BALB/c-3T3 cells or in transfected cells not treated with 5-HT, and are greatly retarded in cells treated with PTX. This result indicates that the 5-HT1A receptor stimulates cell growth and transformation via a PTX-sensitive pathway. Since the 5-HT1A receptor induces the PI turnover pathway in these cells, this stimulatory pathway may initiate enhanced cell proliferation in BALB/c-3T3 cells. However, in NIH-3T3 cells transfected with the 5-HT1A receptor, no stimulation of PI turnover or $[Ca^{2+}]_i$ is detected, yet 5-HT also induces DNA synthesis and causes focus formation and transformation (Varrault et al., 1992). This result suggests that pathways other than PI turnover (perhaps via a tyrosine phosphatase?) mediate 5-HT1A receptor action to enhance cell proliferation. Although the specific pathways involved in the actions of $G_{i/o}$-coupled receptors on cell growth remain to be elucidated (Chambard et al., 1987; Cui et al., 1991; Pouysségur and Seuwen, 1992), the receptors mediate opposite effects on growth depending on the cell type in which they are expressed. In GH cells, an inhibitory phenotype is observed, whereas in fibroblast cells (BALB/c and NIH-3T3 cells), a stimulatory phenotype is described. Whether a switch in the signaling of G protein-coupled receptors plays a role in the normal growth and differentiation processes of a single cell type is unknown. Addressing this problem will require an understanding of the components that determine cell-specific receptor signaling.

B. CELL-SPECIFIC SIGNALING OF G_s-COUPLED RECEPTORS

In addition to $G_{i/o}$-coupled receptors, signal transduction of the G_s-coupled dopamine-D1 receptor expressed in transfected GH4C1 and Ltk$^-$ cells has been examined in detail (Liu et al., 1992a). The actions of G_s-coupled receptors are summarized in Fig. 7. When expressed in

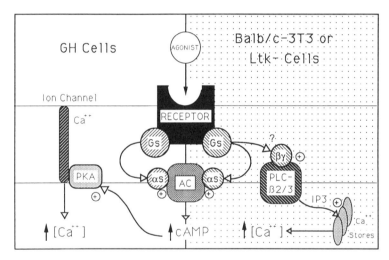

FIG. 7. Cell-specific signaling of G_s-coupled receptors. Agonist binding to receptors, such as VIP or dopamine-D1, initiates the GTP-dependent activation and dissociation of G_s proteins. In GH cells, α_s activates adenylyl cyclase (AC) to increase cAMP levels. Increased cAMP activates protein kinase A (PKA), which catalyzes a phosphorylation-mediated opening of calcium channels to increase [Ca^{2+}]$_i$. In fibroblast BALB/c–3T3 or Ltk$^-$ cells, receptor activation enhances adenylyl cyclase activity via α_s; phosphoinositol (PI) turnover is also enhanced to varying degrees, depending on the receptor, perhaps via a $\beta\gamma$-dependent activation of phospholipase C (PLC)-β2 or β3, as proposed for $G_{i/o}$-coupled receptors. Activation of PLC produces inositol triphosphate (IP$_3$), which releases intracellular calcium stores increasing [CA^{2+}]$_i$. DAG, which activates protein kinase C (PKC; not shown) with feedback, inhibits coupling to PLC. PKA activation enhances D1 receptor coupling to PLC. See references in text. Adapted with permission from Liu and Albert (1991). Copyright © The American Society for Biochemistry & Molecular Biology.

GH4C1 cells, activation of the dopamine-D1 receptor induces a large (10-fold) stimulation of cAMP levels and a small (2-fold), sustained increase in $[Ca^{2+}]_i$ that is maximal 1–2 min after dopamine addition. The D1-induced calcium response is mediated by calcium influx involving the activation of dihydropyridine-sensitive calcium channels. Via activation of endogenous G_s-coupled receptors, VIP also increases cAMP levels and enhances calcium influx in GH cells (Table I). Pretreatment of the cells with cholera toxin (CTX), which ADP-ribosylates G_s causing maximal activation (Gilman, 1987), elevates cAMP levels. No further elevation is elicited by D1 receptor activation. Similarly, the D1 receptor does not induce a calcium signal in CTX-treated cells, implying that the activation of calcium influx utilizes a G_s-coupled pathway. The role of cAMP in the D1-induced calcium influx has been examined by addition of cAMP analogs or forskolin to elevate cAMP levels. These compounds mimic D1-induced elevation of $[Ca^{2+}]_i$ and prevent further increase in $[Ca^{2+}]_i$ by dopamine. This result indicates that D1-induced activation of calcium channels is cAMP dependent, and may be mediated by a cAMP-dependent phosphorylation event at the calcium channel (Armstrong and Eckert, 1987). Although direct coupling of α_s to calcium channels has been proposed to occur in certain tissues (e.g., cardiac cells; Birnbaumer et al., 1990), indirect coupling via cAMP may mediate α_s action even in these tissues (Hartzell and Fischmeister, 1992).

When expressed in L cells, D1 receptor activation increases cAMP levels, as observed in GH cells, and produces a small transient increase in $[Ca^{2+}]_i$ that is mediated by mobilization of intracellular calcium pools (Liu et al., 1992a). The calcium signal is associated with a dopamine-induced increase in [³H]inositol phosphate formation, and is not mimicked by 8-bromo(8-Br)-cAMP or by elevation of cAMP with forskolin or by PGE_1. Instead, these compounds enhance the D1-mediated calcium mobilization rather than preventing it. Pretreatment with CTX abolishes the calcium response, suggesting that a G_s-coupled pathway is involved, since CTX treatment does not alter other $G_{i/o}$- or G_q-coupled receptor signaling pathways (Gilman, 1987). CTX maximally activates G_s, preventing further receptor-mediated activation. Hence, the D1-induced PI and calcium signals are blocked. Since elevation of cAMP using forskolin or 8-Br-cAMP does not mimic CTX-induced blockade of the D1 response, we conclude that the action on PI turnover is not indirect via cAMP-induced actions, but is the result of direct coupling of G_s to PLC (Fig. 7). Ultimately, antisense knockout of G_s could be used to confirm this preliminary conclusion.

Activation of a PI pathway by transfected G_s-coupled receptors has been observed by other researchers in fibroblast or kidney cell lines (Van Sande et al., 1990; Abou-Samra et al., 1992; Gudermann et al., 1992), but not in GH cells (Liu et al., 1992). Thus, as observed for $G_{i/o}$-coupled receptors, GH cells appear to lack receptor-induced coupling to PLC, with the exception of G_q-mediated coupling (e.g., by the TRH or bombesin receptors). A mechanism analogous to that suggested for the $G_{i/o}$-coupled receptors may be occurring. Namely, the βγ dimer of G_s may activate PLC-β2 or -β3 to induce PI turnover in L cells, whereas this enzyme may be lacking in GH cells. Clearly, many gaps must be filled before a more complete understanding of the cell-specific signaling of G_s-coupled receptors is reached. However, for many G_s-coupled receptors, the activation of the PI pathway in target tissues is of profound significance for hormone action. In particular, the cloned thyrotropin (TSH), luteotropin (LH), and parathyroid hormone (PTH) receptors have all been shown to couple strongly to both enhancement of cAMP accumulation and PI turnover (Van Sande et al., 1990; Abou-Samra et al., 1992; Gudermann et al., 1992). A receptor specificity exists in the ability of G_s-coupled receptors to signal PI turnover: TSH, LH, and PTH receptors couple strongly, D1 is intermediate, and vasopressin-V2 and PGE_1 appear to couple weakly, if at all. Putative differences in the G_s proteins to which these receptors couple may explain these differing efficacies. The CTX-sensitivity of the TSH-, LH-, and PTH-mediated PI responses has not been reported, although the G protein which mediates this pathway is proposed to be of the G_q family.

C. THE CELLULAR MILIEU

Previously, cellular response was assumed to be determined solely by the receptor subtype expressed on the cell surface. For example, the D1 subtype was classified as stimulatory, by increasing cAMP, and the D2 receptor subtype was classified as inhibitory, by decreasing cAMP (Civelli et al., 1993). However some cells, such as neurons and pituitary cells, express the inhibitory phenotype of D2 receptor signal transduction, leading to inhibition of spontaneous action potentials, secretion, hormone transcription, and cell proliferation. On the other hand, fibroblast cells respond to the same receptor with a stimulation of cellular metabolism, gene transcription, and cell proliferation. As suggested earlier, differential expression of βγ-responsive PLC isozymes may explain this differential signaling. Similarly, the $α_{2A}$-adrenergic receptor mediates opposite actions on cAMP levels depending on the cell type in which it is expressed (Federman et al., 1992). In this case, the presence or absence of βγ-responsive adenylyl cyclase type II (Tang and Gilman, 1991,1992; Tussig et al., 1993) determines whether this receptor will increase or decrease cAMP levels. The difference between these signaling phenotypes resides not in the receptor, but in discrete intracellular differences in the expression of effector enzymes, such as PLC or adenylyl cyclase, in the two cell types.

Another example of the importance of intracellular response elements in determining receptor function is that of the *trk* oncogene, recently identified as a neurotrophin receptor (Chao, 1992). In neural crest-derived PC-12 pheochromocytoma cells, nerve growth factor (NGF)-induced activation of *trk* initiates a signaling cascade that results in the activation of *ras* and cessation of cell growth, differentiation, and neurite extension (Hagag et al., 1986; Hempstead et al., 1992). When the same receptor is expressed in NIH-3T3 fibroblasts, the receptor initiates a *ras*-dependent stimulation of cell proliferation and transformation, a response opposite to that in PC-12 cells (Chao, 1992).

These examples of cell-specific signaling illustrate the role of the cellular milieu in determining the type of signal or response triggered by a particular receptor. The cellular milieu specifies the response phenotype of the cell: it is determined by the expression of intracellular response elements (e.g., effector enzymes, secreted proteins, transcriptionally active response genes) appropriate to its function. The cellular milieu is transcriptionally determined by the developmental history and lineage of the cell. In several instances, the cellular milieu plays an important role in the signaling of endogenous receptors, and may designate distinct cellular lineages. For example, in hippocampal neurons, the 5-HT1A receptor couples to a CTX-sensitive electrophysiological response that potentiates G_s-coupled receptor actions (Blier et al., 1993; Andrade, 1993). In raphe neurons, the 5-HT1A receptor induces only inhibitory PTX-sensitive responses. In glial cells, this receptor mediates growth stimulatory actions via secretion of growth-promoting S100 protein (Lauder, 1993). Although these observations are complicated by the possibility of multiple 5-HT1A-like receptors *in vivo*, they imply that, depending on the cell in which a receptor is expressed and even on the subcellular location of the receptor, different signaling phenotypes may result.

D. PATHWAY-SELECTIVE MODULATION

The multiplicity of signaling pathways induced by a given receptor provides new targets for the regulation of receptor signaling by intracellular kinases. In studies of signaling in L cells, we have identified a pathway-selective modulation of receptor signaling by PKC and PKA, two protein kinases that are activated or inhibited by $G_{i/o}$- or G_s-coupled receptors (Liu and Albert, 1991; Liu et al., 1992a,b). In particular, both 5-HT1A and D2S receptor-mediated PI turnover and calcium increase are completely abolished by 2-min preactivation of PKC, whereas receptor-mediated inhibition of forskolin-stimulated cAMP accumulation is not altered. Interestingly, the D2L receptor is resistant to PKC-induced uncoupling. Since PKC is activated as a result of receptor-mediated PI turnover and phosphorylates these receptors (Raymond, 1991), this protein appears to play a negative feedback role to regulate activity of this pathway via receptor phosphorylation (Nishizuka, 1988). Whether PKC inactivates the receptor, the G protein, or the PLC enzyme that mediate the PI turnover pathway has yet to be clarified (Ryu et al., 1990). However, PKC does not alter the inhibitory cAMP pathway of the D2 or 5-HT1A receptors, although homologous desensitization of this pathway does occur (Bates et al., 1990). This result implies a pathway-selective modulation of receptor signaling, a concept that is very different from the classical view of desensitization as expounded for the β-adrenergic system, in which the desensitized receptor is regarded as uncoupled or inactive. Acute uncoupling of the β-adrenergic receptor from adenylyl cyclase appears to result from receptor phosphorylation by both PKA and a receptor kinase (BARK), and results in loss of signaling via this pathway (Kobilka, 1992; Lefkowitz, 1993). However, possible effects of these kinases on other pathways have not been examined since the receptor is assume to be uncoupled, that is, nonfunctional. Our results indicate that pathway-selective modulation is observed, so some receptor-mediated pathways are shut off whereas others are left on, and yet others may be potentiated (Fig. 8).

A striking example of such modulation is the action of PKA on D1 receptor signaling (Liu et al., 1992b). In L cells, the dopamine-D1 receptor couples to G_s to increase cAMP (which activates PKA) and to increase PI turnover (which activates PKC). Preactivation of PKA causes a decrease in coupling of the D1 receptor to cAMP (uncoupling, as studied for the β-adrenergic receptor) but induces an enhanced coupling of the receptor to PI turnover and calcium signaling (Liu et al., 1992b; Machida et al., 1992). Preactivation of PKC uncouples the receptor from PI turnover but does not affect the cAMP pathway. This system provides an example of a receptor that displays selective feedback inhibition (by PKC and PKA) and feedforward stimulation (by PKA), as illustrated in Fig. 8.

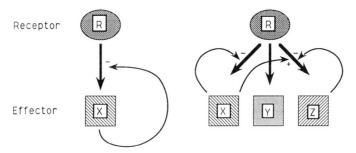

FIG. 8. Pathway-selective modulation of receptor-mediated signals. In classical desensitization (left), such as desensitization of β-adrenergic receptor activation of adenylyl cyclase, only one pathway of the receptor (R) is examined. The product of the effector X (e.g., cAMP) initiates feedback inhibition to block the receptor signaling pathway (e.g., PKA-mediated phosphorylation of the β-adrenergic receptor). In pathway-selective modulation (right), the receptor exhibits multiple signaling to different effector enzymes or channels X, Y, and Z. The products of these effectors differentially regulate signaling pathways. For example, the PLC pathway, but not the cyclase pathway, of the 5-HT1A or dopamine-D2S receptors is inhibited by DAG-mediated activation of PKC within 1–2 min. In some cases, activation of one pathway appears to potentiate another receptor-operated pathway, as for D1-receptor-induced increase in cAMP levels. Elevated cAMP enhances the D1-induced PI turnover and calcium increase several-fold in transfected fibroblast cells, while causing desensitization of the cyclase pathway. See references in text.

The specific site of action (i.e., receptor, G protein, or effector) of these kinases has yet to be determined. Evidence suggests that mutations in the third cytoplasmic loop of the β-adrenergic receptor uncouple it from adenylyl cyclase but not from sodium/proton exchange, suggesting that different coupling domains of the receptor mediate different receptor-mediated actions (Barber et al., 1992). Similarly, mutagenesis of the TSH receptor has localized the PI-coupled and cAMP-coupled domains to distinct portions of the third cytoplasmic loop of this receptor (Kosugi et al., 1992). By analogy, phosphorylation at different sites on the receptor may uncouple one response without altering other responses.

V. Conclusion and Future Prospects

Results from the use of heterologous expression of cloned receptors in diverse cell types are broadening our understanding of receptor signaling mechanisms. Some general conclusions can be drawn.

1. Numerous studies indicate that one receptor generates multiple signals, each of which participates in the action of the receptor on the target cell (Birnbaumer, 1992; Conklin and Bourne, 1993). Thus, $G_{i/o}$-coupled receptors not only inhibit cAMP formation, but also activate serine and tyrosine phosphatases, open potassium channels, and close calcium channels.

2. The routing of these multiple signals is via distinct G proteins. A network of specific interactions between receptor–G protein–effector exists. Distinct G proteins are specialized for certain effectors (e.g., G_o for calcium channel regulation) and certain receptors recognize select G protein combinations (e.g., somatostatin receptor to $\alpha_o 2/\beta 1/\gamma 3$) (Table III).

3. Some receptor-induced pathways are cell specific. Thus, $G_{i/o}$-coupled receptors stimulate PI turnover in several types of fibroblast cells but not in GH4C1 pituitary cells or mature neurons, in which they open potassium channels and close calcium channels. In all cell types, these receptors inhibit stimulation of adenylyl cyclase. Depending on the cell type, the different pathways of receptor action may have opposite actions. Thus, in GH4C1 cells, $G_{i/o}$-coupled receptors inhibit DNA synthesis, whereas the same receptors stimulate DNA synthesis when expressed in fibroblast cells. The response of a target cell depends on the types of receptor-linked pathways present or, additionally, on the types of downstream response elements such as kinases, phosphatases, *trans*-acting factors, and response genes available within the cell. All these response factors constitute the cellular milieu, and determine the response to receptor activation.

4. Finally, these multiple pathways activated by a receptor may interact negatively (feedback inhibition) or synergistically to enhance certain pathways of the receptor. Thus, a receptor is not necessarily turned off following prolonged activation, as in classical desensitization, but may be modulated to send different signals within the cell.

The redefinition of desensitization could have important implications for drug dependence, addiction, and withdrawal. For example, since chronic exposure to a drug may cause an alteration in receptor signaling, rather than an abolition of signaling, the new receptor signal would require maintenance of the drug. On removal of the drug, the new signal is lost and "withdrawal" results.

Detailed mapping of receptor signaling may appear to be esoteric at first glance. However, the definition of specific aspects of receptors may be of importance to understand drug and hormone action. By defining the pathways of receptor action, we will be able to assess the roles of each pathway in receptor signaling. Cells transfected with individual antisense α subunits should provide excellent tools with which to study the role of specific G protein-coupled pathways in complex actions such as control of secretion, phosphatase activity, gene transcription, and cell growth. For example, the roles of G_o-linked pathways in secretion or gene transcription can be assessed using the stable α_o-antisense cell lines. Similarly, the antisense knockout experiments can be extended to downstream effectors. The specificity of G protein activation of effectors could be exploited as a target for drug development. For example, G protein-activating peptides are already proving useful in uncovering new roles for G_O in growth-cone collapse (Igarishi et al., 1993). Similarly, other pathways activated by specific G proteins may serve as drug targets. Ultimately, the full sequence of receptor activation will lead to the discovery of new intracellular drug targets, as the components constituting cellular responsiveness are clarified. The use of antisense technology or pharmaceutical interventions may provide new generations of agents with high specificity for distinct receptor-mediated pathways.

Acknowledgments

I thank all collaborators and members of my laboratory for their support and enthusiasm. In particular, I thank A. H. Tashjian, Jr. for introducing me to GH cells, Olivier Civelli for collaboration and support with the heterologous expression studies, Stephen Morris for critical reading of the manuscript, and Paola Lembo for alignment of G protein sequences. I acknowledge grant support from the National Cancer Institute (Canada), Medical Research Council (Canada), and Chercheur Boursier, Fonds de la Recherche en Santé du Québec.

REFERENCES

Abdel-Baset, H., Bozovic, V., Szyf, M., and Albert, P. R. (1992). Conditional transformation mediated via a pertussis toxin sensitive receptor signalling pathway. *Mol. Endocrinol.* **6,** 730–740.

Abou-Samra, A., Jüppner, H., Force, T., Freeman, M. W., Kong, X., Schipani, E., Urena, P., Richards, J., Bonventre, J. V., Potts, J. T., Jr., Kronenberg, H. M., and Segre, G. V. (1992). Expression cloning of a common receptor for parathyroid hormone and parathyroid hormone-related peptide from rat osteoblast-like cells: A single receptor stimulates intracellular accumulation of both cAMP and inositol trisphosphates and increases intracellular free calcium. *Proc. Natl. Acad. Sci. USA* **89,** 2732–2736.

Aizawa, T., and Hinkle, P. M. (1985). Thyrotropin-releasing hormone rapidly stimulates a biphasic secretion of prolactin and growth hormone in GH4C1 rat pituitary tumor cells. *Endocrinology* **116,** 73–82.

Albert, P. R. (1992). Molecular biology of the 5-HT1A receptor: Low-stringency cloning and eukaryotic expression. *J. Chem. Neuroanat.* **5,** 283–288.

Albert, P. R., and Liston, D. (1993). Deletions of the synenkephalin domain which do not alter cell-specific proteolytic processing or regulated secretion of human proenkephalin. *J. Neurochem.* **60,** 1325–1334.

Albert, P. R., and Tashjian, A. H., Jr. (1984a). Thyrotropin-releasing hormone-induced spike and plateau in cytosolic free Ca^{++} concentrations in pituitary cells: Relation to prolactin release. *J. Biol. Chem.* **259**, 5827–5832.

Albert, P. R., and Tashjian, A. H., Jr. (1984b). Relationship of thyrotropin-releasing hormone-induced spike and plateau phases in cytosolic free Ca^{++} concentrations to hormone secretion: Selective blockade using ionomycin and nifedipine. *J. Biol. Chem.* **259**, 15350–15363.

Albert, P. R., and Tashjian, A. H., Jr. (1985). Dual actions of phorbol esters on cytosolic free Ca^{++} concentrations and reconstitution with ionomycin of acute thyrotropin-releasing hormone responses. *J. Biol. Chem.* **260**, 8746–8759.

Albert, P. R., and Tashjian, A. H., Jr. (1986). Ionomycin acts as an ionophore to release TRH-regulated Ca^{++} stores from GH4C1 cells. *Am. J. Physiol.* **251**, C887–C891.

Albert, P. R., Wolfson, G., and Tashjian, A. H., Jr. (1987). Diacylglycerol increases cytosolic free Ca^{++} concentration in rat pituitary cells. Relationship to thyrotropin-releasing hormone action. *J. Biol. Chem.* **262**, 6577–6581.

Albert, P. R., Neve, K. A., Bunzow, J., and Civelli, O. (1990a). Coupling of a cloned dopamine-D2 receptor to inhibition of adenylyl cyclase and prolactin secretion. *J. Biol. Chem.* **265**, 2098–2104.

Albert, P. R., Zhou, Q.-Y., VanTol, H. H. M., Bunzow, J., and Civelli, O. (1990b). Cloning, functional expression, and mRNA tissue distribution of the rat 5-HT1A receptor gene. *J. Biol. Chem.* **265**, 5825–5832.

Andrade, R. (1983). Enhancement of β-adrenergic responses by Gi-linked receptors in rat hippocampus. *Neuron* **10**, 83–88.

Aragay, A. M., Katz, A., and Simon, M. I. (1992). The $G\alpha_q$ and $G\alpha_{11}$ proteins couple the thyrotropin-releasing hormone receptor to phospholipase C in GH3 rat pituitary cells. *J. Biol. Chem.* **267**, 24983–24988.

Armstrong, D., and Eckert, R. (1987). Voltage-activated calcium channels that must be phosphorylated to respond to membrane depolarization. *Proc. Natl. Acad. Sci. USA* **84**, 2518–2522.

Aub, D. L., Frey, E. A., Sekura, R. D., and Cote, T. E. (1986). Coupling of the thyrotropin-releasing hormone receptor to phospholipase C by a GTP-binding protein distinct from the inhibitory or stimulatory GTP-binding protein. *J. Biol. Chem.* **261**, 9333–9340.

Baertschi, A. J., Audigier, Y., Lledo, P.- M., Israel, J.-M., Bockaert, J., and Vincent, J.-D. (1992). Dialysis of lactotropes with antisense oligonucleotides assigns guanine nucleotide binding protein subtypes to their channel effectors. *Mol. Endocrinol.* **6**, 2257–2265.

Barber, D. L., Ganz, M. B., Bongiorno, P. B., and Strader, C. D. (1992). Mutant constructs of the β-adrenergic receptor that are uncoupled from adenylyl cyclase remain functional in activation of Na-H exchange. *Mol. Pharmacol.* **41**, 1056–1060.

Barros, F., and Kaczorowski, G. J. (1984). Mechanisms of Ca^{2+} transport in plasma membrane vesicles prepared from cultured pituitary cells. II. (CA^{2+} + MG^{2+})-ATPase-Dependent Ca^{2+} transport activity. *J. Biol. Chem.* **259**, 9395–9403.

Barros, F., Katz, G. M., Kaczorowski, G. J., and Vandlen, R. L. (1985). Calcium currents in GH3 cultured pituitary cells under whole-cell voltage-clamp: Inhibition by voltage-dependent potassium currents. *Proc. Natl. Acad. Sci. USA* **82**, 1108–1112.

Bates, M. D., Senogles, S. E., Bunzow, J. R., Liggett, S. B., Civelli, O., and Caron, M. G. (1990). Regulation of responsiveness at D2 dopamine receptors by receptor desensitization and adenylyl cyclase sensitization. *Mol. Pharmacol.* **39**, 55–63.

Berridge, M. J., and Irvine, R. F. (1984). Inositol trisphosphate, a novel second messenger in cellular signal transduction. *Nature (London)* **312,** 315–321.
Berridge, M. J., and Irvine, R. F. (1989). Inositol phosphates and cell signalling. *Nature (London)* **341,** 197–205.
Biden, T. J., Wollheim, C. B., and Schlegel, W. (1986). Inositol 1,4,5-trisphosphate and intracellular Ca^{2+} homeostasis in clonal pituitary cells (GH3). Translocation of CA^{2+} into mitochondria from a functionally discrete portion of the nonmitochondrial store. *J. Biol. Chem.* **261,** 7223–7229.
Birnbaumer, L. (1992). Receptor-to-effector signaling through G proteins: Roles for βγ dimers as well as α subunits. *Cell* **71,** 1069–1072.
Birnbaumer, L., Abramowitz, J., and Brown, A. M. (1990). Receptor-effector coupling by G proteins. *Biochim. Biophys. Acta* **1031,** 163–224.
Bjøro, T., Wiik, P., Opstad, P. K. Gautvik, K. M., and Haug, E. (1987). Binding and degradation of vasoactive intestinal peptide in prolactin-producing cultured rat pituitary cells (GH4C1). *Acta Physiol. Scand.* **130,** 609–618.
Blier, P., Lista, A., and De Montigny, C. (1993). Differential properties of pre- and postsynaptic 5-hydroxytryptamine$_{1A}$ receptors in the dorsal raphe and hippocampus: II. Effect of pertussis and cholera toxins. *J. Pharmacol. Exp. Therapeut.* **265,** 16–23.
Campbell, V., Berrow, N., and Dolphin, A. C. (1993). GABA$_B$ receptor modulation of Ca^{2+} currents in rat sensory neurones by the G protein Go: antisense oligonucleotide studies. *J. Physiol. (London.)* **470,** 1–11.
Camps, M., Carozzi, A., Schnabel, P., Scheer, A., Parker, P. J., and Gierschik, P. (1992). Isozyme-selective stimulation of phospholipase C-β2 by G protein βγ-subunits. *Nature (London)* **360,** 684–686.
Carrillo, A. J., Pool, T. B., and Sharp, Z. D. (1985). Vasoactive intestinal peptide increases prolactin messenger ribonucleic acid content in GH3 cells. *Endocrinology* **116,** 202–206.
Chambard, J. C., Paris, S., L'Allemain, G., and Pouysségur, J. (1987). Two growth factor signalling pathways in fibroblasts distinguished by pertussis toxin. *Nature (London)* **326,** 800–803.
Chao, M. V. (1992). Neurotrophin receptors: A window into neuronal differentiation. *Neuron* **9,** 583–593.
Civelli, O., Bunzow, J. R., and Grandy, D. K. (1993). Molecular diversity of the dopamine receptors. *Annu. Rev. Pharmacol. Toxicol.* **32,** 281–307.
Codina, J., Olate, J., Abramowitz, J., Mattera, R., Cook, R. G., and Birnbaumer, L. (1988). αi-3 cDNA encodes the α subunit of G_k, the stimulatory G protein of receptor-regulated K$^+$ channels. *J. Biol. Chem.* **263,** 6746–6750.
Cohen, C. J., and McCarthy, R. T. (1987). Nimodipine block of calcium channels in rat anterior pituitary cells. *J. Physiol. (London)* **387,** 195–225.
Coleman, T. A., Chomczynski, P., Frohman, L. A., and Kopchik, J. J. (1991). A comparison of transcriptional regulatory element activities in transformed and nontransformed rat anterior pituitary cells. *Mol. Cell. Endocrinol.* **75,** 91–100.
Collins, S., Lohse, M. J., O'Dowd, B., Caron, M. G., and Lefkowitz, R. J. (1991). Structure and regulation of G protein-coupled receptors: The β-adrenergic receptor as a model. *Vit. Horm.* **46,** 1–39.
Conklin, B. R., and Bourne, H. R. (1993). Structural elements of Gα subunits that interact with Gβγ, receptors, and effectors. *Cell* **73,** 631–641.
Crews, C. M., and Erikson, R. L. (1993). Extracellular signals and reversible protein phosphorylation: What to mek of it all. *Cell* **74,** 215–217.

Cui, Z., Zubiaur, M., Bloch, D. B., Michel, T., Seidman, J. G., and Neer, E. J. (1991). Expression of a G protein subunit, αi-1, in BALB/c3T3 cells leads to agonist-specific changes in growth regulation. *J. Biol. Chem.* **266,** 20276–20282.

Davis, R. J. (1993). The mitogen-activated protein kinase signal transduction pathway. *J. Biol. Chem.* **268,** 14553–14556.

Day, R. N., and Maurer, R. A. (1990). Pituitary calcium channel modulation and regulation of prolactin gene expression. *Mol. Endocrinol.* **4,** 736–742.

Dell'Acqua, M. L., Carroll, R. C., and Peralta, E. G. (1993). Transfected m2 muscarinic acetylcholine receptors couple to $G\alpha_i2$ and $G\alpha_i3$ in Chinese hamster ovary cells. *J. Biol. Chem.* **268,** 5676–5685.

Dorflinger, L. J., and Schonbrunn, A. (1983a). Somatostatin inhibits vasoactive intestinal peptide-stimulated cyclic adenosine monophosphate accumulation in GH pituitary cells. *Endocrinology* **113,** 1541–1550.

Dorflinger, L. J., and Schonbrunn, A. (1983b). Somatostatin inhibits basal and vasoactive intestinal peptide-stimulated hormone release by different mechanisms in GH pituitary cells. *Endocrinology* **113,** 1551–1560.

Dorflinger, L. J., and Schonbrunn, A. (1985). Adenosine inhibits prolactin and growth hormone secretion in a clonal pituitary cell line. *Endocrinology* **227,** 2330–2338.

Drummond, A. H. (1985). Bidirectional control of cytosolic free calcium by thyrotropin-releasing hormone. *Nature (London)* **315,** 752–755.

Drust, D. S., and Martin, T. F. J. (1984). Thyrotropin-releasing hormone rapidly activates protein phosphorylation in GH3 pituitary cells by a lipid-linked, protein kinase C-mediated pathway. *J. Biol. Chem.* **259,** 14520–14530.

Drust, D. S., Sutton, C. A., and Martin, T. F. J. (1982). Thyrotropin-releasing hormone and cyclic AMP activate distinctive pathways of protein phosphorylation in GH pituitary cells. *J. Biol. Chem.* **257,** 3306–3312.

Dubinski, J. M., and Oxford, G. S. (1984). Ionic currents in two strains of rat anterior pituitary tumor cells. *J. Gen. Physiol.* **83,** 309–339.

Dubinski, J. M., and Oxford, G. S. (1985). Dual modulation of K channels by thyrotropin-releasing hormone in clonal pituitary cells. *Proc. Natl. Acad. Sci. USA* **82,** 4282–4286.

Dufy, B., and Barker, J. L. (1982). Calcium-activated and voltage-dependent potassium conductances in clonal pituitary cells. *Life Sci.* **30,** 1933–1941.

Elsholtz, H. P., Lew, A. M., Albert, P. R., and Sundmark, V. C. (1991). Inhibitory control of prolactin and PIT-1 gene promoters by dopamine. Dual signaling pathways required for D2 receptor-regulated expression of the prolactin gene. *J. Biol. Chem.* **266,** 22919–22925.

Ewald, D. A., Pang, I., Sternweis, P. C., and Miller, R. J. (1989). Differential G protein-mediated coupling of neurotransmitter receptors to Ca^{2+} channels in rat dorsal root ganglion neurons *in vitro. Neuron* **2,** 1185–1193.

Fargin, A., Raymond, J. R., Lohse, M. J., Kobilka, B. K., Caron, M. G., and Lefkowitz, R. J. (1988). The genomic clone G-21 which resembles a β-adrenergic receptor sequence encodes the 5-HT1A receptor. *Nature (London)* **335,** 358–360.

Fargin, A., Raymond, J. R., Regan, J. W., Cotecchia, S., Lefkowitz, R. J., and Caron, M. G. (1989). Effector coupling mechanisms of the cloned 5-HT1A receptor. *J. Biol. Chem.* **264,** 14848–14852.

Fearon, C. W., and Tashjian, A. H., Jr. (1987). Ionomycin inhibits thyrotropin-releasing hormone-induced translocation of protein kinase C in GH4C1 cells. *J. Biol. Chem.* **262,** 9515–9520.

Federman, A. D., Conklin, B. R., Schrader, K. A., Reed, R. R., and Bourne, H. R. (1992).

Hormonal stimulation of adenylyl cyclase through Gi-protein βγ subunits. *Nature (London)* **356**, 159–162.

Florio, T., Pan, M., Newman, B., Hershberger, R. E., Civelli, O., and Stork, P. J. S. (1992). Dopaminergic inhibition of DNA synthesis in pituitary tumor cells is associated with phosphotyrosine phosphatase activity. *J. Biol. Chem.* **267**, 24169–24172.

Fowler, C. J., Ahlgren, P. C., and Br^annstr^om, G. (1992). GH4ZD10 cells expressing rat 5-HT1A receptors coupled to adenylyl cyclase are a model for the postsynaptic receptors in the rat hippocampus. *Br. J. Pharmacol.* **107**, 141–145.

Frawley, L. S., and Boockfor, F. R. (1991). Mammosomatotropes: Presence and functions in normal and neoplastic pituitary tissue. *Endocrine Rev.* **12**, 337–355.

Gautvik, K. M., and Kriz, M. (1976). Effects of prostaglandins on prolactin and growth hormone synthesis and secretion in cultured rat pituitary cells. *Endocrinology* **98**, 352–358.

Gershengorn, M. C. (1986). Mechanism of thyrotropin releasing hormone stimulation of pituitary hormone secretion. *Annu. Rev. Physiol.* **48**, 515–526.

Gershengorn, M. C., and Thaw, C. (1985). Thyrotropin-releasing hormone (TRH) stimulates biphasic elevation of cytoplasmic free calcium in GH3 cells. Further evidence that TRH mobilizes cellular and extracellular Ca^{2+}. *Endocrinology* **116**, 591–596.

Gilman, A. G. (1987). G proteins: Transducers of receptor-generated signals. *Annu. Rev. Biochem.* **56**, 615–649.

Gourdji, D., Bataille, D., Vauclin, N., Grouselle, D., Rosselin, G., and Tixier-Vidal, A. (1979). Vasoactive intestinal peptide (VIP) stimulates prolactin (PRL) release and cAMP production in a rat pituitary cell line (GH3/B6). Additive effects of VIP and TRH on PRL release. *FEBS Lett.* **104**, 165–168.

Gudermann, T., Birnbaumer, M., and Birnbaumer, L. (1992). Evidence for dual coupling of the murine lutenizing hormone receptor to adenylyl cyclase and phosphoinositide breakdown and Ca^{2+} mobilization. *J. Biol. Chem.* **267**, 4479–4488.

Guild, S., and Drummond, A. H. (1984). Vasoactive-intestinal-polypeptide-stimulated adenosine 3′,5′-cyclic monophosphate accumulation in GH3 pituitary tumour cells. *Biochem. J.* **221**, 789–796.

Gupta, S. K., Gallego, C., Johnson, G. L., and Heasley, L. E. (1992). MAP kinase is constitutively activated in *gip2* and *src* transformed Rat 1a fibroblasts. *J. Biol. Chem.* **267**, 7987–7990.

Hagag, N., Halegoua, S., and Viola, M. (1986). Inhibition of growth factor-induced differentiation of PC12 cells by microinjection of antibody to *ras* p21. *Nature (London)* **319**, 680–682.

Hagiwara, S., and Ohmori, H. (1983). Studies of single calcium channels in rat clonal pituitary cells. *J. Physiol. (London)* **336**, 649–661.

Hallam, T. J., and Tashjian, A. H., Jr. (1987). Thyrotropin-releasing hormone activates Na^+/H^+ exchange in rat pituitary cells. *Biochem. J.* **242**, 411–416.

Hartzell, H. C., and Fischmeister, R. (1992). Direct regulation of cardiac Ca^{2+} channels by G proteins: Neither proven nor necessary? *Trends Pharmacol. Sci.* **13**, 380–385.

Hayes, G., Biden, T. J., Selbie, L. A., and Shine, J. (1992). Structural subtypes of the dopamine D2 receptor are functionally distinct: Expression of the cloned D2A and D2B subtypes in a heterologous cell line. *Mol. Endocrinol.* **6**, 920–926.

Haymes, A. A., Kwan, Y. W., Arena, J. P., Kass, R. S., and Hinkle, P. M. (1992). Activation of protein kinase C reduces L-type calcium channel activity of GH3 pituitary cells. *Am. J. Physiol.* **262**, C1211–C1219.

Hempstead, B. L., Rabin, S. J., Kaplan, L., Reid, S., Parada, L. F., and Kaplan, D. R. (1992). Over-expression of the *trk* tyrosine kinase rapidly accelerates nerve growth factor-induced differentiation. *Neuron* **9**, 883–896.

Herrington, J., Stern, R. C., Evers, A. S., and Lingle, C. J. (1991). Halothane inhibits two components of calcium current in clonal (GH3) cells. *J. Neurosci.* **11**, 2226–2240.
Hescheler, J., Rosenthal, W., Trautwein, W., and Schulz, G. (1987). The GTP-binding protein, Go, regulates neuronal calcium channels. *Nature (London)* **325**, 445–447.
Hinkle, P. M., and Tashjian, A. H., Jr. (1973). Receptors for thyrotropin-releasing hormone in prolactin-producing rat pituitary cells in culture. *J. Biol. Chem.* **248**, 6180–6186.
Hinkle, P. M., Scammell, J. G., and Shanshala, E. D., II (1992). Prolactin and secretogranin-II, a marker for the regulated pathway, are secreted in parallel by pituitary GH4C1 cells. *Endocrinology* **130**, 3503–3511.
Hoyer, D., and Boodeke, H. W. G. M. (1993). Partial agonists, full agonists, antagonists: Dilemnas of definition. *Trends Pharmacol. Sci.* **14**, 270–275.
Igarishi, M., Strittmatter, S. M., Vartanian, T., and Fishman, M. C. (1993). Mediation by G proteins of signals that cause collapse of growth cones. *Science* **259**, 77–79.
Iverson, R. A., Day, K. H., d'Emden, M., Day, R. N., and Maurer, R. A. (1990). Clustered point mutation analysis of the rat prolactin promoter. *Mol. Endocrinol.* **4**, 1564–1571.
Izant, J. G., and Weintraub, H. (1985). Constitutive and conditional suppression of exogenous and endogenous genes by anti-sense RNA. *Science* **229**, 345–352.
Jones, D. T., and Reed, R. R. (1987). Molecular cloning of five GTP-binding protein cDNA species from rat olfactory neuroepithelium. *J. Biol. Chem.* **262**, 14241–14249.
Kaczorowski, G. J., Costello, L., Dethmers, J., Trumble, M. J., and Vandlen, R. L. (1984). Mechanisms of Ca^{2+} transport in plasma membrane vesicles prepared from cultured pituitary cells. I. Characterization of Na^+/Ca^{2+} exchange activity. *J. Biol. Chem.* **259**, 9395–9403.
Katz, A., Wu, D., and Simon, M. I. (1992). Subunits βγ of heterotrimeric G protein activate β2 isoform of phospholipase C. *Nature (London)* **360**, 686–689.
Kleuss, C., Hescheler, J., Ewel, C., Rosenthal, W., Shultz, G., and Wittig, B. (1991). Assignment of G-protein subtypes to specific receptors inducing inhibition of calcium currents. *Nature (London)* **353**, 43–48.
Kleuss, C., Scherübl, H., Hescheler, J., Shultz, G., and Wittig, B. (1992). Different β-subunits determine G-protein interaction with transmembrane receptors. *Nature (London)* **358**, 424–426.
Kleuss, C., Scherübl, H., Hescheler, J., Shultz, G., and Wittig, B. (1993). Selectivity in signal transduction determined by γ subunits of heterotrimeric G proteins. *Science* **259**, 832–834.
Kobilka, B. (1992). Adrenergic receptors as models for G protein-coupled receptors. *Annu. Rev. Neurosci.* **15**, 87–114.
Koch, B. D., and Schonbrunn, A. (1988). Characterization of the cyclic AMP-independent actions of somatostatin in GH cells. II. An increase in potassium conductance initiates somatostatin-induced inhibition of prolactin secretion. *J. Biol. Chem.* **263**, 226–234.
Koch, B. D., Dorflinger, L. J., and Schonbrunn, A. (1985). Pertussis toxin blocks both cyclic AMP-mediated and cyclic AMP-independent actions of somatostatin. Evidence for coupling of Ni to decreases in intracellular free calcium. *J. Biol. Chem.* **260**, 13138–13145.
Koch, B. D., Blalock, J. B., and Schonbrunn, A. (1988). Characterization of the cyclic AMP-independent actions of somatostatin in GH cells. I. An increase in potassium conductance is responsible for both the hyperpolarization and the decrease in intracellular free calcium produced by somatostatin. *J. Biol. Chem.* **263**, 216–225.
Kosugi, S., Okajima, F., Ban, T., Hidaka, A., Shenker, A., and Kohn, L. D. (1992).

Mutation of alanine 623 in the third cytoplasmic loop of the rat thyrotropin (TSH) receptor results in a loss in the phosphoinositide but not cAMP signal induced by TSH and receptor autoantibodies. *J. Biol. Chem.* **267**, 24153–24156.

Kramer, R. H., Kaczmarek, L. K., and Levitan, E. S. (1991). Neuropeptide inhibition of voltage-gated calcium channels mediated by mobilization of intracellular calcium. *Neuron* **6**, 557–563.

Kroll, S. D., Chen, J., De Vivo, M., Carty, D. J., Buku, A., Premont, R. T., and Iyengar, R. (1992). The Q205LGo-α subunit expressed in NIH-3T3 cells induces transformation. *J. Biol. Chem.* **267**, 23183–23188.

LaMorte, V. J., Harootunian, A. T., Spiegel, A. M., Tsien, R. Y., and Feramisco, J. R. (1993). Mediation of growth factor induced DNA synthesis and calcium mobilization by Gq and Gi2. *J. Cell Biol.* **121**, 91–99.

Lang, D. G., and Ritchie, A. K. (1987). Large and small conductance calcium-activated potassium channels in the GH3 anterior pituitary cell line. *Pflugers Arch.* **410**, 614–622.

Lauder, J. M. (1993). Neurotransmitters as growth regulatory signals: Role of receptors and second messengers. *Trends Neuro Sci.* **16**, 233–239.

Law, S. F., and Reisine, T. (1992). Agonist binding to rat brain somatostatin receptors alters the interaction of the receptors with guanine nucleotide-binding regulatory proteins. *Mol. Pharmacol.* **42**, 3988–4002.

Law, S. F., Manning, D., and Reisine, T. (1991). Identification of the subunits of GTP binding proteins coupled to somatostatin receptors. *J. Biol. Chem.* **266**, 17885–17897.

Law, S. F., Yasuda, K., Bell, G. I., and Reisine, T. (1993). Giα3 and Goα selectively associate with the cloned somatostatin receptor subtype SSTR2. *J. Biol. Chem.* **268**, 10721–10727.

Lee, C. H., Park, D., Wu, D., Rhee, S. G., and Simon, M. I. (1992). Members of the Gq α subunit gene family activate phospholipase C β isozymes. *J. Biol. Chem.* **267**, 16044–16047.

Lefkowitz, R. J. (1993). G protein-coupled receptor kinases. *Cell* **74**, 409–412.

Levitan, E. S., and Kramer, R. H. (1990). Neuropeptide modulation of single calcium and potassium channels detected with a new patch clamp configuration. *Nature (London)* **348**, 545–547.

Lew, A. M., and Elsholtz, H. P. (1993). Cell-specific signalling pathways determine the regulation of prolactin gene transcription by dopamine D2 receptors. *Proc. Endocrinol. Soc. Mtg.*, **75**, 265.

Liebow, C., Reilly, C., Serrano, M., and Schally, A. V. (1989). Somatostatin analogues inhibit growth of pancreatic cancer by stimulating tyrosine phosphatase. *Proc. Natl. Acad. Sci. USA* **86**, 2003–2007.

Liu, Y. F., and Albert, P. R. (1991). Cell-specific signalling of the 5-HT1A receptor. Modulation by protein kinases C and A. *J. Biol. Chem.* **266**, 23689–23697.

Liu, Y. F., Civelli, O., Grandy, D. K., and Albert, P. R. (1992a). Differential sensitivity of the short and long human dopamine-D2 receptor subtypes to protein kinase C. *J. Neurochem.* **59**, 2311–2317.

Liu, Y. F., Civelli, O., Zhou, Q.-Y., and Albert, P. R. (1992b). Cholera toxin-sensitive 3′,5′-cyclic adenosine monophosphate and calcium signals of the human dopamine-D1 receptor: Selective potentiation by protein kinase A. *Mol. Endocrinol.* **6**, 1815–1824.

Liu, Y. F., Jakob, K. H., Rasenick, M. M., Albert, P. R. (1994a). G protein specificity in receptor-effector coupling. Analysis of the roles of Go and Gi2 in GH4C1 pituitary cells. *J. Biol. Chem.* **269**, in press.

Lledo, P.-M., Homburger, V., Bockeart, J., and Vincent, J.-D. (1992). Differential G protein-mediated coupling of D2 dopamine receptors to K^+ and Ca^{2+} currents in rat anterior pituitary cells. *Neuron* **8**, 455–463.

Lledo, P.-M., Vernier, P., Vincent, J.-D., Mason, W. T., and Zorec, R. (1993). Inhibition of Rab3B expression attenuates Ca^{2+}-dependent exocytosis in rat anterior pituitary cells. *Nature (London)* **364**, 540–543.

Lo, W. W. Y., and Hughes, J. (1987). A novel cholera toxin-sensitive G-protein (Gc) regulating receptor-mediated phosphoinositide signalling in human pituitary clonal cells. *FEBS Lett.* **220**, 327–331.

Lowndes, J. M., Gupta, S. K., Osawa, S., and Johnson, G. L. (1991). GTPase-deficient $G\alpha_i 2$ oncogene *gip2* inhibits adenylylcyclase and attenuates receptor-stimulated phospholipase A2 activity. *J. Biol. Chem.* **266**, 14193–14197.

Lyons, J., Landis, C. A., Harsh, G., Vallar, L., Grünewald, K., Feichtinger, H., Duh, Q., Clark, O. H., Kawasaki, E., Bourne, H. R., and McCormick, F. (1990). Two G protein oncogenes in human endocrine tumors. *Nature (London)* **249**, 655–659.

Machida, C. A., Searles, R. P., Nipper, V., Brown, J. A., Kozell, L. B., and Neve, K. (1992). Molecular cloning and expression of the rhesus macaque D1 dopamine receptor gene. *Mol. Pharmacol.* **41**, 652–659.

MacPhee, C. H., and Drummond, A. H. (1984). Thyrotropin-releasing hormone stimulates rapid breakdown of phosphatidyl inositol 4,5-biphosphate and phosphatidyl 4-phosphate in GH3 pituitary cells. *Mol. Pharmacol.* **25**, 193–200.

Mariot, P., Dufy, B., Audy, M.-C., and Sartor, P. (1993). Biphasic changes in intracellular pH induced by thyrotropin-releasing hormone in pituitary cells. *Endocrinology* **132**, 846–854.

Martin, T. F. J. (1983). Thyrotropin-releasing hormone rapidly activates the phosphodiester hydrolysis of polyphosphoinositides in GH3 pituitary cells. Evidence for the role of a polyphosphoinositide-specific phospholipase C in hormone action. *J. Biol. Chem.* **258**, 14816–14822.

Martin, T. F. J., and Kowalchyk, J. A. (1984a). Evidence for the role of calcium and diacylglycerol as dual second messengers in thyrotropin-releasing hormone action: Involvement of diacylglycerol. *Endocrinology* **115**, 1517–1526.

Martin, T. F. J., and Kowalchyk, J. A. (1984b). Evidence for the role of calcium and diacylglycerol as dual second messengers in thyrotropin-releasing hormone action: Involvement of Ca^{2+}. *Endocrinology* **115**, 1527–1536.

Martin, T. F. J., Hsieh, K., and Porter, B. W. (1990). The sustained second phase of hormone-stimulated diacylglycerol accumulation does not activate protein kinase C in GH3 cells. *J. Biol. Chem.* **265**, 7623–7631.

Matteson, D. R., and Armstrong, C. M. (1986). Properties of two types of calcium channels in clonal pituitary cells. *J. Gen. Physiol.* **87**, 161–182.

McFadzean, I., Mullaney, I., Brown, D. A., and Milligan, G. (1989). Antibodies to the GTP binding protein, Go, antagonize noradrenaline-induced calcium current inhibition in NG108-15 hybrid cells. *Neuron* **3**, 177–182.

McKenzie, F., and Milligan, G. (1990). Opioid receptor mediated inhibition of adenylyl cyclase is transduced specifically by the G protein $Gi\alpha 2$. *Biochem. J.* **267**, 391–398.

Missale, C., Castelletti, L., Boroni, F., Memo, M., and Spano, P. (1991a). Epidermal growth factor induces the functional expression of dopamine receptors in the GH3 cell line. *Endocrinology* **128**, 13–20.

Missale, C., Boroni, F., Castelletti, L., Dal Toso, R., Gabellini, N., Sigala, S., and Spano, P. (1991b). Lack of coupling of D-2 receptors to adenylate cyclase in GH-3 cells exposed to epidermal growth factor. Possible role of a differential expression of Gi protein subtypes. *J. Biol. Chem.* **266**, 23392–23398.

Mollard, P., Vacher, P., Dufy, B., and Barker, J. L. (1988a). Somatostatin blocks Ca^{2+} action potential activity in prolactin-secreting pituitary tumor cells through coordinate actions of K^+ and Ca^{2+} conductances. *Endocrinology* **123,** 721–732.

Mollard, P., Vacher, P., Dufy, B., Winiger, B. P., and Schlegel, W. (1988b). Thyrotropin-releasing hormone-induced rise in cytosolic calcium and activation of outward K^+ current monitored simultaneously in individual GH3B6 pituitary cells. *J. Biol. Chem.* **263,** 19570–19576.

Mollard, P., Guérineau, N., Chiavaroli, C., Schlegel, W., and Cooper, D. M. F. (1991). Adenosine A1 receptor-induced inhibition of Ca^{2+} transients linked to action potentials in clonal pituitary cells. *Eur. J. Pharmacol.* **206,** 271–277.

Montmayeur, J.-P., Guiramand, J., and Borrelli, E. (1993). Preferential coupling between dopamine D2 receptors and G-proteins. *Mol. Endocrinol.* **7,** 161–170.

Moxham, C. M., Hod, Y., and Malbon, C. C. (1993). Induction of Giα2-specific antisense RNA in vivo inhibits neonatal growth. *Science* **260,** 991–995.

Murdoch, G. H., Waterman, M., Evans, R. M., and Rosenfeld, M. G. (1985). Molecular mechanisms of phorbol ester, thyrotropin-releasing hormone, and growth factor stimulation of prolactin gene transcription. *J. Biol. Chem.* **260,** 11852–11858.

Murray-Whelan, R., and Schegel, W. (1992). Brain somatostatin receptor-G protein interaction. *J. Biol. Chem.* **267,** 2960–2965.

Nishimoto, I., Okamoto, T., Matsuura, Y., Takahashi, S., Okamoto, T., Murayama, Y., and Ogata, E. (1993). Alzheimer amyloid protein precursor complexes with brain GTP-binding protein Go. *Nature (London)* **362,** 75–79.

Nishizuka, Y. (1988). The molecular heterogeneity of protein kinase C and its implications for cellular regulation. *Nature (London)* **334,** 661–665.

Osborne, R., and Tashjian, A. H., Jr. (1981). Tumor-promoting phorbol esters affect production of prolactin and growth hormone by rat pituitary cells. *Endocrinology* **108,** 1164–1170.

Ostrowski, J., Kjelsberg, M. A., Caron, M. G., and Lefkowitz, R. J. (1992). Mutagenesis of the β_2-adrenergic receptor: How structure elucidates function. *Annu. Rev. Pharmacol. Toxicol.* **32,** 167–183.

Ozawa, S. (1981). Biphasic effect of thyrotropin-releasing hormone on membrane K^+ permeability in rat clonal pituitary cells. *Brain Res.* **209,** 240–244.

Park, D., Jhon, D., Lee, C., Lee, K., and Rhee, S. G. (1993). Activation of phospholipase C isozymes by G protein $\beta\gamma$ subunits. *J. Biol. Chem.* **268,** 4573–4576.

Paulssen, E. J., Paulssen, R. H., Haugen, T. B., Gautvik, K. M., and Gordeladze, J. O. (1991). Cell specific distribution of guanine nucleotide-binding regulatory proteins in rat pituitary tumour cells. *Mol. Cell. Endocrinol.* **76,** 45–53.

Perez-Reyes, E., Kim, J. S., Lacerda, A. E., Horne, W., Wei, X., Rampe, D., Campbell, K., Brown, A. M., and Birnbaumer, L. (1990). Induction of calcium currents by the expression of the α1-subunit of the dihydropyridine receptor from skeletal muscle. *Nature (London)* **340,** 233–236.

Pinkas-Kramarski, R., Edelman, R., and Stein, R. (1990). Indications for selective coupling to phosphoinositide hydrolysis or to adenylate cyclase inhibition by endogenous muscarinic receptor subtypes M3 and M4 but not by M2 in tumor cell lines. *Neurosci. Lett.* **335,** 335–340.

Pouysségur, J., and Seuwen, K. (1992). Transmembrane receptors and intracellular pathways that control cell proliferation. *Annu. Rev. Physiol.* **54,** 195–210.

Ramsdell, J. S. (1991). Transforming growth factor-α and -β are potent and effective inhibitors of GH4 pituitary tumor cell proliferation. *Endocrinology* **128,** 1981–1990.

Raymond, J. R. (1991). Protein kinase C induces phosphorylation and desensitization of the human 5-HT1A receptor. *J. Biol. Chem.* **266,** 14747–14753.

Rebecchi, M. J., Kolesnick, R. N., and Gershengorn, M. C. (1983). Thyrotropin-releasing hormone stimulates rapid loss of phosphatidylinositol and its conversion to 1,2-diacylglycerol and phosphatidic acid in rat mammotropic cells. Association with calcium mobilization and prolactin secretion. *J. Biol. Chem.* **258**, 227–234.

Rhee, S. G., Suh, P., Ryu, S., and Lee, S. Y. (1988). Studies of inositol phospholipid-specific phospholipase C. *Science* **244**, 546–550.

Ritchie, A. K. (1987). Thyrotropin-releasing hormone stimulates a calcium-activated potassium current in a rat anterior pituitary cell line. *J. Physiol. (London)* **385**, 611–625.

Rogawski, M. A. (1988). Transient outward current (IA) in clonal pituitary cells: Blockade by aminopyridine analogs. *Naunyn-Schmied. Arch. Pharmacol.* **338**, 125–132.

Rogawski, M. A. (1989). Aminopyridines enhance opening of calcium activated potassium channels in GH3 anterior pituitary cells. *Mol. Pharmacol.* **35**, 458–468.

Rogawski, M. A., Inoue, K., Suzuki, S., and Barker, J. L. (1988). A slow calcium-dependent chloride conductance in clonal anterior pituitary cell line. *J. Neurophysiol.* **59**, 1854–1870.

Ross, E. M. (1989). Signal sorting and amplification through G protein-coupled receptors. *Neuron* **3**, 141–152.

Ryu, S. H., Kim, U., Wahl, M. I., Brown, A. B., Carpenter, G., Huang, K., and Rhee, S. G. (1990). Feedback regulation of phospholipase C-β by protein kinase C. *J. Biol. Chem.* **265**, 17941–17945.

Scammell, J., and Dannies, P. S. (1983). Veratridine and ouabain stimulate calcium-dependent prolactin release. *Endocrinology* **113**, 1228–1235.

Schlegel, W., Roduit, C., and Zahnd, G. R. (1984). Polyphosphoinositide hydrolysis by phospholipase C is accelerated by thyrotropin releasing hormone (TRH) in clonal rat pituitary cells (GH cells). *FEBS Lett.* **168**, 5459.

Schlegel, W., Wuarin, F., Zbaren, C., Wollheim, C. B., and Zahnd, G. R. (1985). Pertussis toxin selectively abolishes hormone induced lowering of cytosolic calcium in GH3 cells. *FEBS Lett.* **189**, 27–32.

Schlegel, W., Winiger, B. P., Mollard, P., Vacher, P., Wuarin, F., Zahnd, G. R., Wollheim, C. B., and Dufy, B. (1987). Oscillations of cytosolic Ca^{2+} in pituitary cells due to action potentials. *Nature (London)* **329**, 719–721.

Schonbrunn, A., and Loose-Mitchell, D. (1993). Somatostatin receptor 1 (SSR1) is expressed in GH4C1 pituitary cells. *Proc. Endocrinol. Soc.* **75**, 476.

Schonbrunn, A., and Tashjian, A. H., Jr. (1978). Characterization of functional receptors for somatostatin in rat pituitary cells in culture. *J. Biol. Chem.* **253**, 6473–6483.

Schonbrunn, A., Krasnoff, M., Westendorf, J., and Tashjian, A. H., Jr. (1980). Epidermal growth factor and thyrotropin releasing hormone act similarly on a clonal pituitary cell strain. *J. Cell Biol.* **85**, 786–797.

Sellar, R. E., Taylor, P. L., Lamb, R. F., Zabavnik, J., Anderson, L., and Eidne, K. A. (1993). Functional expression and molecular characterization, of the thyrotropin-releasing hormone receptor from the rat anterior pituitary gland. *J. Endocrinol.* **10**, 199–206.

Simasko, S. M. (1991a). Evidence for a delayed rectifier-like potassium current in the clonal rat pituitary cell line GH3. *Am. J. Physiol.* **261**, E66–E75.

Simasko, S. M. (1991b). Reevaluation of the electrophysiological actions of thyrotropin-releasing hormone in a rat pituitary cell line (GH3). *Endocrinology* **128**, 2015–2026.

Simmonds, W., Goldsmith, P., Codina, J., Unson, C., and Spiegel, A. M. (1989). Giα2 mediates *alpha2*-adrenergic inhibition of adenylyl cyclase in platelet membranes: *In situ* identification of Gα C-terminal antibodies. *Proc. Natl. Acad. Sci. USA* **86**, 7809–7813.

Simon, M., Strathmann, M. P., and Gautam, N. (1991). Diversity of G proteins in signal transduction. *Science* **252**, 802–808.
Solomon, M. J. (1993). Activation of the various cyclin/cdc2 protein kinases. *Curr. Opin. Cell Biol.* **5**, 180–186.
Spedding, M., and Kenny, B. (1992). Voltage-dependent calcium channels: Structures and drug-binding sites. *Biochem. Soc. Trans.* **20**, 147–153.
Straub, R. E., and Gershengorn, M. C. (1986). Thyrotropin-releasing hormone and GTP activate inositol trisphosphate formation in membranes isolated from rat pituitary cells. *J. Biol. Chem.* **261**, 2712–2717.
Strittmatter, S. M., Valenzuela, D., Kennedy, T. E., Neer, E. J., and Fishman, M. C. (1990). Go is a major growth cone protein subject to regulation by GAP-43. *Nature (London)* **344**, 836–841.
Strittmatter, S. M., Valenzuela, D., Sudo, Y., Linder, M. E., and Fishman, M. C. (1991). An intracellular guanine nucleotide release protein for Go. GAP-43 stimulates isolated α subunits by a novel mechanism. *J. Biol. Chem.* **266**, 22465–22471.
Tallent, M., and Reisine, T. (1992). Giα1 selectively couples somatostatin receptors to adenylyl cyclase in pituitary-derived AtT-20 cells. *Mol. Pharmacol.* **41**, 452–455.
Tan, K.-N., and Tashjian, A. H., Jr. (1981). Receptor-mediated release of plasma membrane-associated calcium and stimulation of calcium uptake by thyrotropin-releasing hormone in pituitary cells in culture. *J. Biol. Chem.* **256**, 8094–9002.
Tan, K.-N., and Tashjian, A. H., Jr. (1984a). Voltage-dependent calcium channels in pituitary cells in culture. I. Characterization by $^{45}Ca^{2+}$ fluxes. *J. Biol. Chem.* **259**, 418–426.
Tan, K.-N., and Tashjian, A. H., Jr. (1984b). Voltage-dependent calcium channels in pituitary cells in culture. II. Participation in thyrotropin-releasing hormone action on prolactin release. *J. Biol. Chem.* **259**, 427–434.
Tang, W., and Gilman, A. G. (1991). Type-specific regulation of adenylyl cyclase by G protein βγ subunits. *Science* **254**, 1500–1503.
Tang, W., and Gilman, A. G. (1992). Adenylyl cyclases. *Cell* **70**, 869–870.
Tashjian, A. H., Jr. (1979). Clonal strains of hormone-producing pituitary cells. *Meth. Enzymol.* **58**, 527–535.
Tashjian, A. H., Jr., Yasumura, Y., Levine, L., Sato, G. H., and Parker, M. L. (1968). Establishment of clonal strains of rat pituitary tumor cells that secrete growth hormone. *Endocrinology* **82**, 342–352.
Tashjian, A. H., Jr. Heslop, J. P., and Berridge, M. J. (1987). Subsecond and second changes in inositol polyphosphates in GH4C1 cells induced by thyrotropin-releasing hormone. *Biochem. J.* **243**, 305–308.
Taussig, R., Sanchez, S., Rifo, M., Gilman, A. G., and Belardetti, F. (1992). Inhibition of the ω-conotoxin-sensitive calcium current by distinct G proteins. *Neuron* **8**, 799–809.
Taussig, R., Inigues-Lluhi, J. A., and Gilman, A. G. (1993). Inhibition of adenylyl cyclase by Giα. *Science* **261**, 218–221.
Tornquist, K., and Tashjian, A. H., Jr. (1992). pH homeostasis in pituitary GH4C1 cells: Basal intracellular pH is regulated by cytosolic free Ca^{2+} concentration. *Endocrinology* **130**, 717–725.
Vallar, L., Muca, C., Magni, M., Albert, P., Bunzow, J., Meldolesi, J., and Civelli, O. (1990). Differential coupling of dopaminergic D2 receptors expressed in different cell types. *J. Biol. Chem.* **265**, 10320–10326.
Van Sande, J., Raspé E., Perret, J., Lejeune, C., Maenhaut, C., Vassart, G., and Dumont, J. (1990). Thyrotropin activates both the cyclic AMP and the PIP2 cascades in CHO

cells expressing the human cDNA of TSH receptor. *Mol. Cell. Endocrinol.* **74,** R1–R6.

Varrault, A., Bockaert, J., and Waeber, C. (1992). Activation of 5-HT1A receptors expressed in NIH-3T3 cells induces focus formation and potentiates EGF effect of DNA synthesis. *Mol. Biol. Cell* **3,** 961–969.

Watkins, D. C., Johnson, G. L., and Malbon, C. C. (1992). Regulation of differentiation of teratocarcinoma cells into primitive endoderm by $G\alpha_i2$. *Science* **258,** 1373–1375.

Westendorf, J. M., and Schonbrunn, A. (1982). Bombesin stimulates prolactin and growth hormone release by pituitary cells in culture. *Endocrinology* **110,** 352–358.

Westendorf, J. M., and Schonbrunn, A. (1983). Characterization of bombesin receptors in a rat pituitary cell line. *J. Biol. Chem.* **258,** 7527–7535.

White, R. E., Schonbrunn, A., and Armstrong, D. L. (1991). Somatostatin stimulates Ca^{2+}-activated K^+ channels through protein dephosphorylation. *Nature (London)* **351,** 570–573.

White, R. E., Lee, A. B., Shcherbatko, A. D., Lincoln, T. M., Schonbrunn, A., and Armstrong, D. L. (1993). Potassium channel stimulation by natriuretic peptides through cGMP-dependent dephosphorylation. *Nature (London)* **361,** 263–266.

Wojcikiewicz, R. J. H., Dobson, P. R. M., and Brown, B. L. (1984). Muscarinic acetylcholine receptor activation causes inhibition of cyclic AMP accumulation, prolactin and growth hormone secretion in GH3 rat anterior pituitary tumour cells. *Biochim. Biophys. Acta* **805,** 25–29.

Wong, Y. H., Federman, A., Pace, A. M., Zachary, I., Evans, T., Pouysségur, J., and Bourne, H. R. (1991). Mutant α subunits of Gi2 inhibit cyclic AMP accumulation. *Nature (London)* **351,** 63–65.

Yajima, Y., Akita, Y., and Saito, T. (1986). Pertussis toxin blocks the inhibitory effects of somatostatin on cAMP-dependent vasoactive intestinal peptide and cAMP-independent thyrotropin releasing hormone-stimulated prolactin secretion of GH_3 cells. *J. Biol. Chem.* **261,** 2684–2689.

Yan, G., Pan, W. T., and Bancroft, C. (1991). Thyrotropin-releasing hormone action on the prolactin promoter is mediated by the POU protein Pit-1. *Mol. Endocrinol.* **5,** 535–541.

Yatani, A., Codina, J., Sekura, R. D., Birnbaumer, L., and Brown, A. M. (1987). reconstitution of somatostatin and muscarinic receptor mediated stimulation of K^+ channels by isolated GK protein in clonal rat anterior pituitary membranes. *Mol. Endocrinol.* **1,** 283–289.

Yatani, A., Mattera, R., Codina, J., Graf, R., Okabe, K., Padrell, E., Iyengar, R., Brown, A. M., and Birnbaumer, L. (1988). The G protein-gated atrial K^+ channel is stimulated by three distinct Giα subunits. *Nature (London)* **336,** 680–682.

Yatani, A., Okabe, K., Polakis, P., Halenbeck, R., McCormick, F., and Brown, A. M. (1990a). ras p21 and GAP inhibit coupling of muscarinic receptors to atrial K^+ channels. *Cell* **61,** 769–776.

Yatani, A., Quilliam, L. A., Brown, A. M., and Bokoch, G. M. (1990b). Rap1A antagonizes the ability of *ras* and *ras*-Gap to inhibit muscarinic K^+ channels. *J. Biol. Chem.* **266,** 22222–22226.

Zhao, D., Yang, J., Jones, K. E., Gerald, C., Suzuki, Y., Hogan, P. G., Chin, W. W., and Tashjian, A. H., Jr. (1992). Molecular cloning of a complementary deoxyribonucleic acid encoding the thyrotropin-releasing hormone receptor and regulation of its messenger ribonucleic acid in rat GH cells. *Endocrinology* **130,** 3529–3536.

Receptors for the TGF-β Ligand Family

CRAIG H. BASSING, JONATHAN M. YINGLING, AND XIAO-FAN WANG

Department of Pharmacology
Duke University Medical Center
Durham, North Carolina 27710

I. Introduction
II. TGF-β Ligands
   A. TGF-β Family
   B. TGF-β Superfamily
   C. TGF-β Structure
   D. Structural and Functional Relationship of TGF-β
III. TGF-β Receptors
   A. Identification of Cell Surface TGF-β Binding Proteins
   B. Structural Features of the TGF-β Receptors
IV. TGF-β Signal Transduction
   A. TGF-β Signaling Receptors
   B. Genetic Evidence for a Signaling Complex
   C. Signaling through a Heteromeric Receptor Complex
   D. Identification of a Functional Type I Receptor
   E. Mechanisms of TGF-β Receptor Activation
   F. Two Receptor Pathways and Signaling Thresholds
   G. Other Signaling Mechanisms
   H. Type III Receptor/Betaglycan Presents TGF-β to the Type II Receptor
   I. Role of Endoglin
V. Perspectives and Future Directions
References

I. INTRODUCTION

The transforming growth factor βs (TGF-βs) are a group of multifunctional peptide hormones that regulate many aspects of cellular function (Lyons and Moses, 1990; Massague, 1990; Roberts and Sporn, 1990). TGF-β was originally identified as a protein that induced normal rat kidney fibroblasts to proliferate in soft agar in the presence of epidermal growth factor (EGF). TGF-β can inhibit the differentiation of certain mesodermal cells, induce the differentiation of others, and inhibit proliferation of cells derived from epithelial, endothelial, neuronal, hematopoietic, lymphoid, and fibroblastic origins. In addition to its effects on individual cells, TGF-β is also important in many biological processes (Lyons and Moses, 1990; Massague, 1990; Roberts and

Sporn, 1990). For example, TGF-β is an important regulator of immune responses, wound healing, cell adhesion, cell–cell recognition, and extracellular matrix deposition. Deregulation of TGF-β function is implicated in the pathological processes of several diseases including arthritis, atheosclerosis, and glomerulonephritis. In addition, cells that lose the ability to respond to TGF-β may be more likely to exhibit uncontrolled proliferation and become tumorigenic. Therefore, a deeper understanding of the mechanisms of TGF-β signaling may improve our comprehension of several diseases and ultimately may lead to improved treatments.

Because of the widespread importance of TGF-β in many normal biological processes and the implication of TGF-β involvement in several diseases, tremendous effort has been invested in studying the molecular mechanisms of TGF-β action. Most studies have concentrated on two objectives: (1) identification of the TGF-β receptors and (2) elucidation of their intracellular signaling pathways. Initial efforts involving affinity labeling of cells with radio-iodinated TGF-β helped identify several cell-surface proteins that could specifically bind TGF-β with high affinity (Massague, 1990; Segarini, 1993). These binding proteins were biochemically characterized subsequently, but the molecular nature of receptors capable of signaling TGF-β responses remained unknown until recently. Elucidation of the intracellular TGF-β signaling pathways has been hindered by the lack of knowledge about the TGF-β receptors. None of the known intracellular signaling pathways, including tyrosine phosphorylation, phosphatidyl inositol turnover, and modulation of intracellular cAMP and Ca^{2+} levels has been convincingly demonstrated to be directly linked to the actions of TGF-β (Massague, 1990; Segarini, 1993).

Over the past few years, cDNAs for the most widely distributed and possibly the most biologically relevant receptors for the ligands of the TGF-β superfamily have been cloned. At least three of these receptors belong to an emerging family of transmembrane serine/threonine receptor kinases. The kinases of these receptors are able to autophosphorylate on serine/threonine residues; this catalytic activity is essential for signal transduction. The TGF-β receptors utilize this novel activity to transduce signals through largely undefined pathways. This functional feature of the TGF-β receptors distinguishes them from the transmembrane receptors with intrinsic tyrosine kinase activity, many of which act to transduce mitogenic or trophic signals into the cell in response to their respective growth factors. In contrast, the TGF-β receptors transduce signals that result in TGF-β-induced gene responses and growth inhibition.

This chapter summarizes the current status of research in the TGF-β receptor field. Most of the discussion focuses on the type I, II, and III receptors, since they are widely expressed and appear to be the most biologically relevant mediators of TGF-β signal transduction. Throughout the discussion, we attempt to address current questions and present the challenges that await researchers studying the molecular mechanisms of TGF-β action.

II. TGF-β Ligands

A. TGF-β Family

Any detailed discussion concerning a group of receptors must start with the ligands that interact with the receptors to initiate biological responses. Five structurally related TGF-β molecules (TGF-β1, 2, 3, 4, and 5) constitute the TGF-β family (Massague, 1990; Roberts and Sporn, 1990). The cDNAs for TGF-β1, TGF-β2, and TGF-β3 have all bone cloned from mammalian sources, whereas the cDNAs for TGF-β4 and TGF-β5 have only been isolated from chicken and *Xenopus,* respectively. The biologically active polypeptide TGF-β hormones are all 25-kDa disulfide-linked dimers. The TGF-β monomers are peptides of 112 amino acid residues, except for the TGF-β4 isoform, which has 114 residues. The five isoforms exhibit approximately 64–82% identities at the amino acid level. The structural similarity includes nine invariant cysteine residues, suggesting that these strictly conserved cysteines may play important structural or functional roles.

The individual TGF-β isoforms exhibit a startling degree of conservation (97%) among different mammalian species (Massague, 1990; Roberts and Sporn, 1990). This conservation indicates a strong evolutionary pressure to maintain the molecular nature of separate TGF-β isoforms, suggesting that these isoforms may mediate distinct and evolutionarily conserved processes. The observation that genetically manipulated mice with homozygous defects in TGF-β1 expression develop into adult animals without gross developmental abnormalities indicates the existence of a certain degree of functional redundancy for the TGF-β isoforms (Schull *et al.,* 1992). However, the same animals that lack TGF-β1 expression develop a wasting syndrome characterized by multifocal inflammatory cell responses and tissue necrosis that lead to organ failure and death of the animal. Therefore, available evidence indicates that the three mammalian TGF-β isoforms play overlapping but sometimes distinct roles, a notion supported by the observation that the proteins display common but not identical expression patterns during development.

The TGF-βs are synthesized as precursors of approximately 400 amino acids and secreted as 100-kDa latent complexes (Massague, 1990; Roberts and Sporn, 1990). These complexes consist of the active TGF-β dimer noncovalently associated with an N-terminal dimer of the remainder of the precursor peptide. Prior to secretion, the precursor is cleaved at a 4-basic-amino-acid sequence directly preceeding the active dimer without affecting the complex, resulting in the maintenance of TGF-β in a latent form. Dissociation of the latent complex and, thus, release of the active form of TGF-β appears to be regulated, since only the active dimer is capable of binding TGF-β receptors and initiating a biological response.

All three mammalian TGF-β isoforms appear to bind the same signaling type II receptor and elicit identical biological responses (Wrana *et al.*, 1992). However, whether these different isoforms signal through distinct subsets of type I receptors to signal unique responses remains to be seen.

B. TGF-β SUPERFAMILY

The TGF-β family belongs to a larger, extended superfamily of peptide hormones (Massague, 1990; Roberts and Sporn, 1990; Segarini 1993). The TGF-β superfamily includes Mullerian inhibiting substance (MIS), the *Drosophila* morphogen decapentaplegic factor (DPP), the bone morphogenic factors (BMP-2, -3, -4, -5, -6, and -7), the *Xenopus* sequences Vg-1 and Vgr-1, the growth differentiation factors (GDF-1, -3, and -5), the activins (A, B, and AB), and the inhibins (alpha and beta). These polypeptide factors are similar to the TGF-βs in overall structure, but share less than 40% amino acid identity with members of the TGF-β family. The superfamily members contain seven of the nine invariant cysteine residues found in the TGF-βs, with the exception of the inhibin β chain, which contains all nine of these invariant cysteines. Members of the TGF-β superfamily play important roles in modulating a variety of different biological processes, including cell growth, differentiation, and development.

Like TGF-β, members of the superfamily initiate their biological responses by binding to cell-surface receptors. Activin A and BMP-4 selectively bind with high affinities to 55-kDa and 75- to 85-kDa receptors. These receptors have been designated, respectively, as type I and type II receptors and are analogous to the TGF-β receptors of the same designation. The activin type II receptor has been cloned and is a putative serine/threonine receptor kinase (Mathews and Vale, 1991;

Attisano *et al.*, 1992; Mathews *et al.*, 1992). In addition, a *Caenorhabditis elegans* gene, *daf*-1, that also encodes a serine/threonine receptor kinase has been cloned (Georgi *et al.*, 1990). The ligand of *daf*-1 has not yet been identified, but the *daf*-1 signaling pathway is known to be important in cell differentiation. Therefore, members of the TGF-β superfamily are likely to transduce their signals by a common mechanism initiated by the binding of ligands to signaling receptor molecules, the transmembrane serine/threonine receptor kinases. In this chapter, we occasionally use results generated from the study of receptors for the other members of the TGF-β superfamily to gain insight into the mechanisms by which the TGF-β receptors transduce signals.

C. TGF-β STRUCTURE

The crystal structure of TGF-β2 has been solved independently by two different groups (Daopin *et al.*, 1992; Schlunegger and Grutter, 1992). The monomer is approximately 60 Å × 20 Å × 15 Å. The secondary structure of the molecule includes three α helices and nine β strands. The overall architecture of the monomer resembles that of an outstretched hand. The protein contains a pair of slightly twisted, extended antiparallel β strands and a large α helix with an axis that is separated from the strands. The 3-turn α helix forms the "heel" and the β strands form the "fingers" of the TGF-β2 "hand." The monomer contains an extensive network of hydrogen bonds that hold together fingers 1 and 2 and fingers 3 and 4.

The nine strictly conserved cysteines of the TGF-βs are all involved in disulfide bridges. Eight of these cysteines make four intrachain disulfide bridges, whereas the ninth cysteine (Cys 77) forms an interchain disulfide bond. Three of the intramolecular disulfides form an unusual structural feature referred to as the TGF-β "knot." In the TGF-β knot, two of the disulfide bridges form a compact 8-amino-acid ring through which a third disulfide bridge passes. These last two cysteines are involved in a disulfide bond that connects the N-terminal α helix of TGF-β2 to the first finger. The TGF-β knot, in conjunction with the fourth disulfide bridge, forms the highly compact core ("palm") of the TGF-β2 monomer.

The TGF-β2 dimer has a most unusual feature. Unlike other dimeric proteins, the interface of the monomers is not continuous but contains a large water-filled cavity. A total of eight water molecules fills the cavities between the hydrophobic interfaces and the interchain disulfide bridge. These water molecules are involved in an extended hydrogen bonding network between the dimers, which may stabilize the structure. In the active TGF-β2 dimer, the monomers pack in an antiparallel configuration. The monomers assemble "heels to fingers" with the palms of the subunits oriented toward the dimer interface. The monomers are held together by hydrophobic interfaces between residues of the long α helix and several of the β strands. The interchain disulfide bond also stabilizes the dimer.

As described earlier, all members of the TGF-β superfamily share at least seven of the nine invariant cysteines as well as a significant degree of sequence homology. In the crystal structure of TGF-β2, these invariant cysteine residues are involved in disulfide bridges that constitute the TGF-β knot. Most of the amino acids involved in the dimer interface and the hydrogen bonding network also belong to the invariant group of residues conserved among members of the TGF-β superfamily. This observation led to the proposal that the TGF-β knot, the monomer fold, and the orientation of the monomers in the dimer would be characteristic for all members of the TGF-β superfamily.

Although the overall structures of the five TGF-β isoforms are very similar, some minor but important structural differences in these molecules may contribute to the specific biological activity of each isoform. For instance, variant residues located in the solvent-accessible regions of the TGF-β dimer, including the N-terminal helix, the C-terminal end of the heel α helix, the loop connecting fingers 1 and 2, and the turn connecting fingers 3 and 4 may play important roles in determining the binding affinity and specificity of each isoform for the signaling receptors.

The crystal structure of TGF-β2 and the NMR-determined structure of TGF-β1 in solution (Archer et al., 1993) reveal that the two isoforms contain the same monomeric fold and dimeric composition. This result is not surprising because of the high degree (70%) of sequence similarity between the two molecules. However, in two regions the structures differ from one another: the loop between fingers 3 and 4 and the amino acids stretches connecting the long α helix to the β sheets. Both these regions are in solvent-accessible areas of the TGF-β dimer structure. Researchers proposed that these secondary structures could form the interacting area with the TGF-β receptors. Whether the biological and biochemical differences of the two TGF-β isoforms are actually derived from these differences in their molecular structures remains to be seen.

D. STRUCTURAL AND FUNCTIONAL RELATIONSHIP OF TGF-β

As discussed earlier, the structures of two TGF-β isoforms, TGF-β1 and TGF-β2, have been solved. These molecules share a common tertiary structure and interact with the same receptors, although with different affinities. However, whether the observed functional variety of the isoforms results from their structural differences is not clear. To address this question, Qian et al. (1992) created a chimeric TGF-β molecule from the two isoforms with different biological potencies.

TGF-β1 and TGF-β2 differ in their ability to inhibit growth of certain cell lines. Whereas TGF-β1 and TGF-β2 are equally potent in inhibiting the growth of mink lung epithelial (Mv1Lu) cells, TGF-β1 is 100 times more potent in causing growth arrest of fetal bovine heart endothelial (FBHE) cells (Cheifetz et al., 1990). This difference in biological potency may derive from a difference in binding affinity for the signaling receptors by the two isoforms. To test this hypothesis, a chimeric molecule was constructed from TGF-β1 and TGF-β2 in an attempt to map the ligand binding domain(s) of TGF-β (Qian et al., 1992). The chimeric monomer consists of amino acids 1–39 and 83–112 of TGF-β2 and amino acids 40–82 of TGF-β1. The chimeric TGF-β(2-1-2) was expressed in NIH-3T3 cells and tested for its biological potency on the Mv1Lu and FBHE cell lines. TGF-β(2-1-2) was found to be as effective as TGF-β1 in inhibiting the growth of both Mv1Lu cells and FBHE cells. This result indicates that the chimeric TGF-β(2-1-2) retains a biological activity that resembles that of TGF-β1 and that the region constructed of amino acids 40–82 is likely to contain a binding domain for the receptors. In this region, TGF-β1 and TGF-β2 differ only in 14 amino acids among the 42 residues.

Based on this result of functional study of the chimeric TGF-β, additional structural analysis of TGF-β1 and TGF-β2 molecules reveals that the region of amino acids 40–82 contains a specific domain with a large 3-turn α helix (heel of the monomer) and a stretch with residues connecting the domain to the rest of the monomer. This type of structural analysis predicts that this particular region of the TGF-β molecule constitutes the potential ligand-binding site, a conclusion that is consistent with the results of functional analysis of the chimeric ligand. The relatively small difference in the peptide sequences (14 aa) and the minor differences in structural features of TGF-β1 and TGF-β2 suggest that subtle amino acid substitutions in a region involved in ligand–receptor contact could significantly influence the biological potencies of the TGF-β isoforms.

III. TGF-β Receptors

A. Identification of Cell Surface TGF-β Binding Proteins

Like other polypeptide hormones, TGF-β initiates its signaling cascade by interacting with specific cell-surface receptors on its target cells. Available experimental evidence suggests that almost all cell types are responsive to TGF-β. Consequently, cell-surface TGF-β binding proteins have been detected in most types of cells and tissues (Massague et al., 1990). A classical biochemical assay has been widely used to identify cell surface TGF-β binding proteins. The assay consists of chemically cross-linking radio-iodinated TGF-β to cell-surface proteins using a bifunctional reagent DSS. The cross-linked proteins are subsequently analyzed by gel electrophoresis and characterized according to their molecular weight. Using this assay, at least nine membrane-associated proteins have been identified that specifically bind TGF-β with relatively high affinities.

The type I, type II, and type III TGF-β receptors were the first TGF-β binding proteins identified by this method, primarily because of their near ubiquitous distribution in most mammalian cell lines of various origins (Massague et al., 1990). The type III receptor is the most abundant TGF-β binding protein expressed on the surface of many types of cells. Almost all TGF-β responsive cell lines express both type I and type II receptors, although the levels of their expression may vary among different cell types. The type I and type II receptors bind to TGF-β1 and TGF-β3 with a higher affinity than they bind TGF-β2, whereas the type III receptor binds all three isoforms with lower but equivalent affinity.

Several other TGF-β binding proteins with a more limited distribution have been identified through the ligand-binding and cross-linking assay. The type IV receptor is a 70- to 74-kDa protein only found in pituitary tumor cells (Cheifetz et al., 1988b). In addition to binding TGF-β, the type IV receptor binds activin and inhibin with an equally high affinity. This cross-reactivity of ligand binding for the type IV receptor may explain the observation that these ligands are functionally equipotent when applied to pituitary cells (Cheifetz et al., 1988b). In addition, certain receptors of the TGF-β superfamily may interact with different members of the ligand superfamily simply because of their shared sequence and structural similarities. Although it may occur for the type IV receptor, whether this cross-reactivity in ligand binding can always translate into functional consequences remains to be seen. We discuss this issue in more detail in the context of the type I receptors.

The type VI receptor is a 180-kDa protein found in most types of cells. The type VI receptor shows an unusual ligand-binding characteristic: it binds TGF-β2 with very low affinity but its binding to TGF-β1 is dependent on the presence of a high concentration of TGF-β2 (Segarini et al., 1992). The mechanism of this cooperative binding is unknown. BHE and MG-63 human osteosarcoma cells express a TGF-β1 binding protein of 180 kDa and TGF-β2 binding proteins of 60 and 140 kDa (Cheifetz and Massague, 1991). These receptors are attached to the plasma membrane through phosphatidyl inositol anchors and could be secreted by the actions of phospholipase.

The only two TGF-β binding proteins that were not initially characterized by the classic binding assay were the type V receptor and endoglin. The type V receptor is a 400-kDa protein originally identified in an attempt to purify the type I, type II, and type III receptors (O'Grady et al., 1991a). Except in some epithelial tumor lines, this receptor is widely expressed (O'Grady et al., 1991b). Endoglin is a cell-surface protein composed of two 95-kDa subunits found in vascular endothelial cells, erythroblasts, macrophages, and various myeloid and lymphoid leukemias (Gougos and Letarte, 1990). Endoglin binds TGF-β1 and TGF-β3 with much higher affinity than it binds TGF-β2 (Cheifetz et al., 1992). Cells that express endoglin typically do not express the type III TGF-β receptor.

Currently, only the type I, type II, and type III receptors and endoglin have been cloned, whereas the type V receptor has been characterized only by peptide analysis. The molecular identity of other TGF-β binding proteins and the characterization of their functional roles in TGF-β action are the subject of current research.

B. STRUCTURAL FEATURES OF THE TGF-β RECEPTORS

To understand fully the molecular mechanisms of TGF-β signaling, researchers had to clone the receptors responsible for mediating these actions. Initial attempts to clone the TGF-β receptors relied on conventional approaches, but were largely unsuccessful. Therefore, an expression cloning strategy was developed based on the success of this method in cloning other receptors for growth factors and cytokines. Expression cloning has two attractive advantages over conventional cloning strategies. First, the method requires no biochemical characterization of the receptor since the screening is based solely on the specific interaction between the receptor and its ligand. Second, the signal-to-noise ratio is not a major factor, since visual screening is used to identify positive clones. The type II and type III TGF-β receptors were eventually cloned by this approach (Wang et al., 1991; Lin et al., 1992). The type III receptor was also cloned independently through a polymerase chain reaction (PCR) cloning strategy that utilized all the possible oligonucleotides encoding peptide sequences derived from several tryptic fragments of the purified receptor (Lopez-Casillas et al., 1991).

The molecular characterization of the type III receptor resulted in the identification of an additional TGF-β binding protein, endoglin, which shares structural and sequence homology with the type III receptor and was originally characterized as a transmembrane protein without a known function (Gougos and Letarte, 1990).

The type I receptor cannot bind TGF-β in the absence of a type II receptor (Wrana et al., 1992; C. H. Bassing et al., 1994). Consequently, isolating cDNAs encoding type I receptors by expression cloning, which requires the detecting of positive clones through ligand binding, was not possible. Several proteins of the transmembrane serine/threonine kinase family have been cloned using a PCR technique based on protein sequence information derived from the type II receptors for activin, TGF-β, and the product of the *C. elegans* gene *daf*-1 (Attisano et al., 1993; Ebner et al., 1993; Franzen et al., 1993; He et al., 1993; Matsuzaki et al., 1993; ten Dijke et al., 1993; Tsuchida et al., 1993). One of these molecules was identified as a type I TGF-β receptor based on its ability to restore a TGF-β-induced gene response in a mutant cell line lacking the expression of endogenous type I receptor (Franzen et al., 1993; C. H. Bassing et al., 1994).

The TGF-β type V receptor has not been cloned. The sequence of a peptide fragment from the purified receptor indicates that the cytoplasmic domain of the type V receptor may also be a member of the emerging class of serine/threonine receptor kinases (O'Grady et al., 1992). The structural features of these receptors will be discussed further in the order in which they were cloned and/or characterized.

1. *Type III Receptor*

The type III receptor, alternatively referred to as β-glycan, is a transmembrane proteoglycan with a molecular mass varying between 280 and 330 kDa, primarily because of its heavily modified and heterogeneous glycosaminoglycan (GAG) chains (Cheifetz et al., 1988a; Segarini and Seyedin, 1988). The protein core of the type III receptor is approximately 100 kDa and can bind TGF-β directly without the modification of its GAG chains. A soluble form of the type III receptor, lacking the transmembrane and cytoplasmic domains, has been found in the serum and extracellular matrix of cultured fibroblasts (Andres et al., 1989).

As mentioned earlier, the type III TGF-β receptor was cloned using a COS cell-transfection- and ligand-binding-based expression cloning strategy (Wang et al., 1991). A cDNA clone, derived from the rat vascular smooth muscle cell line A10, was shown to cause increased radiolabeled TGF-β1 binding when expressed in COS cells. This cDNA clone was found to lack the coding region for the cytoplasmic portion of the molecule. Subsequently, the cDNA was used as a probe to isolate a full-length cDNA clone from a rat fibroblast cDNA library. When expressed in COS cells, this new clone gave rise to a significant increase in TGF-β1 binding to a group of cross-linked proteins with the typical characteristics of proteoglycans, ranging from 100 to 280 kDa. The binding of radiolabeled ligand to this species of protein was specific, since the binding could be significantly reduced when excess unlabeled ligand was included in the binding assay. Additional evidence suggesting that the cDNA clone encodes the type III receptor derived from an assay in which the transfected COS cells were treated with deglycosylating enzymes that specifically remove the GAG chains. This treatment reduced the molecular mass of the heterogeneous TGF-β-binding species to a single component of approximately 100 kDa, a molecular mass characteristic of the type III receptor protein core. This result is also consistent with earlier biochemical results that showed that TGF-β is able to bind the type III receptor when it is stripped of its GAG chains. Stable expression of the isolated cDNA clone in L6 rat skeletal myoblasts, which lack endogenous type III TGF-β receptor expression, led to the same conclusion derived from expression in COS cells. Based on these characteristics, the isolated clone was confirmed to be the type III TGF-β receptor.

The type III receptor was cloned independently by another group using synthesized oligonucleotides for peptide sequences deduced from several tryptic fragments of the purified receptor (Lopez-Casillas et al., 1991). A DNA fragment was generated in a PCR reaction using these oligonucleotides as primers, and was used as a probe to isolate a cDNA clone from a Rat-1 fetal rat fibroblast cDNA library. Affinity labeling of COS cells transfected with the isolated full-length cDNA gave rise to an increased binding of TGF-β to a protein that exhibited biochemical properties characteristic of the type III TGF-β receptor.

The type III receptor cDNA encodes a protein of 853 amino acids (Lopez-Casillas et al., 1991; Wang et al., 1991). The protein contains a predicted 23-residue signal peptide, a 752-amino-acid hydrophilic extracellular domain, a 27-amino-acid hydrophobic transmembrane domain, and a very short cytoplasmic domain of only 41 amino acids. The extremely large extracellular domain, which composes nearly the entire molecular topology of the receptor, is a most unusual feature for cytokine receptors. The extracellular domain of the receptor contains six consensus N-linked glycosylation sites and 15 cysteine residues. This domain also contains a polyproline stretch (PIPPPP) at the base of the extracellular domain that has been proposed to act as a flexible hinge (Lopez-Casillas et al., 1991). The cytoplasmic domain contains no known signaling motifs. The most notable feature of this short stretch of amino acids is that it contains 42% serine and threonine residues, mostly located in clusters. The cytoplasmic domain also contains a potential protein kinase C (PKC) phosphorylation site.

Shortly after the rat type III TGF-β receptor was cloned, sequences for the porcine and human receptors were obtained from uterine and placental libraries, respectively (Moren et al., 1992). The human receptor exhibits 83 and 81% amino acid identity to the porcine and rat receptors, respectively. The human type III receptor contains six potential N-linked glycosylation sites and five potential GAG attachment sites. Five of the six glycosylation sites and three of the possible GAG attachment sites are conserved among the three receptors. Among the extracellular domains, 15 cysteine residues are strictly conserved. The polyproline hinge region is also present in all three proteins. The transmembrane and cytoplasmic domains share great homology and display very minor sequence differences.

Most recently, a chick type III receptor was cloned from a 17-day *in ovo* chick brain cDNA library (J. V. Barnett, A. Moustakos, W. Lin, X. F. Wang, H. Y. Lin, G. B. Galper, R. A. Weinberg, H. F. Lodish, and R. L. Maas, unpublished data). The chick clone contains an open reading frame of 841 amino acids. Sequence comparison with the 853-amino-acid rat type III receptor shows that the proteins share 78% sequence similarity and 67% identity. The chick receptor contains the 15 invariant cysteines found in the mature extracellular domain of the other type III receptors. This receptor also contains the conserved six N-linked glycosylation sites and three potential GAG attachment sites. The polyproline sequence near the transmembrane domain is not present in the chick sequence. The transmembrane and cytoplasmic domains of the chick type III receptor also exhibit an extremely high degree of conservation. The chick, porcine, and human receptors all lack the consensus PKC phosphorylation site that was found in the cytoplasmic domain of the rat type III receptor.

The type III receptor has been found in serum and extracellular matrices as a soluble protein lacking a transmembrane anchor and the cytoplasmic domain (Andres et al., 1989). To determine whether this soluble form of the receptor is encoded by the same mRNA transcript, the type III receptor cDNA was transfected into COS cells and the cultured medium was analyzed for the presence of the soluble receptor (Lopez-Casillas et al., 1991). This form of the type III receptor was found to be secreted from these transfected cells. Therefore, a single transcript is likely to encode both the membrane-anchored and the soluble form of the type III receptor. The soluble receptor derives from the membrane-anchored receptor by specific proteolytic cleavage. Indeed, sequence analysis of rat β-glycan revealed the presence of two potential proteolytic cleavage sites near the base of the extracellular domain (Lopez-Casillas et al., 1991). One of these is a dibasic site (KK) susceptible to cleavage by trypsin-like proteases. The other site contains an elastase-like cleavage sequence (LAVV) that is identical to the cleavage site of the membrane bound TGF-α precursor. Of these two potential proteolytic sites, only the dibasic cleavage site is conserved among the other three species. This observation has led to the suggestion that cleavage at the dibasic site may be sufficient to generate the soluble type III receptor. The regulation of soluble receptor production and the functional role of its release to the matrix and serum remain to be determined.

The stringent conservation of peptide sequences in the transmembrane and cytoplasmic domains of the type III receptors of different species suggests that these domains serve important, but unknown, functional roles in the modulation of TGF-β signal transduction. For instance, the transmembrane domain may interact with other transmembrane proteins to form complexes of a higher order in the plane of the membrane. Likewise, the cytoplasmic domain of the type III receptor could play a role in TGF-β signal transduction by interacting with other signaling components. Several other cytokine receptors that lack signaling motifs in their intracellular domains are known to transduce signals through other signaling components. These molecules are non-covalently associated with the receptors and contain their own intrinsic catalytic activity. Alternatively, type III receptors may not directly participate in TGF-β signal transduction, but may modulate signal transduction initiated by the other signaling receptors.

Sequence analysis also reveals that the TGF-β type III receptor contains regions of significant homology with several other proteins. The cytoplasmic domain, transmembrane domain, and parts of the extracellular domain are highly related to the corresponding regions of endoglin (discussed in the next section). A pattern-based sequence comparison indicates that the type III receptor contains a conserved domain that is located between amino acids 432 and 730 of the extracellular domain (Bork and Sander, 1992). This homology is shared by several proteins with various functions, including the urine protein uromodulin, the zymogen granule membrane protein GP-2, and two sperm receptors of the zona pellucida (ZP2, 3). All these proteins are heavily glycosylated and are secreted as soluble proteins. Interestingly, three of these proteins, like the type III receptor, have also been found in the form of integral membrane proteins. The significance of this shared conserved domain remains unknown.

2. Endoglin

Endoglin is expressed at low levels in erythroblasts, macrophages, and various myeloid and lymphoid leukemic cells, and at high levels in vascular endothelial cells and syncytiotrophoblasts. Endoglin was originally identified as an antigen recognized by a monoclonal antibody 44G4 generated against a pre-B cell leukemic cell line (Gougos and Letarte, 1990). This molecule was subsequently characterized as a glycoprotein of two 95-kDa disulfide-linked homodimers. The endoglin gene was isolated from a human umbilical vein endothelial (HUVE) cell cDNA library and encodes a 633-amino-acid protein with a predicted molecular mass of 68 kDa. The protein contains a 561-amino-acid extracellular domain, a 27-residue hydrophobic transmembrane domain, and a very short cytoplasmic domain of only 47 amino acids. The extracellular domain contains four potential N-linked glycosylation sites and 16 cysteine residues. An Arg–Gly–Asp (RGD) sequence is located in a region of the extracellular domain that is predicted to be exposed to solvent and, therefore, may be involved in the binding of integrins. Like the type III receptor, the intracellular domain of endoglin contains a high proportion of serine and threonine residues.

As previously discussed, endoglin shares 27% amino acid identity with the type III TGF-β receptor. The overall structures of the two molecules are similar, both being large transmembrane glycoproteins with very short cytoplasmic domains. The transmembrane and cytoplasmic domains of endoglin and the type III receptor are highly homologous, with 63% identity. Less sequence homology is seen between the two extracellular domains, although two small stretches of highly conserved regions with identities of 41 and 48% are seen. The positions of eight cysteine residues are also highly conserved in the extracellular domain.

Although endoglin was predicted to play a role in cell adhesion and/or cell–cell recognition based on its structural features (Gougos and Letarte, 199), such functions have not been demonstrated. Once the structural similarity between the type III TGF-β receptor and endoglin was revealed, the ability of endoglin to bind TGF-β was tested (Cheifetz et al., 1992). The profile of cell-surface TGF-β binding proteins was re-examined in HUVE cells, from which endoglin was originally isolated. Researchers found that HUVE cells display the classic type I and type II receptors in addition to a disulfide-linked 180-kDa TGF-β binding protein on the cell surface. Levels of type III receptor were undetectable. This 180-kDa protein ran at approximately 98 kDa under reducing conditions, suggesting that the molecule consisted of two subunits linked by disulfide bonds. Lysates from HUVE cells affinity-labeled with radioactive TGF-β1 were immunoprecipitated with the endoglin-specific monoclonal antibody 44G4 and analyzed. The resulting profile of precipitated proteins indicated that the 180-kDa disulfide-linked dimer was the main species affinity-labeled by TGF-β. Additional analysis in COS cells confirmed that endoglin was indeed the protein labeled by TGF-β. Subsequently, researchers showed that, although endoglin binds TGF-β1 and TGF-β3 with equally high affinity, it cannot bind TGF-β2. This difference in binding capability of different TGF-β isoforms to endoglin correlates closely with their ability to inhibit the growth of endothelial cells. For example, investigators demonstrated that TGF-β1 and TGR-β3 are 100 times more effective than TGF-β2 in inhibiting the proliferation of certain endothelial cells. This result suggests that endoglin may be directly involved in modulating TGF-β induction of growth arrest in endothelial cells.

As described earlier, the transmembrane and cytoplasmic domains are stringently conserved in the type III receptor and endoglin. Perhaps these domains play important roles in modulation of the initiation of TGF-β signal transduction, such as presenting the ligand to the signaling receptors, or in direct mediation of a TGF-β signal to specific downstream pathways, in parallel with the signaling receptors. Endoglin and the type III TGF-β receptor bind TGF-β with high affinities although their extracellular domains share only limited sequence homology. These conserved cysteine residues and the two homologous stretches are likely to form a TGF-β binding pocket. The lack of higher structural similarity in the regions beyond these homologous patches may explain the different binding affinities of endoglin and the type III receptor for the TGF-β isoforms.

3. Type II Receptor

The type II TGF-β receptor is a membrane-associated glycoprotein with a molecular mass of 75–85 kDa, depending on its state of glycosylation (Massague et al., 1990). The receptor is distributed ubiquitously in most cells and tissues, except for a few tumor cell lines. Based on studies with mutant mink lung epithelial cells, researchers concluded that the type II receptor functions as a *bona fide* signaling molecule that is directly involved in transducing TGF-β signals across the membrane (Laiho et al., 1990).

Because of the functional importance of this receptor, a great deal of effort was made to clone the gene encoding the receptor. The gene was eventually cloned using the same expression cloning strategy that yielded the type III receptor cDNA (Lin et al., 1992). Initially, a cDNA encoding a truncated type II receptor was isolated from a porcine renal epithelial cell line (LLC-PK). This cDNA was then used to isolate a full-length cDNA clone from a human hepatoma cDNA library. To verify that the isolated clone encoded a TGF-β receptor, affinity labeling experiments were performed in transiently transfected COS cells. Expression of this clone led to increased TGF-β binding to a protein that comigrated with the endogenous type II receptor. This increased ligand-binding could be competed with specifically by an excess of unlabeled TGF-β1 in the binding reaction. Under the same assaying conditions, unlabeled TGF-β2 competed much less efficiently. This differential ligand-binding ability is characteristic of the type II TGF-β receptor. Based on these observations, the isolated cDNA clone was concluded to encode the type II receptor.

The isolated full-length human cDNA is 4.7 kb in length and contains an open reading frame of 565 amino acids (Lin et al., 1992). The amino acid sequence of the type II receptor predicts a protein with a signal sequence of 23 amino acids, a hydrophilic extracellular domain of 136 amino acids, a hydrophobic transmembrane domain of 30 amino acids, and an intracellular domain of 376 amino acids. The extracellular domain of the type II receptor contains three consensus N-linked glycosylation sites. The majority of the cytoplasmic domain contains the conserved features of a kinase. The sequences that do not constitute the kinase consist of a 56-amino-acid juxtamembrane domain, two short kinase inserts between subdomains VIII–IX and X–XI, and a 22-amino-acid serine/threonine rich C-terminal tail. Comparison of the type II receptor kinase with other known kinases reveals sequence homology to both tyrosine and serine/threonine kinases. However, the type II receptor kinase contains 18 of the 21 invariant residues found in the serine/threonine kinases; the sequence of subdomains VI-B and VIII shows a higher homology to the serine/threonine family of kinases.

To examine the catalytic properties of the type II receptor kinase, the cytoplasmic domain was fused in frame to the glutathione-S-transferase (GST) gene (Lin et al., 1992). Bacterially purified fusion protein was shown to be capable of autophosphorylation, mainly on serine residues, with some threonine activity. Therefore, the cytoplasmic domain of the type II receptor contains an intrinsic kinase activity that has since been shown to be essential for TGF-β signal transduction.

Sequence comparison of the human type II receptor with the cloned portion of the porcine receptor reveals an 88% amino acid identity between the two receptors (85% in the extracellular domain). Since the initial cloning of human and porcine type II TGF-β receptors, cDNAs for the mink and chicken type II receptors have been isolated and sequenced (Wrana et al., 1992; J. V. Barnett, A Moustakas, W. Lin, X. F. Wang, H. Y. Lin, J. B. Galper, R. A. Weinberg, H. F. Lodish, and R. L. Maas, unpublished data). The mink type II receptor shows 94% amino acid identity to the human receptor, with 99% identity in the intracellular domain. Two forms of avian type II TGF-β receptor have been cloned from a chick brain cDNA library (J. V. Barnett, A. Moustakas, W. Lin, X. F. Wang, H. Y. Lin, J. B. Galper, R. A. Weinberg, H. F. Lodish, and R. L. Maas, unpublished data). The first isolated clone contains sequences encoding the extracellular and transmembrane domains, as well as 53 amino acids of the cytoplasmic domain. This truncated receptor contains a stop codon located just before the start of the kinase domain. The second clone encodes a full-length receptor of 554 amino acids. The truncated and full-length receptors are 66 and 83% similar and 44 and 70% identical to the human type II TGF-β receptor. The truncated receptor is less homologous to the human type II receptor because it lacks the highly conserved kinase domain.

An alignment of the human, mink, porcine, and chick type II receptors reveals some interesting structural features. With the exception of the first 40 amino acids of the chick receptor, the extracellular domain and the transmembrane domains exhibit a high degree of similarity. Located within the extracellular domains of the type II receptors are 12 invariant cysteine residues. The first 39 amino acids of the cytoplasmic domain exhibit the same degree of conservation as the extracellular domains. All four receptors, including the truncated chick type II receptor, contain a completely conserved stretch of 15 amino acids (SDISSTCANNINHNT) just upstream of the kinase domain. The kinase domains of the four receptors exhibit more than 97% amino acid identity, including the kinase inserts and the C-terminal tail.

Note that the extracellular domains of the type II receptors from different species do not have a very high level of sequence homology. The three mammalian TGF-β isoforms exhibit greater than 97% homology across species and all bind to the same type II TGF-β receptor. Perhaps the molecular structures of the TGF-β ligands have been subjected to a much greater evolutionary restraint than the receptors, simply because the ligands must interact simultaneously with multiple binding proteins, including the receptors.

The invariant 15 amino acids located immediately on the N-terminal side of the kinase domain may serve an important function in signal transduction and/or modulation of the TGF-β receptor kinase activity. Researchers have shown that phosphorylation of specific tyrosine residues in the juxtamembrane region and in inserts between subdomains of tyrosine kinase receptors serve as specific binding sites for substrates and effectors of the receptor. By analogy, this conserved region may interact with a specific substrate and/or effector of the type II receptor kinase. This type of interaction may be dynamically regulated by the event of ligand binding to the extracellular domain of the receptor. Alternatively, this region may be involved in specific interactions between different partners in the receptor complex, such as type II–type I or type II–type III receptor interactions. Also conceivable is that this region forms a contact point with the C-terminal tail of the receptor, which in turn may serve as a pseudosubstrate that regulates the kinase activity.

Analysis of the truncated chick type II TGF-β receptor indicates that it is generated by a splicing error occurring at an atypical donor site. RNase protection assays fail to detect this transcript in messenger RNAs isolated from cultured cells, suggesting that the truncated receptor is the result of an incompletely processed cDNA. Naturally occurring truncated versions of the activin type II receptor (Nishimatsu et al., 1992), the erythropoietin receptor (Nakamura et al., 1992), and several nerve growth factor tyrosine kinase (NGF/Trk) receptors (Glass and Yancopoulos, 1993) have been characterized. The truncated erythropoietin receptor is unable to prevent apoptosis of erythroid cells, although it retains the capability of initiating the erythropoietin mitogenic signal. Although no biological proof exists, investigators have proposed that other forms of truncated receptors could act similarly as dominant negative receptors to interfere with the functions of their wild-type receptor counterparts. Whether naturally produced truncated TGF-β receptors exist and act to modulate TGF-β signal transduction through a similar mechanism remains to be seen.

4. Type V Receptor

The type V TGF-β receptor is a widely distributed high molecular weight protein originally identified in an attempt to purify the other types of TGF-β receptor from bovine liver (O'Grady et al., 1991a). Using a ligand-based affinity-labeling assay, a specific cross-linked product of 400 kDa was initially detected in cultured cells that comigrated with a 400 kDa protein that was purified from bovine liver to near homogeneity. Subsequent purification of this protein, termed type V receptor, from plasma membrane tissue fractions demonstrated that it was widely expressed. Biochemical studies indicated that the type V receptor could be retained on a lectin column, suggesting that the receptor is a glycoprotein. This receptor was shown not to be modified by GAG chains, thus ruling out the possibility that it is a proteoglycan like the type III TGF-β receptor. The type V receptor can specifically bind TGF-β and is internalized on exposure to the ligand.

Additional biochemical studies with the purified protein also showed that, like the type I and type II TGF-β receptors, the type V receptor may act as a functional serine/threonine kinase (O'Grady et al., 1992). The purified receptor is capable of autophosphorylation and can be stimulated on addition of TGF-β, even at concentrations as low as 4 nM. One-dimensional phosphoamino acid analysis of the autophosphorylated receptor demonstrated that the majority of autophosphorylated residues consisted of phosphoserine. Addition of TGF-β increased the total radioactive phosphate incorporation into the receptor without altering the relative ratios of the phosphoamino acids. Collectively, these results suggest that the type V receptor acts as an intrinsic serine/threonine kinase.

Subsequent affinity cross-linking studies with conditions optimized to detect the type V receptor indicate that this receptor is widely distributed among mammals (O'Grady et al., 1991b). The receptor is expressed in a wide variety of cell types, including fibroblasts, chondrocytes, and cells of epithelial and endothelial origins. The widespread expression of the type V receptor suggests that it plays an important role in mediating TGF-β responses. In support of this hypothesis, expression of the type V receptor was found to be absent in five of the six screened epithelial tumor lines (O'Grady et al., 1991b). Unlike their normal counterparts, these tumor lines are nonresponsive to the growth inhibitory effects of TGF-β. Typically, the loss of TGF-β responsiveness has been correlated with a loss of the type I, the type II, or both type I and type II TGF-β receptors. All six of these tumor lines express both type I and type II TGF-β receptors, indicating that loss of the type V receptor may play a role in tumorigenesis. However, the importance of the type V receptor in TGF-β signal transduction relative to the demonstrated signaling capabilities of the type I and type II receptors, remains to be established.

5. *Type I Receptor*

A type I TGF-β receptor has been identified using a functional assay (Franzen et al., 1993; C. H. Bassing et al., 1994), the details of which are discussed in Section IV,D). At this time, we introduce only the structural characteristics of this type I receptor, referred to by its original name R4. The R4 clone was originally isolated with three other structurally related molecules (R1–R3) from a rat urogenital ridge cDNA library, using a PCR-based cloning strategy designed to isolate serine/threonine kinase receptors that share structural homology with the type II receptors (He *et al.*, 1993). ALK-5, the human homolog of R4, was also isolated using a similar cloning strategy (Franzen *et al.*, 1993).

The R4 receptor is a 53-kDa glycoprotein containing a putative 22-amino-acid signal sequence, a 101-amino-acid hydrophilic extracellular domain, a 23-amino-acid transmembrane domain, and a 355-amino-acid intracellular domain (He *et al.*, 1993). This protein is modified by approximately 5 kDa of N-linked carbohydrates in the extracellular domain. Structurally similar to the type II receptors, the cytoplasmic portion of this receptor is occupied by a kinase that is more closely related to the known serine/threonine kinases based on sequence analysis. In addition, the kinase domain also contains two inserts between certain subdomains that are characteristic of kinases of the transmembrane serine/threonine receptor kinase family.

Additional sequence comparison of R4 to the other type II receptor-like molecules reveals that this receptor belongs to a subgroup in the serine/threonine receptor kinase family. This subgroup contains the other three members (R1–R3), which were isolated in the same cDNA library screening as the R4 receptor. The rat R1 clone homologs from other mammalian sources have also been isolated, including the murine (tsk 7L), rat (ActXIR), and the human (SKR1) and (ActR-1) clones (Attisano *et al.*, 1990; Ebner *et al.*, 1993a; Matsuzaki *et al.*, 1993; Tsuchida *et al.*, 1993). Members of this subgroup share overall structural similarities with the type II receptor subgroup, but display features that distinguish them from those original members of the serine/threonine receptor kinase family. For example, the type II receptor subgroup contains highly conserved and characteristic 10 cysteines in the extracellular domain. The R4 receptor subgroup, on the other hand, has 10–12 cysteines in the extracellular domain but their positions are not highly conserved. This subgroup also has smaller extracellular domains and shorter C-terminal tails than their counterparts in the other subgroup. Based on these structural features, members of the R4 subgroup of serine/threonine receptor kinases may all be type I receptors for ligands of the TGF-β superfamily.

The kinase domain of R4 receptor exhibits 41% similarity to the type II receptor kinase. As discussed earlier, the type II TGF-β receptor was shown to be an active kinase capable of autophosphorylation mainly on serine residues, with a small amount of phosphorylation on threonines. To address the issue of whether R4 could act similarly as an intrinsic serine/threonine kinase *in vitro,* a fusion protein between GST and the cytoplasmic domain of R4 was generated and used in an *in vitro* kinase assay (C. H. Bassing *et al.,* 1994). The GST–R4 kinase was demonstrated to be capable of autophosphorylation primarily on threonine residues, with detectable serine phosphorylation. The kinase activity of the GST-R4 molecule is comparable to that of the GST type II kinase (C. H. Bassing *et al.,* 1994). Therefore, this type I TGF-β receptor is also a functional transmembrane serine/threonine kinase.

Tsk 7L, the murine homolog of R1, was originally reported to be a TGF-β type I receptor based mainly on its biochemical properties (Ebner *et al.,* 1993a). More recently, tsk 7L and its human homolog, ActR-1, have been shown to bind to activin when coexpressed with the activin type II receptor (Ebner *et al.,* 1993b; Attisano *et al.,* 1993). However, this receptor was demonstrated to mediate only activin-dependent gene responses at physiological concentrations of ligand and therefore concluded to be an activin type I receptor (Attisano *et al.,* 1993). TSR-1, the human homolog of R3, was demonstrated to bind both TGF-β and activin, however, no signaling ability has been reported for this receptor (Attisano *et al.,* 1993). Therefore, identification of putative ligands and signaling capabilities of the members of this emerging subgroup of type I serine-threonine receptor kinases remains to be accomplished. In addition, the issue of whether these receptors can act as type I receptors and signal for more than one ligand of the TGF-β superfamily must be addressed.

IV. TGF-β Signal Transduction

A. TGF-β Signaling Receptors

As described in earlier sections, affinity cross-linking cells with radiolabeled TGF-β allows identification of a number of TGF-β binding proteins. However, this biochemical assay was not designed to identify receptors responsible for signaling the actions of TGF-β. Initially, the type I, type II, and type III TGF-β receptors were considered likely candidates solely because of their wide distributions in different cell types and tissues.

Subsequent studies generated several lines of evidence that supported the role of the type I and type II receptors in transduction of TGF-β initiated signals (Massague *et al.,* 1990). First, these two proteins are ubiquitously expressed except in a few tumor cell lines, unlike the other TGF-β binding proteins which are expressed in limited cell types. Second, the biological effects of TGF-β can be achieved by addition of concentrations of the ligand corresponding to the known binding constants for the type I and type II receptors (K_d, 5–20 pM). Finally, several tumor cell lines, such as retinoblastoma, pheochromocytoma, neuroblastoma, and breast carcinoma cells, lack detectable expression of either one or both of these receptors and show no growth inhibition by TGF-β.

B. Genetic Evidence for a Signaling Complex

In an attempt to determine which of the TGF-β receptors are important for TGF-β signaling, mutant cell lines resistant to the growth inhibitory effects of TGF-β were generated. Mink lung epithelial (Mv1Lu) cells that express the type I, type II, and type III TGF-β receptors and are very responsive to the growth inhibitory effects of TGF-β were subjected to mutagenesis (Boyd and Massague, 1989).

Boyd and Massague (1989) used (EMS) to mutagenize Mv1Lu cells chemically and grew the cells in the presence of high concentration of TGF-β1. After further amplification, six independent TGF-β-resistant colonies were isolated and shown to maintain this phenotype for at least 6 mo. In addition to being refractory to both TGF-β1 and TGF-β2 growth inhibition, the mutant cell lines also lacked TGF-β-induced morphological changes and fibronectin production. Affinity-labeling analysis of those TGF-β resistant Mv1Lu cell lines revealed that four of the six mutant cell lines selectively lost the type I TGF-β receptor (R mutants). The other two mutants showed wild-type receptor profiles and apparently were defective in some downstream signaling pathways (S mutants). Observations with the R mutant cells suggest that the type I receptor is an essential component of the TGF-β signal transduction pathway in Mv1Lu cells. Expression of the type I receptor is required for mediating both TGF-β induced gene responses and growth inhibition.

Although this initial screening of Mv1Lu mutant cells generated fruitful results, the lack of any mutants that displayed defective expression of the other TGF-β receptors suggested that the original design of this mutagenesis and selection strategy may have been less than optimal. In theory, mutations in more than one of the cell surface components would have caused a resistant phenotype because of the suggested importance of all three types of TGF-β receptors. An additional surprising result of the study was the limited number of TGF-β resistant clones defective in the signaling pathways isolated. Conventional wisdom predicts that signaling pathways originating at the plasma membrane and ending inside the nucleus require a cascade of consecutive signaling events mediated by multiple cellular components. Defects in any one of these molecules could affect or even eliminate the transduction of signal, unless the TGF-β signal is transduced through multiple parallel pathways. Therefore, more potential signaling-defective mutants could be generated.

To address these concerns, a second round of mutagenesis and selection experiments was conducted with more cells and both TGF-β1 and TGF-β2 as selective agents (Laiho et al., 1990). After 3–4 wk selective growth in the presence of TGF-β, 71 mutant Mv1Lu cell lines resistant to TGF-β growth inhibition were isolated. The receptor profiles of the resistant clones were categorized by radio-iodinated TGF-β cross-linking analysis. Of the isolated clones, 35 had normal receptor profiles (S mutants) and 10 of the clones lacked the type I receptor (R mutants). The remaining third of the mutants had novel receptor-defective phenotypes. These phenotypes included mutants with low levels of type I receptor (LR mutants), low levels of type I and type II receptors (DRb mutants), no detectable type I and type II receptors (DRa mutants), and low levels of type I and type II receptors that had faster electrophoretic mobilities (DRC mutants). Not a single mutant displaying only the type I receptor without the type II receptor was isolated. With the exception of one isolated DRb clone, the mutants were completely nonresponsive to both TGF-β1 and TGF-β2.

These results indicate that the type I and type II TGF-β receptors act together in a common signaling pathway. A complete loss, or low level expression, of either of these receptors results in the TGF-β non-responsive phenotype. In addition, all the isolated mutants defective in the type II receptor were also defective in the type I receptor. The high frequency of these DR mutants suggests that the phenotype arises from a single mutational event. These observations indicate that expression of the type I receptor on the cell surface and its binding to TGF-β are in some manner linked to expression of and binding by the type II receptor. Researchers proposed that the type I receptor requires an interaction with the type II receptor to be stably expressed on the cell surface.

Subsequent hybrid cell fusion experiments with mutant Mv1Lu cell lines from different categories provided additional evidence that expression of both type I and type II receptors is necessary for TGF-β-mediated signal transduction (Laiho et al., 1991). Somatic cell hybrids of the R and DR mutants expressed normal levels of the type I and type II receptors and were fully responsive to TGF-β. Evidently, the DR mutants maintained the ability to express a functional type I receptor, although the presence of the receptor was undetectable by ligand binding. Three explanations are possible: (1) the type I receptor requires the type II receptor to be expressed on cell surface, (2) the ligand-binding ability of the type I receptor is induced through association with the type II receptor, and (3) the type I receptor is not a *bona fide* receptor for TGF-β, but a signaling partner in the receptor complex that is accidently cross-linked to the radiolabeled ligand bound to the type II receptor. Whatever the reason, TGF-β-resistant colonies that lack expression of only the type II receptor were never isolated.

In summary, the improved mutagenesis strategy resulted in the isolation of 71 TGF-β resistant colonies, two-thirds of which were defective in TGF-β receptors. Subsequent studies with these mutants demonstrated that the type I and type II receptors are likely to participate in a common signaling complex. The remainder of the TGF-β-resistant colonies had normal receptor profiles and were assumed to be defective in signaling pathways downstream from the receptor complex. Note that the number of signaling mutants isolated was almost equivalent to the number with mutations in the receptor components. This observation again argues for the possibility of more than one signaling pathway initiated from the receptor complex, all of which act to transduce the same signals. Thus those distinct pathways can functionally compensate for each other when one pathway is mutated.

C. Signaling through a Heteromeric Receptor Complex

The type II serine/threonine receptor kinases and the tyrosine receptor kinases are structurally similar, both containing extracellular domains that bind ligand and cytoplasmic domains with intrinsic kinase activity. Since the active form of TGF-β is a dimeric molecule, ligand-binding-induced dimerization of the type II receptors could represent the initial event in TGF-β-mediated signal transduction. However, previous genetic studies with the Mv1Lu mutant cells indicated that both type I and type II receptors are required for TGF-β signaling. To determine which of these models might be correct, Wrana *et al.* (1992) initiated a series of experiments by introducing the cDNA encoding the human type II TGF-β receptor into the DR Mv1Lu mutant cells lacking expression of the endogenous TGF-β type II receptors.

To facilitate these studies, a construct containing a TGF-β-inducible promoter-controlled reporter luciferase gene (p3TP-Lux) was generated to study TGF-β induction of gene responses in Mv1Lu cells (Wrana *et al.*, 1992). In a transient assay, transfection of p3TP-Lux into normal Mv1Lu cells, but not into the nonresponsive mutants, allowed the measurement of TGF-β-dependent induction of luciferase activity. Transient co-transfection of p3TP-Lux and the human type II TGF-β receptor cDNA into the mutant Mv1Lu cell lines resulted in restoration of a TGF-β-dependent gene response in two of the DR mutants. This phenomenon was not observed when the plasmids were transfected into the R or the S mutants, suggesting that the type II receptor is the only missing component in DR mutant cells that is critical for this restoration of gene response.

Subsequently, Mv1Lu DR and R mutant stable lines stably expressing the human type II receptor were generated (Wrana et al., 1992). In agreement with results derived from the previously described transient experiments, the DR cell lines that stably expressed the type II receptor were shown to increase both fibronectin and plasminogen activator inhibitor type I (PAI-1) production in a TGF-β-dependent fashion. Expression of the type II receptor also restored TGF-β growth inhibition in these DR mutants. In contrast, the R mutants stably expressing the type II receptor remained nonresponsive to TGF-β.

Affinity labeling of these stable lines with radio-iodinated TGF-β showed an interesting result. The transfected type II receptor was expressed in both the DR and R stable lines; however, the DR stable lines also displayed cell-surface expression of the type I TGF-β receptor. This observation was consistent with earlier findings that somatic cell hybrids between DR mutants (no detectable type II and type I receptors) and R mutants (no type I receptor expression) rescued the expression of the type I receptor on cell surface, consequently restoring TGF-β responsiveness. Collectively, these results again demonstrated that the DR mutant lines maintained the capacity to express a functional type I receptor, but the receptor could not be detected on the cell surface without the presence of a type II receptor.

The observation that expression of the type II receptor in the R mutant cells failed to restore any TGF-β responses implies that the type II receptor requires the expression of a functional type I receptor to signal, probably by forming a heteromeric receptor signaling complex. To test this hypothesis, a clonal antibody epitope tag was attached to the type II receptor and used in immunoprecipitation studies (Wrana et al., 1992). The results of these immunoprecipitations demonstrated that the type I and type II receptors could indeed form a complex in the presence of TGF-β. This complex was not observed in cells lacking the expression of a type I TGF-β receptor.

Similar phenomena were observed in a TGF-β-responsive human hepatoma Hep3B cell line and a mutant derivative line (Hep3B-TR) that lacked the expression of type II TGF-β receptor (Inagaki et al.,

1993). Affinity-labeling experiments revealed that the mutant line had lost TGF-β binding to both the type I and the type II receptor. Stable expression of the human type II receptor cDNA in these lines restored TGF-β growth inhibition and ligand binding to the two receptors. Immunoprecipitations of cell lysates with an antibody specific for the type II receptor recovered both receptors, but only in the presence of TGF-β. Heating the lysates in SDS prior to the immunoprecipitation resulted in recovery of only the type II receptor.

The lack of TGF-β binding to the type I receptor in the DR mutants suggests that the type I receptor requires expression of a functional type II receptor to bind ligand. To investigate this issue further, the type II TGF-β receptor cDNAs from two of the DR mutants were cloned and sequenced (Wrana et al., 1992). Not surprisingly, these genes contained point mutations that would be expected to occur from the EMS mutagenesis treatment that generated the mutant lines. In the DR-27 cell line, a mutation in the type II receptor gene caused a cysteine-to-tyrosine substitution. This altered cysteine was one of the conserved residues in the extracellular domain of the TGF-β type II receptor. This substitution is likely to have disrupted the three-dimensional structure of the ligand-binding extracellular domain, since these cysteines are believed to be important in forming the structural backbone of the receptor. Therefore, this altered receptor is unlikely to bind TGF-β and transduce a signal, even if expressed on the cell surface.

The cytoplasmic domain of the type II receptor is a functional serine/threonine kinase. To determine whether this catalytic activity is essential for TGF-β signal transduction, a receptor with a null kinase was created through the substitution of an arginine residue for the Lys 277 residue in the ATP-binding domain (Wrana et al., 1992). Stable expression of this mutant type II TGF-β receptor in the DR mutant cells resulted in the restoration of TGF-β binding to both type I and type II receptors. Although the TGF-β-binding ability of this receptor was unaltered, the signaling capacity of the kinase-negative receptor was completely eliminated, since TGF-β-induced growth inhibition and gene responses remained defective in these stable lines. Therefore, a functional type II receptor kinase is essential for TGF-β signal transduction.

Based on their findings, Wrana et al. (1992) proposed that TGF-β signals through a heteromeric receptor complex. The components of this complex are interdependent, since the type I receptor requires the type II receptor to bind ligand and the type II receptor requires the presence of type I receptor to signal. In addition, the kinase activity of the type II receptor is essential for TGF-β signal transduction.

D. Identification of a Functional Type I Receptor

In an attempt to clone the type I receptor, a murine epithelial cell line (NMS 90) hyperexpressing the type I TGF-β receptor was isolated (Ebner et al., 1993a). A cloning strategy was designed based on the assumption that, like the type II TGF-β receptor, the type I receptor is also a transmembrane serine/threonine kinase. Consequently, a DNA fragment was generated through PCR amplification using degenerate oligonucleotides representing conserved peptide sequences of the activin type II receptor and the *daf*-1 serine/threonine kinases. This DNA fragment was used as a probe in low stringency hybridization of a λ cDNA library prepared from messenger RNAs isolated from the NMS 90 cell line. This strategy yielded several serine–threonine receptors that share sequence homology with the type II activin and TGF-β receptors. Subsequently, the full-length cDNA representing the most abundant receptor was isolated and named *tsk* 7L (Ebner et al., 1993a).

The protein encoded by *tsk* 7L was concluded to be a type I TGF-β receptor based primarily on the observations that it has a molecular weight of 53 kDa when translated in a reticulocyte lysate and that it requires the presence of the type II TGF-β receptor to be expressed on the cell surface and to bind TGF-β (Ebner et al., 1993a). These characteristics are all known biochemical properties of the type I TGF-β receptor (Boyd and Massague, 1989; Wrana et al., 1992). Although the protein encoded by *tsk* 7L possesses biochemical properties of the type I TGF-β receptor, the ability of the receptor to complex with the type II receptor and mediate the biological responses of TGF-β remains undetermined.

In a separate attempt to isolate type II receptors for other members of the TGF-β superfamily ligands, such as MIS, the same PCR-based cloning strategy was employed (He et al., 1993). The structural similarity between the activin and TGF-β type II receptors implied that type II receptors for ligands of the entire TGF-β superfamily may belong to the same class of transmembrane serine/threonine kinases. To explore this possibility, a set of oligonucleotide primers was designed based on the conserved peptide sequences of the activin and TGF-β type II receptors. Four clones (termed R1–R4) encoding structurally related 501- to 509-amino-acid receptors of this family were isolated by PCR from 14.5–15 day fetal Sprague–Dawley rat urogenital ridges (He et al., 1993). Subsequent sequence comparison indicated that the R1 clone represents the rat homolog of the murine transmembrane serine/threonine kinase *tsk* 7L. Since the R1–R4 clones were highly homologous and the functional nature of R1 remained unknown, the clones were assayed for their ability to serve as functional TGF-β type I receptors.

As previously mentioned, co-transfection of the human type II TGF-β receptor cDNA with the reporter gene p3TP-Lux into the Mv1Lu DR mutants restored TGF-β-induced gene responses (Wrana et al., 1992). The same experimental design was used to screen the ability of R1–R4 to restore a functional TGF-β gene response to the TGF-β-nonresponsive Mv1Lu R mutant cell line R1B. The results from these experiments clearly demonstrated that R4 is the only one of these proteins capable of restoring a TGF-β-inducible gene response (C. H. Bassing et al., 1994). The magnitude of induction in luciferase activity in the R4-transfected R1B mutant cells was equivalent to that of the wild-type Mv1Lu cells, indicating that a full TGF-β-induced gene response was restored by transfection with the R4 clone. The observation that R4 expression fully restored a TGF-β-induced gene response to cells with no endogenous type I TGF-β receptor expression strongly suggests that R4 is a functional type I receptor.

As discussed previously, the type I and type II receptors are both required for TGF-β signaling through a heteromeric receptor complex (Wrana et al., 1992). Therefore, R4 was tested for its ability to induce a TGF-β gene response when transfected into the DR mutants, which lack expression of endogenous type II TGF-β receptor. The results of the experiment indicate that R4 cannot act alone and does require the presence of a functional type II TGF-β receptor to transduce the TGF-β signal (C. H. Bassing et al., 1994). This observation further establishes the molecular nature of R4 as a functional type I TGF-β receptor.

Studies with the Mv1Lu mutants also suggested that TGF-β binding to the type I receptor required the presence of the type II receptor in the same cell, whereas ligand binding to the type II receptor could occur independently (Wrana et al., 1992). To determine whether R4 represents a classic type I receptor based on this standard, the ligand-binding properties of R4 were investigated by affinity cross-linking in transfected COS cells (C. H. Bassing et al., 1994). Affinity labeling of COS cells transfected with only the R4 cDNA showed no increased TGF-β1 binding. The type II cDNA transfected alone only resulted in an increase of TGF-β1 binding to a protein migrating as a 85- to 97-kDa band. When the two cDNAs were co-transfected, a dramatic increase in TGF-β1 binding to proteins migrating as 63- to 65-kDa and 85- to 97-kDa bands was observed. These bands correspond to the classic type I and type II TGF-β receptors. The dramatic increase in binding of TGF-β1 to the type II receptor when the type I receptor is coexpressed suggests that the two receptors may cooperatively bind TGF-β. This observation is in agreement with the earlier findings that overexpression of the type II receptor enhanced ligand binding to the type I receptor (X. F. Wang, unpublished data), and with the model that the type II receptor gains affinity for TGF-β1 when it binds in concert with the type I receptor (Lopez-Casillas *et al.*, 1993).

Researchers have demonstrated that the kinase activity of the type II receptor is essential for TGF-β signaling through the heteromeric receptor complex (Wrana *et al.*, 1992). Expression of a kinase-negative type II receptor in the DR mutant cells failed to restore TGF-β-dependent growth inhibition and gene responses. To determine whether the kinase activity of the type I receptor is also required for TGF-β signaling, a kinase negative R4 receptor was transiently co-transfected with p3TP-Lux into R mutants. The kinase-negative R4 failed to restore a TGF-β gene response in cells lacking an endogenous type I receptor (C. H. Bassing *et al.*, 1994). Therefore, an active type I receptor kinase is essential for signaling TGF-β gene responses.

Currently, we have not been able to determine clearly whether R4 can restore TGF-β growth inhibition to the Mv1Lu R mutants. We attempted to address this issue by establishing stable cell lines expressing R4 in the R mutant cells. Preliminary results with one of these stable clones indicated that R4 was able to restore TGF-β growth inhibition in these mutant R1B cells (C. H. Bassing *et al.*, unpublished data). However, this result could not be repeated because the high level of expression of R4 in those cells appears to cause a growth disadvantage. Consequently, the introduced R4 cDNA was lost by the transfected cells during the process of drug selection and clonal expansion. We are currently in the process of establishing stable Mv1Lu R lines expressing either the wild-type R4 or the kinase-negative R4 using low expression or inducible vectors. We anticipate that we will be able to address the question of the functional role of R4 in signaling TGF-β growth inhibition in the near future.

Despite the clear demonstration that the R4 receptor could fully restore the TGF-β-induced gene response in the R Mv1Lu cells, the formal possibility exists that R4 represents only one of the functional type I receptors for TGF-β. The functional assay used may have limited our ability to test the functional nature of the other members of this receptor subfamily, since the method only tested TGF-β induction of a specific promoter in a particular cell type. In addition, only one isoform of TGF-β, TGF-β1, was used in the functional assay to test the ability of these clones to mediate TGF-β responses. This experimental feature may have further limited our ability to detect other type I TGF-β receptors that may have different profiles of response to different ligand isoforms.

Developmental studies of TGF-β inductive fields indicate that, during chick organogenesis, the pattern of expression of TGF-β2 is restricted while that of the type II and type III receptors is more generalized (J. V. Barnett, A. Moustakas, W. Lin, X. F. Wang, H. Y. Lin, J. B. Galper, R. A. Weinberg, H. F. Lodish, and R. L. Maas, unpublished data). This finding contradicts the result obtained with other ligand–receptor pairs and leads to the speculation that the inductive effects of TGF-β2 are potentially mediated by another type of receptor. Although no direct evidence exists currently, the expression of various type I TGF-β receptors may be developmentally and tissue-specifically regulated. To extend this model, different type I receptors may mediate distinct TGF-β responses in different cell contexts. Therefore, differential expression of the type I receptors may be an important factor contributing to the significant diversity of TGF-β responses in different cell types and tissues.

E. MECHANISMS OF TGF-β RECEPTOR ACTIVATION

Researchers have demonstrated that the actions of TGF-β are transduced through a heteromeric signaling complex of the type I and type II receptor kinases (Wrana et al., 1992; C. H. Bassing et al., 1994). However, the precise molecular mechanisms of this process have yet to be resolved. The TGF-β serine/threonine receptor kinases display molecular topologies that are very similar to those of the tyrosine receptor kinases. Therefore, a detailed analysis of the structure–function relationship of these receptors may greatly facilitate elucidation of the activation mechanism of the TGF-β receptors. We will use this opportunity to draw insights into the potential molecular mechanism and to speculate on the actions of signal transduction by the TGF-β receptor complex.

1. *Activation through Dimerization*

Signaling by the receptor tyrosine kinases (RTKs) occurs through ligand-induced homodimerization mediated through three mechanisms: (1) monomeric ligand inducing a receptor conformational change and subsequent association, (2) ligand inducing a conformational change in a disulfide-linked receptor dimer, and (3) dimeric ligand recruiting two receptor subunits together (Ullrich and Schlessinger, 1990).

The existence of the TGF-β receptor complex has been demonstrated by immunoprecipitation of the type II receptor from cells previously cross-linked with TGF-β (Wrana et al., 1992; Franzen, et al., 1993; Inagaki et al., 1993). In these experiments, a cross-linked band migrating as the type I receptor was recovered unless the lysate had been exposed to denaturing conditions prior to precipitation. This result proved the formation of a noncovalently linked ternary complex in the presence of TGF-β. However, this approach was incapable of determining whether this complex formed before or after the addition of TGF-β. The type I and type II receptors may exist on the cell surface as disulfide-linked heterodimers; TGF-β binding may merely cause a conformational change that activates the receptors. With the availability of antibodies against the type I and type II receptors, experiments can now be done to address this important issue.

Dimerization of the RTKs causes aggregation of the cytoplasmic domains, resulting in intermolecular phosphorylation events that activate the intrinsic kinase activity of the receptors (Ullrich and Schlessinger, 1990). Although the kinase activities of both type I (C. H. Bassing et al., 1994) and type II (Wrana et al., 1992) receptors have been shown to be required to transduce TGF-β signals, whether and how the activities of these serine/threonine kinases are regulated remains largely unknown. Immunoprecipitation of the type II receptor in the presence or absence of TGF-β showed no significant difference in the overall activity of the serine/threonine kinase (J. M. Yingling, unpublished data). However, one main concern with the results derived from these experiments is that the tests were performed with cells in which only the type II receptor was overexpressed. Under such circumstances, the level of functional receptor complex formed between the type I and type II receptors may remain very low, whereas the overexpressed type II receptor may not be induced to become active in the proper context. With cDNAs for both receptors cloned, we may now induce the proper formation of TGF-β receptor complex in cells by overexpressing both type I and type II receptors. Then, a ligand-induced change in the intrinsic kinase activity of these receptors, reflecting either the activation or the inactivation of the kinases, may be observed. Alternatively, ligand binding could cause a conformational change in the cytoplasmic domains that allows substrate(s) to gain access to the active site of the kinases.

A heteromeric signaling complex is required for TGF-β signal transduction, but the molecular composition of the ligand–receptor complex has not been determined. For unknown reasons, the type I receptor is only capable of binding TGF-β that is tethered to the type II receptor. Therefore, the simplest interpretation of the available data is that TGF-β binds to the type II receptor and, subsequent to this interaction, the type I receptor joins the complex and increases the affinity of the type II receptor for TGF-β. Analogous to the behavior of RTKs, this heterodimerization and consequent aggregation of the cytoplasmic domains of the type I and type II receptors could activate TGF-β signaling. However, the dimeric nature of TGF-β suggests that formation of a heterotetrameric complex is equally possible. In this model, receptor activation could be initiated through homodimeric or heterodimeric interactions of the cytoplasmic kinase domains.

Another issue that must be resolved is the mechanism by which the type II receptor gains affinity for TGF-β when it binds in concert with the type I receptor (Lopez-Casillas *et al.*, 1993; C. H. Bassing *et al.*, unpublished data). In this respect, the TGF-β receptors are most analogous to the interleukin 2 (IL-2) receptor system. The IL-2 receptor consists of three subunits; the βγ dimer is essential for signaling, but requires the α subunit to gain affinity for IL-2 (Taniguchi and Minami, 1993). In this process, IL-2 binding is believed to cause conformational changes in its receptor components that allow them to interact and form a high-affinity binding complex (Voss *et al.*, 1993). In a similar fashion, TGF-β may induce a conformational change in the type II receptor that allows it to assemble with the type I receptor in a high-affinity complex. Alternatively, the type I receptor may play a more active role in initiating the high-affinity interaction.

2. *Receptor Structure and Function*

The tyrosine and serine/threonine receptor kinases can be considered enzymes that are activated through allosteric interactions with their respective ligands. This activating event causes conformational changes that may affect the intrinsic kinase activity of these receptors. For example, activation of the EGF receptors causes autophosphorylation that is believed to result in a phosphorylated state of the kinase that is capable of interacting with its substrates (Ullrich and Schlessinger, 1990). Analogous events may occur during activation of the serine/threonine receptor kinases. In fact, the kinase of the TGF-β type II receptor has been repeatedly observed to be capable of tyrosine autophosphorylation *in vitro* when certain point mutations and deletions are introduced into the kinase domain (J. M. Yingling, unpublished data), suggesting that tyrosine autophosphorylation may occur as a result of conformational changes. In support of this possibility, purified type II activin receptor is able to autophosphorylate on serine or threonine and tyrosine residues (T. Nakamura *et al.*, 1992). Although no direct experimental evidence is available currently, TGF-β binding to the receptor complex may result in autophosphorylation on tyrosine residues, thereby creating a conformation of the kinase that is capable of interacting with substrate. Alternatively, TGF-β binding may induce a conformational change in the receptor complex, resulting in the activation of tyrosine autophosphorylation and/or phosphorylation of substrates on tyrosine residues.

Most enzymes have regulatory and effector domains. The juxtamembrane domain, the small inserts between domains, and the C-terminal tail have been shown to serve these functions for the RTKs (Ullrich and Schlessinger, 1990). The juxtamembrane domain modulates the kinase activity through modifications mediated by a variety of reagants. The kinase inserts mainly serve as effector domains. Autophosphorylation of tyrosine residues within these regions creates high-affinity binding sites for substrates and effectors of the receptor. Tyrosine residues in the kinase subdomains and the juxtamembrane domain may also serve as high-affinity binding sites. The C-terminal tail of the EGF receptor acts as a negative regulator of the kinase. Deletion of this region leads to an increased oncogenic capacity of the EGF receptor (Ullrich and Schlessinger, 1990). The molecular basis of this inhibition is attributed to the ability of the tail to interact with the substrate-binding site of the kinase.

The analogous regions of the serine/threonine receptor kinase may serve similar functional roles. We have observed that deletion of either the C-terminal kinase tail or residues in the juxtamembrane domain leads to an increase in the intrinsic activity of the receptor kinase (J. M. Yingling, unpublished data). The biological significance of these manipulations has yet to be determined. Evidence supporting this notion derives from a study in which a truncated activin receptor lacking the C-terminal subdomains of the kinase caused increased signal transduction *in vivo* (Nishimatsu *et al.*, 1992). The invariant kinase inserts unique to the serine/threonine receptor kinases contain significant numbers of serine and threonine residues. Consequently, those inserts may serve as ideal targets for autophosphorylation and may play important functional roles.

F. TWO RECEPTOR PATHWAYS AND SIGNALING THRESHOLDS

Growth factor-dependent activation of tyrosine kinase receptors usually results in the induction of multiple cellular responses, many of which are transduced by different intracellular pathways that originate at different levels of the signal cascade. The TGF-β signal appears to be transmitted through at least two distinct intracellular pathways, because of its two major effects on cell growth and extracellular matrix induction. At which point the TGF-β signal diverges into the two main pathways in the cell is still unclear. As mentioned earlier, the formation of a heteromeric TGF-β receptor signaling complex is essential for signal transduction to both pathways. The simplest model would propose that each receptor type in the complex is responsible for activation of a distinct pathway. Several attempts have been made to test this hypothesis.

In a previous study, researchers demonstrated that certain TGF-β responses are mediated, at least in part, by different molecules in the nucleus (Laiho et al., 1991). The product of the retinoblastoma susceptibility gene (RB) plays an important regulatory role in the G_1–S transition of the mammalian cell cycle. TGF-β treatment of Mv1Lu cells retains RB in its active hypophosphorylated form, thereby preventing cells from entering S phase. The simian virus (SV40) large T antigen binds the hypophosphorylated form of RB and inactivates its regulatory effects. Expression of a transfected large T antigen in Mv1Lu cells completely blocks TGF-β-mediated growth inhibition, but has no effect on the induction of TGF-β-responsive genes. These results suggest that RB is a component only of the TGF-β growth inhibition signaling pathway(s) and not of the pathway(s) that mediates gene responses.

Overexpression of truncated tyrosine kinase receptors that lack the kinase domain resulted in a dominant negative phenotype because of the formation of inactive heterodimers with endogenous receptors. Likewise, a truncated type II activin receptor was shown to inhibit activation of the activin signaling pathways (Hemmati-Brivanlou and Melton, 1992). Chen et al., (1993) utilized a similar dominant negative, truncated type II TGF-β receptor to demonstrate that the divergence of the two TGF-β pathways occurs at the level of the receptor complex. These investigators demonstrated that Mv1Lu cells stably expressing the truncated type II receptor, as determined by affinity labeling, were nonresponsive to the growth inhibitory effects of TGF-β. TGF-β treatment of these cells showed no effect on DNA synthesis or the regulation of RB phosphorylation. However, TGF-β could still cause induction of fibronectin and PAI-1. The induction pattern of these genes in cells expressing the dominant negative type II receptor was very similar to that in the parental Mv1Lu cells.

Based on these results, Chen et al. (1993) concluded that the type I receptor does not require a functional type II receptor to mediate certain TGF-β responses. This conclusion implies that TGF-β binding to a complex containing the truncated type II receptor and the full-length type I receptor may be sufficient to transduce signals through the type I receptor kinase. In contrast, earlier findings using mutant Mv1Lu cells that completely lacked expression of endogenous type II receptor demonstrated that, even in the presence of a functional type I receptor, the absence of an active type II receptor kinase could eliminate the mediation of any TGF-β responses (Wrana et al., 1992). Therefore, in the experiments using the truncated type II receptor, heterodimerization of the type I receptor with a limited number of functional endogenous type II receptor could be sufficient to cause selective activation of the type I receptor. To fully address this issue, the effects of the truncated receptor should be assayed in the mink lung epithelial mutants that lack any normal endogenous type II receptors.

Although the results obtained by Chen et al. (1993) did not clearly define the requirements of the type II receptor in signaling TGF-β gene responses, they did suggest an inherent difference in the TGF-β signaling mechanisms that mediate gene responses and growth inhibition. These findings imply that a reduced expression level of the type II receptor may cause selective abrogation of the TGF-β growth inhibitory pathway. Evidence supporting this tentative conclusion exists in certain epithelial tumor lines (Geiser et al., 1992). For example, the human EJ bladder carcinoma and SW480 colon adenocarcinoma cell lines are nonresponsive to the growth inhibitory effects of TGF-β and are tumorigenic in nude mice, but they maintain TGF-β gene responses. These cell lines exhibit low level expression of the type I receptor and a much lower level of the type II receptor. Hybrid cell lines of the two carcinomas were generated and characterized for tumorigenicity in athymic nude mice and for TGF-β responsiveness in culture (Geiser et al., 1992). Five of the six hybrids were nontumorigenic, whereas the sixth line (LT-HYB) had a reduced level of tumorigenicity relative to the parental lines. TGF-β growth inhibition was restored in all the hybrid clones, albeit to a lesser extent in the LT-HYB clone. Affinity labeling indicated that the expression level of the type II receptor in those hybrid cell lines was increased to a level comparable to that of the type I receptor. This increase in type II receptor correlated closely with the restoration of TGF-β responsiveness and the loss of tumorigenicity in animals. Interestingly, the LT-HYB hybrid did not show an increase in its type II TGF-β receptor expression. These studies support the hypothesis that restoration of TGF-β growth inhibition to the hybrids is a direct result of the increased expression of type II receptor.

Collectively, these experiments suggest that the level of type II receptor expression is a key factor in determining activation of the signaling pathway by which TGF-β inhibits cell proliferation. The signaling pathway involved in induction of gene responses, however, may be less sensitive to the level of type II receptor expression. In other words, the presence of a relatively low level of type II receptor may still be sufficient for TGF-β activation of the extracellular matrix gene-induction pathway. Therefore, a decrease in the level of the type II TGF-β receptor may allow cells to escape from the negative growth influences of TGF-β. This event may be a causative step in the progression of human malignancies. In contrast to the human retinoblastoma cell lines that have lost all three types of TGF-β receptors, this model may have broader implications for other types of human tumor cell lines that express variable levels of all the TGF-β receptors, yet still fail to respond to the inhibitory effects of TGF-β.

G. OTHER SIGNALING MECHANISMS

The mink lung epithelial cell line has been the model system used to study the mechanisms of TGF-β receptor action. However, accumulating evidence suggests that TGF-β may signal in different fashions in different cellular contexts. In contrast to the heteromeric signaling complex, the type I and type II receptors may be capable of mediating TGF-β responses independently of each other. The hematopoietic progenitor cell lines B6SUtA and 32D-C13 (Ohta et al., 1987) and the human gastric carcinoma cell line TMK-1 (Ito et al., 1992) only express one type of TGF-β receptor, but all are growth inhibited in the presence of TGF-β. In addition, the human embryonic palate mesenchymal (HEPM) cell line proliferates in response to TGF-β, but expresses only one type of receptor (Linask et al., 1991). In all these cell lines, the one type of TGF-β receptor expressed has the molecular weight of the type I receptor. Investigators claimed that the type II receptor was either absent or at levels too low to be detected by the ligand-binding assay.

These observations raise an important issue. Classically, the differences in molecular weight of affinity cross-linked TGF-β receptors were used to classify them into subtypes. Since the structural differences between the type I and type II receptors are very subtle, we suggest the following molecular characteristics of the receptors as the criteria for subtype classification: the presence of a cytoplasmic kinase domain and the capability of independent ligand binding by the receptor. By this convention, any TGF-β binding protein that signals through its kinase domain and independently binds TGF-β should be referred to as a type II TGF-β receptor, irrespective of its molecular weight. A type I receptor contains a cytoplasmic kinase domain, but is unable to bind ligands alone. Therefore, the TGF-β-responsive cell lines just mentioned that only express one TGF-β binding protein would be considered to express only the type II receptor. If those cells indeed express only one type of receptor, the mechanism of receptor activation is likely to be ligand-induced homodimerization of the type II receptor. Evidence supporting this model derives from *in vitro* studies of the type II receptor kinase in which the kinase was shown to be capable of *trans*-phosphorylation, suggesting that kinase activity can be regulated through homodimerization (J. M. Yingling, unpublished data).

H. Type III Receptor/Betaglycan Presents TGF-β to the Type II Receptor

The type III TGF-β receptor is the most highly expressed TGF-β binding protein with a wide distribution. These characteristics suggest that the type III receptor plays a biologically significant role in TGF-β function. Several lines of evidence indicate that the type III receptor is not essential for TGF-β signal transduction. First, the type III receptor is absent in a number of TGF-β-responsive cell lines. Second, the expression level of type III receptor remains unaltered in all the TGF-β-nonresponsive Mv1Lu mutants. Finally, the cytoplasmic domain of the type III receptor lacks any known signaling motif. However, this last observation does not exclude a signaling role for the type III receptor, since several cytokine receptors that lack intracellular signaling domains are capable of signal transduction by association with another signaling molecule with catalytic capability.

Several growth factor receptors that do not appear to participate directly in signal transduction have been postulated to regulate the access of ligand to the actual signaling receptors (Lopez-Casillas *et al.*, 1993). These accessory receptors usually have the following properties: they bind ligand with lower affinity than the signaling receptors, they are expressed at higher levels than the signaling receptor, and they lack a cytoplasmic signaling motif. The type III receptor has all three of these characteristics and could, therefore, regulate the access of TGF-β to its signaling receptors. The observation that *de novo* expression of the type III receptor in L6 cells, which express only the type I and type II TGF-β receptors, increases TGF-β1 binding to the type II receptor 2.5-fold is consistent with this prediction (Wang *et al.*, 1992).

This increased binding of TGF-β to the type II receptor was originally attributed to an increased expression level, an altered affinity, or an altered configuration of the receptor. Lopez-Casillas *et al.* (1993) attempted to discriminate among these possibilities and to determine the effect of type III receptor expression on TGF-β responses. These researchers utilized the same L6E9 skeletal muscle myoblast cell line in which the phenomenon was originally observed. Since type III receptor overexpression in the L6 cells caused an increased binding of TGF-β to the type II receptor, the simplest model proposed to explain the phenomenon was that a complex could be formed between the two receptors. This physical association could result in the unloading of ligand bound to the type III receptor to the type II receptor or in a conformational change for the type II receptor to increase its ligand-binding affinity.

To address the issue of complex formation, a type III receptor containing an immunological epitope for a monoclonal antibody was created for use in immunoprecipitation studies of transiently transfected L6 myoblasts (Lopez-Casillas et al., 1993). Immunoprecipitation from transfectants cross-linked with TGF-β1 recovered the type III receptor and a second protein that was confirmed to be the type II TGF-β receptor. Additional experiments revealed that the formation of this complex was dependent on the presence of TGF-β, and that the receptors were not covalently linked. Collectively, these results indicate that the type II and type III receptors form a complex in the presence of TGF-β. These complexes were shown to occur between endogenously expressed receptors in several different cell lines, indicating that they are not the result of type III receptor overexpression. Further, the observations that this complex exists in the mutant Mv1Lu R1B cells, which lack type I receptor expression, and that the type I receptor is never recovered during immunoprecipitation of the type III receptor indicate that this complex forms independent of the type I receptor.

Although the existence of a complex between the type II and type III receptors was demonstrated, the mechanism for the increased binding to the type II receptor had to be addressed. Northern blots revealed that type II receptor mRNA levels were unchanged in the type III-transfected L6 cells (Lopez-Casillas et al., 1993). The cell-surface expression levels of the type II receptor were also demonstrated to be unaltered. These results suggested that the increased binding of the TGF-β to the type II receptor was the result of an intrinsic increase in the binding affinity of the receptor. Consequently, the molecular basis for this increased affinity was studied. Affinity labeling normal L6E9 cells with biologically active concentrations of TGF-β1 revealed that only 10% of the total population of type II receptors bound the ligand with high affinity (Lopez-Casillas et al., 1993). When the same experiment was repeated with the L6 cells overexpressing the type III receptor, a dramatic increase in the number of type II receptors binding TGF-β1 with high affinity was observed. In both experiments, the amount of TGF-β1 bound to the type I receptor was determined to be unchanged. Therefore, the apparent increased binding of TGF-β1 to the type II receptor is the result of increased availability of receptor complexes with a higher binding affinity.

Similar to the endothelial and hemopoietic cells that lack the type III TGF-β receptor, L6E9 cells are much more responsive to TGF-β1 and TGF-β3 than they are to TGF-β2 (Lopez-Casillas et al., 1993). The type I and type II receptors bind TGF-β1 and TGF-β3 with higher affinity than they bind TGF-β2 (Cheifetz et al., 1987,1990; Segarini et al., 1989). Therefore, the biological responses of the L6E9 cells to the individual isoforms correlate closely with the affinity of these receptors for the ligands. In agreement with this observation, the type III TGF-β receptor binds all three isoforms with equal affinity; most cultured cell lines that respond equally to the three TGF-β isoforms, express the type III receptor at high levels (Massague et al., 1990). These observations imply that the type III receptor could regulate the biological activity of TGF-β2 by affecting the interaction between TGF-β2 and the signaling receptor complex.

To have a better understanding of the mechanism of this regulation, the TGF-β2 binding characteristics of normal L6E9 cells and of those overexpressing the type III receptor were examined (Lopez-Casillas et al., 1993). TGF-β2 is virtually incapable of binding normal L6E9 cells, but binds the type III receptor-expressing L6E9 lines with a very high affinity, comparable to the binding affinity of TGF-β1 on the same cells. Affinity labeling revealed that TGF-β2 binding to the type II receptors was increased in parallel with its binding to the type I receptors in the type III receptor-expressing cells. Expression of a type III receptor mutant incapable of binding TGF-β failed to increase TGF-β2 binding to the type I and type II receptors, indicating that the increased binding of TGF-β2 was dependent on the ability of the type III receptor to bind this isoform. These results demonstrate that the type III receptor increases the amount of TGF-β2 that can interact with the type I and type II receptors. Presumably, the effect of increased TGF-β2 binding to the type I receptor is mediated by ligand binding to the type II receptor, since only the type II and type III receptors were shown to interact directly.

Since TGF-β signal transduction occurs through a heteromeric complex between the type I and type II receptors, determining whether L6E9 cells expressing the type III receptor become more responsive to TGF-β2 was necessary. Normal L6E9 cells are growth inhibited by TGF-β1 and TGF-β2; however, TGF-β1 is 10 times more potent. In contrast, L6 cells stably expressing the type III receptor showed equivalent growth inhibition when treated with TGF-β1 and TGF-β2 (Lopez-Casillas et al., 1993). These results clearly indicate that expression of the type III receptor increases the potency of TGF-β2 through selective up-regulation of the interaction between TGF-β2 and the signaling complex.

These results obtained by Lopez-Casillas et al. (1993) demonstrate that the type III TGF-β receptor acts as a modulator of TGF-β function. The type III receptor binds TGF-β and presents is to the type II receptor, which binds the tethered ligand with higher affinity than the soluble TGF-β. This presentation has no effect on TGF-β1 binding, since the type II receptor has an intrinsic high affinity for soluble TGF-β1. In contrast, the type III receptor can significantly increase the affinity of the type II receptor for TGF-β2. Consequently, more signaling receptor complexes can be formed and/or activated, leading to the similar pattern of responses of the cells to TGF-β2 and TGF-β1.

This model raises several interesting questions concerning the assembly and disassembly of the type II and type III receptor complex. Like the models for other ligand-induced dimerizations, TGF-β binding to the type III receptor may cause a conformational change in the receptor that makes it competent to interact with the type II receptor. Alternatively, the type III receptor may simply bind and display TGF-β so the type II receptor binding pocket of the ligand becomes available for the higher affinity interaction with the type II receptor. In this way, ligand binding by the type III receptor may cause a conformational change in the TGF-β ligand isoforms so they can all subsequently bind the type II receptor with higher affinity. Another interesting phenomenon observed during the presentation process is that the TGF-β molecule is never found in the same complex with all three receptors. The simplest explanation is that the type I and type III receptors compete for the same binding epitope on the ligand. Since type I receptor has a higher ligand affinity than the type III receptor, the net consequence of this competition is the formation of a signaling complex and the separation of the type II receptor from TGF-β. A second possibility is that the binding of the type I receptor causes a conformational change in the ligand so the type III receptor is no longer able to bind. A third possibility is that phosphorylation of the serine/threonine-rich cytoplasmic domain of the type III receptor by the type I and/or type II receptor kinase causes a conformational change of the type III receptor, resulting in the dissociation of the type III receptor from the complex.

I. ROLE OF ENDOGLIN

Endoglin shares several structural characteristics with the type III receptor. As suggested by its nonoverlapping expression pattern with the type III receptor, endoglin may serve a similar function by presenting TGF-β to the signaling receptor complex in cells that do not express the type III receptor. However, endoglin and the type III receptor may also have distinct biological functions. For example, endoglin has an intrinsic affinity for TGF-β2 that is 100 times lower than its affinity for the other isoforms (Cheifetz et al., 1992). This distinction would be expected to result in a functional consequence for the effects of TGF-β2 in cells expressing only endoglin. In fact, endoglin is predominantly expressed in endothelial cells, which are not ordinarily targets of TGF-β2 actions.

The extracellular domain of endoglin contains an RGD sequence in a region of the structure predicted to be solvent accessible (Gougos and Letarte, 1990). Therefore, endoglin may bind extracellular matrix proteins and serve a functional role in cell adhesion or cell–cell recognition. The type III receptor lacks a similar peptide sequence, but its extracellular domain is heavily modified with GAG chains of both chondroitin sulfate and heparin sulfate. Several extracellular matrix proteins contain GAG-binding domains; membrane-bound proteoglycans have been shown to participate in cell adhesion. Endoglin and the type III receptor may interact with extracellular matrix to constitute a signaling event. TGF-β is a major regulator of extracellular matrix protein deposition and cellular adhesion processes. An attractive, but purely speculative, model is that endoglin and the type III receptor may act, through increased interactions with the proteins up-regulated by TGF-β, to send a signal in a negative feedback loop to down-regulate this particular TGF-β signaling pathway.

V. Perspectives and Future Directions

Research in the TGF-β receptor field is currently at an exciting stage. The main molecular players that transduce TGF-β signals across the plasma membrane have all been isolated. However, the mechanism of action by which these receptor transduce the signal and the downstream signaling pathways which they activate remain largely unknown. The functional relationship between the type I and type II receptors must be investigated further to address the issue of why a heteromeric receptor complex is essential for signal transduction. In addition, the formation of this heteromeric complex results in the aggregation of distinct cytoplasmic kinases, each of which may potentially interact with seperate signaling molecules to activate different downstream signaling pathways. These downstream pathways must be identified so their interaction with each other and with the other known signaling pathways may be deduced.

In addition to the pursuits centering on the type I–type II heteromeric receptor complex and its signaling mechanism, other signaling mechanisms must be addressed. Sufficient evidence suggests that TGF-β may signal through type II homodimers in some cellular contexts, but this possibility remains to be confirmed. Whether the type III receptor and endoglin can signal TGF-β responses or whether they act only to modulate signaling of the type I and type II heteromeric complex by presenting the ligand remains unknown. Finally, the biochemical classification of a potentially novel TGF-β serine/threonine receptor kinase, the type V receptor, raises many intriguing questions. Therefore, the molecular structure of this receptor must be determined.

To better address many of the remaining issues, more definitive biological assays for TGF-β responses must be established. The conventional assays for TGF-β growth inhibition and gene responses all look at changes in events late in the TGF-β signaling pathway. Therefore, assays must be developed that detect earlier cellular responses that may distinguish signals derived from the type I receptor from those produced by the type II receptor.

In closing, many interesting and unexpected lessons have been learned through our studies of the TGF-β receptors; many more probably lie ahead. Hopefully, using the knowledge derived from studies of other classes of receptors, the molecular means by which TGF-β signals will soon be discovered.

ACKNOWLEDGMENTS

We thank M. Datto and D. Howe for technical assistance and helpful discussions of the manuscript. This work was supported by Grant DK45746 from the National Institutes of Health and Grant 3613 from the Council for Tobacco Research to X.-F. Wang. C. H. Bassing was supported by a National Science Foundation Graduate Research Fellowship. X.-F. Wang is a Leukemia Society Scholar.

REFERENCES

Andres, J. L., Stanley, K., Cheifetz, S., and Massague, J. (1989). Membrane-anchored and soluble forms of betaglycan, a polymorphic proteoglycan that binds transforming growth factor-beta. *J. Cell Biol.* **109,** 3137–3145.

Archer, S. J., Bax, A. Roberts, A. B., Sporn, M. B., Ogawa, Y., Piez, K. A., Weatherbee, J. A., Tsang, M. L., Lucas, R., Zheng, B. L., Wenker, J., and Torchia, D. A. (1993). Transforming growth factor beta 1: Secondary structure as determined by heteronuclear magnetic resonance spectroscopy. *Biochemistry* **32,** 1164–1171.

Attisano, L., Wrana, J. L., Cheifetz, S., and Massague, J. (1992). Novel activin receptors: Distinct genes and alternative mRNA splicing generate a repertoire of serine/threonine kinase receptors. *Cell* **68,** 97–108.

Attisano, L., Carcamo, J., Ventura, F., Weis, F.M.B., Massague, J., and Wrana, J. L. (1993). Identification of human activin and TGFβ type I receptors that form heteromeric kinase complexes with type II receptors. *Cell* **75**, 671–680.
Barnett, J. V., Moustakas, A., Lin, W., Wang, X. F., Lin, H. Y., Galper, J. B., Weinberg, R. A., Lodish, H. F., and Maas, R. L. (1993). Expression of the type II and type III TGFβ receptors during chick organogenesis: Comparison with TGF-β expression. Submitted.
Bassing, C. H., Yingling, J. M., Howe, D. H., Wang, T., He, W. W., Gustafson, M.L., Shah, P., Donahoe, P. K., and Wang, X.-F. (1994). A transforming growth factor-β type I receptor that signals to activate gene expression. *Science*, in press.
Bork, P., and Sander, C. (1992). A large domain common to sperm receptors (Zp2 and Zp3) and TGF-beta type III receptor. *FEBS Lett.* **300**, 237–240.
Boyd, F. T., and Massagué, J. (1989). Transforming growth factor-β inhibition of epithelial cell proliferation linked to the expression of a 53-kD membrane protein. *J. Biol. Chem.* **264**, 2272–2278.
Cheifetz, S., and Massague, J. (1991). Isoform-specific transforming growth factor-beta binding proteins with membrane attachments sensitive to phosphatidylinositol-specific phospholipase C. *J. Biol. Chem.* **266**, 20767–20772.
Cheifetz, S., Like, B., and Massagué, J. (1986). Cellular distribution of type I and type II receptors for transforming growth factor-β. *J. Biol. Chem.* **261**, 9972–9978.
Cheifetz, S., Weatherbee, J. A., Tsang, M. L.-S., Anderson, J. K., Mole, J. E., Lucas, R., and Massague, J. (1987). The transforming growth factor-β system, a complex pattern of cross-reactive ligands and receptors. *Cell* **48**, 409–415.
Cheifetz, S., Andres, J. L., and Massagué, J. (1988a). The transforming growth factor-β receptor type III is a membrane proteoglycan. *J. Biol. Chem.* **263**, 16984–16991.
Cheifetz, S., Ling, N., and Guillemin, R. (1988b). A surface component of GH3 pituatory cells that recognizes transforming growth factor β, activin, and inhibin. *J. Biol. Chem.* **263**, 17225–17228.
Cheifetz, S., Hernandez, H., Laiho, M., Dijke, P., Iwata, K. K., and Massague, J. (1990). Distinct transforming growth factor-β (TGF-β) receptor subsets as determinants of cellular responsiveness to three TGF-beta isoforms. *J. Biol. Chem.* **265**, 20533–20538.
Cheifetz, S., Bellon, T., Cales, C., Vera, S., Bernabeu, C., Massague, J., and Letarte, M. (1992). Endoglin is a component of the transforming growth factor-beta receptor system in human endothelial cells. *J. Biol. Chem.* **267**, 19027–19030.
Chen, R. H., Ebner, R., and Derynck, R. (1993). Inactivation of the type II receptor reveals two receptor pathways for the diverse TGF-β activities *Science* **260**, 1335–1338.
Daopin, S., Piez, K. A., Ogawa, Y., and Davies, D. R. (1992). Crystal structure of transforming growth factor-β2: An unusual fold for the superfamily. *Science* **257**, 369–373.
Ebner, R., Chen, R. H., Shum, L., Lawler, S., Zioncheck, T. F., Lee, A., Lopez, A. R., and Derynck, R. (1993a). Cloning of a type I TGF-β receptor and its effect on TGF-β binding to the type II receptor. *Science* **260**, 1344–1348.
Ebner, R., Chen, R. H., Lawler, S., Zionchick, T., and Dorynek, R. (1993b). Determination of receptor specificity by the type II receptors for TGFβ or activin. *Science* **262**, 900–902.
Franzen, P., ten Dijke, P., Iohijo, H., Yamashita, H., Schulz, P., Heldin, C.-H., and Miyazono, K. (1993). Cloning of a TGF-β type I receptor that forms a heteromeric complex with the TGF-β type II receptor. *Cell* **75**, 681–692.
Geiser, A. G., Burmester, J. K., Webbink, R., Roberts, A. B., and Sporn, M. B. (1992).

Inhibition of growth by transforming growth factor-β following fusion of two nonresponsive human carcinoma cell lines: Implication of the type II receptor in growth-inhibitory responses. *J. Cell Biol.* **267;** 2588–2593.

Georgi, L. L., Albert, P. S., and Riddle, D. L. (1990). *daf*-1, a *C. elegans* gene controlling dauer larva development, encodes a novel receptor protein kinase. *Cell* **61,** 635–645.

Glass, D. J., and Yancopulos, G. D. (1993). The neurotrophins and their receptors. *Trends Cell Biol.* **3,** 262–267.

Gougos, A., and Letarte, M. (1990). Primary structure of endoglin, an RGD-containing glycoprotein of human endothelial cells. *J. Biol. Chem.* **265,** 8361–8364.

He, W. W., Gustafson, M. L., Hirobe, S., and Donahoe, P. K. (1993). Developmental expression of four novel serine/threonine kinase receptors homologous to the activin/transforming growth factor-β type II receptor family. *Dev. Dynam.* **196,** 133–142.

Hemmati-Brivanlou, A., and Melton, D. A. (1992). A truncated activin receptor inhibits mesoderm induction and formation of axial structures in *Xenopus* embryos. *Nature (London)* **359,** 609–614.

Inagaki, M., Moustakas, A., Lin, H. Y., Lodish, H. F., and Carr, B. I. (1993). Growth inhibition by transforming growth factor beta (TGF-β) type I is restored in TGF-β-resistant hepatoma cells after expression of TGF-beta receptor type II cDNA. *Proc. Natl. Acad. Sci. USA* **90,** 5359–5363.

Ito, M., Yasui, W., Kyo, E., Yokozaki, H., Nakayama, H., Ito, H., and Tahara, E. (1992). Growth inhibition of transforming growth factor beta on human gastric carcinoma cells: Receptor and postreceptor signaling. *Cancer Res.* **52,** 295–300.

Laiho, M., Weis, F.M.B., and Massagué, J. (1990). Concomitant loss of transforming growth factor (TGF)-β receptor types I and II in TGF-β-resistant cell mutants implicates both receptor types in signal transduction. *J. Biol. Chem.* **265,** 18518–18524.

Laiho, M., Weis, F.M.B., Boyd, F., Ignotz, R. A., and Massagué, J. (1991). Responsiveness to transforming growth factor-β (TGF-β) restored by genetic complementation between cells defective in TGF-β receptors I and II. *J. Biol. Chem.* **266,** 9108–9112.

Lin, H. Y., and Lodish, H. F. (1993). Receptors for the TGF-β superfamily: Multiple polypeptides and serine/threonine kinases. *Trends Cell Biol.* **3,** 14–19.

Lin, H. Y., Wang, X. F., Ng-Eaton, E., Weinberg, R. A., and Lodish, H. F. (1992). Expression cloning of the TGF-β type II receptor, a functional transmembrane serine/threonine kinase. *Cell* **68,** 775–785.

Linask, K. K., D'Angelo, M., Gehris, A. L., and Greene, R. M. (1991). Transforming growth factor-β receptor profiles of human and murine embryonic palate mesenchymal cells. *Exp. Cell Res.* **192,** 1–9.

Lopez-Casillas, F., Cheifetz, S., Doody, J., Andres, J. L., Lane, W. S., and Massague, J. (1991). Structure and expression of the membrane proteoglycan betaglycan, a component of the TGF-β receptor system. *Cell* **67,** 785–795.

Lopez-Casillas, F., Wrana, J. L., and Massague, J. (1993). Betaglycan presents ligand to the TGFβ signaling receptor. *Cell* **73,** 1435–1444.

Lyons, R. M., and Moses, H. L. (1990). TGF-βs and the regulation of cell proliferation. *Eur. J. Biochem.* **187,** 467–473.

Massague, J. (1990). The transforming growth factor-β family. *Annu. Rev. Cell Biol.* **6,** 597–641.

Massague, J. (1992). Receptors for the TGF-beta family. *Cell* **69,** 1067–1070.

Massague, J., Cheifetz, S., Boyd, F. T., and Andres, J. L. (1990). TGF-β receptors and TGF-β binding proteoglycans: Recent progress in identifying their functional properties. *Ann. N.Y. Acad. Sci.* **593,** 59–72.

Mathews, L. S., and Vale, W. W. (1991). Expression cloning of an activin receptor, a predicted transmembrane serine kinase. *Cell* **65,** 973–982.

Mathews, L. S., Vale, W. W., and Kintner, C. R. (1992). Cloning of a second type of activin receptor and functional characterization in *Xenopus* embryos. *Science* **255,** 1702–1705.
Matsuzaki, K., Xu, J., Wang, F., McKeehan, W. L., Krummen, L., and Kan, M. (1993). A widely expressed transmembrane serine/threonine kinase that does not bind activin, inhibin, transforming growth factor B, or bone morphogenic factor. *J. Biol. Chem.* **268,** 12719–12723.
Moren, A., Ichijo, H., and Miyazono, K. (1992). Molecular cloning and characterization of the human and porcine transforming growth factor-β type III receptors. *Biochem. Biophys. Res. Commun.* **189,** 356–362.
Nakamura, T., Sugino, K., Kurosawa, N., Sawai, M., Takio, K., Eto, Y., Iwashita, S., Muramatsu, M., Titani, K., and Sugino, H. (1992). Isolation and characterization of activin receptor from mouse embryonal carcinoma cells. Identification of its serine/threonine/tyrosine protein kinase activity. *J. Biol. Chem.* **267,** 18924–18928.
Nakamura, Y., Kamatsu, N., and Nakauchi, H. (1992). A truncated erythropoietin receptor that fails to prevent programmed cell death of erythroid cells. *Science* **257,** 1138–1141.
Nishimatsu, S., Iwao, M., Nagai, T., Oda, S., Suzuki, A., Asashima, M., Murakami, K., and Veno, N. (1992). A carboxy-terminal truncated version of the activin receptor mediates activin signals in early *Xenopus* embryos. *FEBS Lett.* **312,** 169–173.
O'Grady, P., Huang, S. S., and Huang, J. S. (1991a). Expression of a new type high molecular weight receptor (type V receptor) of transforming growth factor β in normal and transformed cells. *Biochem. Biophys. Res. Commun.* **179,** 378–385.
O'Grady, P., Kuo, M. D., Baldassare, J. J., Huang, S. S., and Huang, J. S. (1991b). Purification of a new type high molecular weight receptor (type V receptor) of transforming growth factor β (TGF-β) from bovine liver. Identification of the type V TGF-β receptor in cultured cells. *J. Biol. Chem.* **266,** 8583–8589.
O'Grady, P., Liu, Q., Huang, S. S., and Huang, J. S. (1992). Transforming growth factor beta (TGF-β) type V receptor has a TGF-β-stimulated serine/threonine-specific autophosphorylation activity. *J. Biol. Chem.* **267,** 21033–21037.
Ohta, M., Greenberger, J. S., Anklesaria, P., Bassols, A., and Massague, J. (1987). Two forms of transforming growth factor-β distinguished by multipotential hematopoietic progenitor cells. *Nature (London)* **329,** 539–541.
Qian, S. W., Burmester, J. K., Merwin, J. R., Madri, J. A., Sporn, M. B., and Roberts, A. B. (1992). Identification of a structural domain that distinguishes the actions of the type 1 and 2 isoforms of transforming growth factor β on endothelial cells. *Proc. Natl. Acad. Sci. USA* **89,** 6290–6294.
Roberts, A. B., and Sporn, M. B. (1990). The transforming growth factor-βs. *In* "Peptide Growth Factors and Their Receptors" (M. Sporn and A. B. Roberts, eds.), pp. 421–472. Springer-Verlag, Heidelberg.
Schlunegger, M. P., and Grutter, M. G. (1992). An unusual feature revealed by the crystal structure at 2.2 Å resolution of human transforming growth factor-β2. *Nature (London)* **358,** 430–434.
Segarini, P. R. (1993). TGF-β receptors: A complicated system of multiple binding proteins. *Biochim. Biophys. Acta* (in *press*).
Segarini, P. R., and Seyedin, S. M. (1988). The high molecular weight receptor to transforming growth factor-β contains glycosaminoglycans chains. *J. Biol. Chem.* **263,** 8366–8370.
Segarini, P. R., Rosen, D. M., and Seyedin, S. M. (1989). Binding of transforming growth factor-β to cell surface proteins varies with cell type. *Mol. Endocrinol.* **3,** 261–272.
Segaraini, P. R., Ziman, J. M., Kane, C. J., and Dasch, J. R. (1992). Two novel patterns of

transforming growth factor β (TGF-β) binding to cell surface proteins are dependent upon the binding of TGF-β1 and indicate a mechanism of positive cooperativity. *J. Biol. Chem.* **267,** 1048–1053.

Shull, M. M., Ormsby, I., Kier, A. B., Pawlowski, S., Diebold, R. J., Yin, M., Allen, R., Sidman, C., Proetzel, G., Calvin, D., *et al.* (1992). Targeted disruption of the mouse transforming growth factor-β1 gene results in multifocal inflammatory disease. *Nature (London)* **359,** 693–699.

Sporn, M. B., and Roberts, A. B. (1992). Transforming growth factor-β: Recent progress and new challenges. *J. Cell Biol.* **119,** 1017–1021.

Taniguchi, T., and Minami, Y. (1993). The IL-2/IL-2 receptor system: A current overview. *Cell* **73,** 5–8.

Tsuchida, K., Mathews, L. S., and Vale, W. W. (1993). Cloning and characterization of a transmembrane serine kinase that acts as an activin type I receptor. *PNAS* **90,** 11242–11246.

ten Dijke, P., Iohijo, H., Franzen, P., Schults, P., Saras, J., Toyoshima, H., Heldin, C.-H., and Miyazono, K. (1993). Activin receptor-like kinases: A novel subclass of cell surface receptors with predicted serine/threonine kinase activity. *Oncogene* **8,** 2879–2887.

Ullrich, A., and Schlessinger, J. (1990). Signal transduction by receptors with tyrosine kinase activity. *Cell* **61,** 203–212.

Voss, S. D., Leary, T. P., Sondel, P. M., and Robb, R. J. (1993). Identification of a direct interaction between interleukin 2 and the p64 interleukin 2 receptor gamma chain. *Proc. Natl. Acad. Sci. USA* **90,** 2428–2432.

Wang, E. A., Rosen, V., Cordes, P., Hewick, R. M., Kriz, M. J., Luxenberg, D. P., Sibley, B. S., and Wozney, J. M. (1988). Purification and characterization of other distinct bone-inducing factors. *Proc. Natl. Acad. Sci. USA* **85,** 9484–9488.

Wang, X. F., Lin, H. Y., Ng-Eaton, E., Downward, J., Lodish, H. F., and Weinberg, R. A. (1991). Expression cloning and characterization of the TGF-β type III receptor. *Cell* **67,** 797–805.

Wrana, J. L., Attisano. L., Carcamo, J., Zentella, A. Doody, J., Laiho, M., Wang, X.-F., Massague, J. (1992). TGF-β signals through a heteromeric protein kinase receptor complex. *Cell* **71,** 1003–1014.

Biological Actions of Endothelin

KATHERINE STEPHENSON,* CHANDRASHEKHAR R. GANDHI,[†] AND MERLE S. OLSON*

*Department of Biochemistry
The University of Texas Health Science Center at San Antonio
San Antonio, Texas 78284
†Department of Anesthesiology and Critical Care Medicine
University of Pittsburgh
Pittsburgh, Pennsylvania 15261

I. Introduction
II. Discovery of Endothelin
III. Endothelin Structure
IV. Endothelin Structure–Activity Relationships
V. Processing of Endothelin
VI. Genes of the Endothelin Family
VII. Structural Organization of the Endothelin Gene
VIII. Factors that Stimulate Endothelin Production
IX. Endothelin Receptors
X. Localization of Endothelin
XI. Endothelin-Activated Transmembrane Signaling Systems
XII. Biological Actions of Endothelin
XIII. Pathophysiology of Endothelin
XIV. Endothelin and the Liver
XV. Conclusions
References

I. Introduction

The metabolic and hemodynamic function of various tissues is regulated by a variety of humoral, paracrine, and autacrine signaling mechanisms. For decades, a major portion of the experimental consideration of the regulation of macromolecular synthesis and degradation and of the regulation of blood flow in the circulatory system concentrated on humoral factors, including but not limited to the pancreatic peptides, insulin and glucagon, and adrenal-generated catecholamines. More recently, it has become apparent and of crucial interest to characterize a variety of mediators and cytokines of both lipid and peptide derivations which exhibit potent modulating or signaling responses in various tissues and organs. Consideration of paracrine and autacrine factors that are the signaling molecules involved in numerous intercellular regulatory mechanisms in physiological and pathophysiological situations has opened a new era of intense research into how tissues and organs respond to the numerous imperatives of maintaining an organism in an otherwise inhospitable environment. This chapter summarizes, somewhat selectively, the growing amount of information that has developed in the past 5 years on, arguably, the most potent vasoactive peptide agonist yet discovered. This discussion attempts to pursuade the reader that endothelin participates in a variety of important signaling mechanisms in normal and disease states.

II. Discovery of Endothelin

The discovery that endothelial cells synthesize and secrete potent vasoactive factors distinguished the vascular endothelium as an integral regulatory element involved in maintaining vascular smooth muscle tone. The elucidation of the vasorelaxant prostaglandin I_2 (Moncada et al., 1980) and the documentation of a short-lived diffusable substance named "endothelium-derived relaxing factor" (EDRF; Furchgott and Zawadski, 1980) heralded a period of intense investigation of endothelial-derived vasoactive factors (Furchgott, 1984). The nonprostanoid factor EDRF synthesized from L-arginine was later identified as nitric oxide (NO) (Palmer et al., 1987,1988; Moncada et al., 1988). In addition to mediating vasodilatory responses, research has indicated that the intact endothelium is essential for the vasoconstrictive responses stimulated by various chemical and mechanical factors (DeMey and Vanhoutte, 1982; Holden and McCall, 1983,1984; O'Brien and McMurtry, 1984; Rubanyi and Vanhoutte, 1984,1985; Vanhoutte et al., 1986; Harder, 1987; Katusic et al., 1987). In the quest for the identity of a nonprostanoid vasodilatory agent, researchers observed serendipitously that the addition of endothelial cell-conditioned medium to isolated rings of canine, porcine, or bovine coronary arteries provoked a prolonged vasoconstrictive response (Hickey et al., 1985). This work, in conjunction with other studies, indicated the existence of an endothelial cell-derived constricting factor that was peptidergic in nature (Agricola et al., 1984; Hickey et al., 1985; Gillespie et al., 1986; O'Brien et al., 1987). This series of important investigations led Yanagisawa and his colleagues to isolate and clone endothelin and to determine the primary structure of this peptide. Endothelin is the most potent vasoconstrictor peptide yet discovered and was isolated from the serum-free conditioned medium of porcine endothelial cells cultured in confluent monolayers (Yanagisawa et al., 1988a). This peptide was purified by anion-exchange column chromatography and reversed-phase high performance liquid chromatography (HPLC). A bioassay of vasoconstrictor activity on porcine coronary artery strips was used to detect the presence of the peptide in fractions collected during purification.

III. Endothelin Structure

Endothelin is a 21-amino-acid peptide found in at least four isoforms: endothelin-1, endothelin-2, endothelin-3, and endothelin-β or vasoactive intestinal contractor (VIC) (Itoh et al., 1988; Yanagisawa et al., 1988a,b; Inoue et al., 1989a; Saida et al., 1989). Endothelin-2 differs from endothelin-1 by the amino acid substitutions of Trp^6–Leu^7 and shares 90% homology with endothelin-1. Endothelin-3 contains 6 amino acid substitutions relative to endothelin-1—Thr^2, Phe^4–Thr^5–Tyr^6–Lys^7, and Tyr^{14}—and shares 71% sequence homology with endothelin-1 and -2. VIC differs from endothelin-1 by three amino acid substitutions: Asn^4, Trp^6, and Lys^7.

Endothelin peptides share 67% homology and compare in bioactivity with the sarafotoxins (STX), rare peptide toxins isolated from the venom of the Israeli burrowing asp *Atractapsis engaddensis* (Takasaki et al., 1988a,b; Bdolah et al., 1989; Kloog et al., 1988,1989a). Four STXs have been described: STX-S6a, -b, -c, and -d. The STXs exhibit potent coronary constrictor activity, which accounts for their cardiac toxicity and potential lethality (Lee et al., 1986). When conservative substitutions are considered, approximately 80% homology exists among endothelin-1,-2,-3, VIC, and STX-S6b (Takasaki et al., 1988a). The similarities among the STXs and the endothelins suggest that the families share a common evolutionary origin (Kloog et al., 1989a). The primary structures for the endothelin family are shown in Fig. 1.

The members of the endothelin family of peptides share several structural motifs. The endothelin isopeptides contain two disulfide linkages occurring at Cys^1–Cys^{15} and Cys^3–Cys^{11} that provide a tightly constrained peptide structure (Yanagisawa et al., 1988b; Hirata et al., 1989a). The specific location of each disulfide pair was confirmed by synthesis in which selective protection–deprotection was implemented to verify unambiguous coupling between the four cysteine residues (Immer et al., 1988; Liu et al., 1990; Nomizu et al., 1990). Other distinctive structural features of the endothelin family are a cluster of three perfectly conserved polar sidechains (Asp^8–Lys^9–Glu^{10}) re-

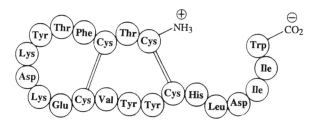

Endothelin-3

Endothelin-1	C	S	C	S	S	L	M	D	K	E	C	V	Y	F	C	H	L	D	I	I	W
Endothelin-2	C	S	C	S	S	W	L	D	K	E	C	V	Y	F	C	H	L	D	I	I	W
Endothelin-3	C	T	C	F	T	Y	K	D	K	E	C	V	Y	Y	C	H	L	D	I	I	W
VIC	C	S	C	N	S	W	L	D	K	E	C	V	Y	F	C	H	L	D	I	I	W
STX S6b	C	S	C	K	D	M	T	D	K	E	C	L	Y	F	C	H	Q	D	V	I	W

FIG. 1. Tertiary structure for endothelin-3 and primary structures for other endothelin isopeptides. The isopeptides of the endothelin family share several structural motifs. Each contains two disulfide linkages occurring between Cys 1–Cys 15 and Cys 3–Cys 11, a cluster of three perfectly conserved polar-charged sidechains at Asp 8–Lys 9–Glu 10 that reside within a hairpin loop structure containing amino acids 6–10, and a hydrophobic C terminus in residues 16–21 that contains the aromatic indole side chain at Trp 21.

siding in a hairpin loop structure containing amino acids 6–10 and a hydrophobic C terminus in residues 16–21 that contains the aromatic indole side chain at Trp[21] (Itoh et al., 1988; Yanagisawa et al., 1988a,b).

The residues at positions 4–7 represent the variable region specific for each endothelin isopeptide (Kloog et al., 1989a). The lack of sequence homology in these regions contributes to differences in the overall net charge of the isopeptides. STX-a, -b, and -d, with two positive and three negative charges, possess an overall charge of -1 in the loop structure, as do endothelin-1 and -2. STX-c, with four negative charges, has a net charge of -4 whereas endothelin-3 has a net charge of zero, with two positive and two negative charges.

Analysis of endothelin tertiary structure by nuclear magnetic resonance (NMR) suggests a helical conformation between Lys 9 and His 16 with an extended conformationally flexible C terminus (Endo et al., 1989; Hiley, 1989; Nakajima et al., 1989b; Saudek et al., 1989; Brown et al., 1990). Circular dichroism (CD) studies have indicated that the helical formation may be stabilized by electrostatic interactions (Brown et al., 1990). Further, the His[16] and Asp[18] residues are located in close proximity (1.95 Å) so hydrogen bonding may occur between these functional groups. The aromatic substituent Trp[21] lies in close proximity to the His[16], Asp[18] pair and may function to stabilize the donation transfer of a proton between these two amino acids (Topiol, 1987; Spinella et al., 1989).

IV. Endothelin Structure–Activity Relationships

The perfect conservation of the two disulfide linkages present in each of the endothelin isopeptides indicates the importance of these bonds in maintaining the structural and functional integrity of the peptide. The significance of these structures is indicated by the observation that reduction and alkylation of the two disulfide linkages increases the EC_{50} value for vascular smooth muscle contraction by 3 orders of magnitude (Kimura et al., 1988). Further, the topology of the disulfide bonds is also necessary for full activity, since two isomers of endothelin—[Cys[1]–Cys[11]/Cys[3]–Cys[15]]-endothelin and [Cys[1]–Cys[3]/Cys[11]–Cys[15]]-endothelin—are at least 100 times less active than native endothelin (Kumagaye et al., 1988). Various functional roles have been hypothesized for the Cys[3]–Cys[11] disulfide linkage. During peptide synthesis, Cys[3] and Cys[11] may react to aid in the initial folding process since these residues are in closer proximity. This occurrence would allow disulfide bond formation between the more distal Cys[1] and Cys[15]. Alternatively, since disulfide linkages are susceptible to nucleophilic attack, the Cys[3]–Cys[11] disulfide bond could participate actively in the receptor–ligand binding process if a cysteine residue is present at the receptor surface. The latter possibility may explain the extremely tight binding that has been observed between the receptor and endothelin itself. The destruction of the hairpin loop structure by hydrolysis at Lys[9] with lysyl endopeptidase, or reduction and alkylation of the disulfide linkages, attenuates endothelin-induced contractile activity by three orders of magnitude, underscoring the importance of the loop conformation in maintaining full bioactivity.

The C terminus, which is referred to as the hydrophobic tail, is crucial for endothelin activity. Deletion of Trp[21] displaces the ED_{50} for smooth muscle contraction 3–4 log units to the right whereas deletion of residues 17–21 results in total loss of activity (Kimura et al., 1988). Additional investigations in which the C terminus was elongated by one amino acid demonstrated the significance of the terminal Trp[21] residue in maintaining full agonist activity (Nishikori et al., 1991). Structure–activity studies of the C-terminal region performed on rabbit pulmonary artery and rat left atria confirmed that the C-terminal carboxylate, the indole nitrogen, and the aromatic residue are important for receptor recognition and binding (Doherty et al., 1991). Nakajima et al. (1989a) demonstrated that the amino and carboxyl groups of Asp[8] and Glu[10] as well as Phe[14] were crucial for receptor–ligand interactions. In addition, the alternating polar–nonpolar scheme represented in the C terminus may participate in a charge relay system, as mentioned previously. The His[16], Asp[18] and Trp[21] substituents have formed a close triad that is likely to be internally hydrogen bonded to form the proton relay system that may interact with a carboxyl group in the receptor pocket (Topiol, 1987; Spinella et al., 1989).

V. Processing of Endothelin

The sequences of cDNAs isolated from porcine aortic endothelial cell and human placenta cDNA libraries reveal that the 21-amino-acid isopeptides are proteolytic fragments of a prepropeptide (Itoh et al., 1988; Yanagisawa et al., 1988a; Inoue et al., 1989a; Saida et al., 1989). The endothelin precursors are large polypeptides of approximately 200 residues that possess distinct isopeptide sequences (Fig. 2). The first 19 residues of preproendothelin are characteristic of a secretory signal sequence with a hydrophobic core, followed by amino acid residues with small polar side chains (Perlman and Halvorson, 1983). Further, preproendothelin contains a sequence of residues, located from Cys[110] to Cys[124], that shows a remarkable similarity to the 15 N-terminal residues of mature endothelin; the relative positions of the four cysteine residues are intact in this fragment. Interestingly, this endothelin-like peptide sequence is located between two dibasic pairs of residues, Arg[106]–Arg[107] and Lys[139]–Arg[140], and could be secreted during proteolytic processing (Yanagisawa et al., 1988a). Additional analysis has revealed the endothelin-like sequence to be biologically inactive (Cade et al., 1990).

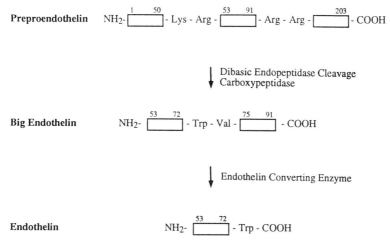

FIG. 2. Proteolytic processing pathway for the conversion of the prepro form of endothelin to mature endothelin. In this processing scheme, preproendothelin is converted to big endothelin via dibasic endopeptidase and carboxypeptidase cleavages. Big endothelin is processed to mature endothelin by a specific endopeptidase, endothelin converging enzyme.

The existence of the prepropeptides suggests a putative biosynthetic processing pathway for the endothelins that involves proteolytic processing of isopeptide-specific preprohormones, analogous to that for various other hormones and neuropeptides. The mature human endothelin sequence is found in residues Cys^{53}–Trp^{73}. The proposed proteolytic processing scheme of the prepropeptide includes two N-terminal cleavages of amino acids by dibasic pair-specific endopeptidases followed by subsequent carboxypeptidase cleavage to yield proendothelin, referred to as "big endothelin." The porcine isoform of big endothelin-1 contains 39 amino acids whereas the human isoform contains 38 amino acids. Big endothelin-2 and big endothelin-3 contain 37 and 41 amino acids, respectively. The processing of endothelin is completed by a novel protease referred to as "endothelin-converting enzyme," which cleaves between Trp^{21} and Ile^{22} of big endothelin-3 or Trp^{21} and Val^{22} of the remaining big endothelin isoforms to form the mature peptides (Yanagisawa et al., 1988a; Rubanyi and Parker-Botelho, 1991). The endothelin-converting enzyme was proposed to be an endopeptidase with chymotrypsin-like activity (Yanagisawa et al., 1988a). Additional investigations into the conversion of big endothelin-1 to endothelin-1 in porcine aortic endothelial cells and in the cultured bovine carotid artery indicate that endothelin-converting enzyme may be a membrane-bound neutral metalloprotease since phosphoramidon, a metalloprotease inhibitor, selectively blocks the conversion (Ikegawa et al., 1990a; Okada, et al., 1990). Aspartic, serine, and cysteine protease inhibitors have no appreciable effect on this cleavage reaction. In addition, a cathepsin D-like aspartic protease was found, at low pH, to participate in the conversion of big endothelin-1 to the mature peptide in porcine aortic endothelial cells (Ikegawa et al., 1990b). However, the physiological significance of this converting enzyme is in question since pepstatin A had no inhibitory effect on the *in vivo* conversion of endothelin (McMahon and Palomo, 1990). Current information indicates that multiple types of endothelin-converting enzyme may participate in the activation process.

VI. Genes of the Endothelin Family

Low hybridization stringency Southern blot analyses of human, porcine, and rat genomic DNA indicate that the endothelin isopeptides are encoded by at least three distinct genes (Inoue et al., 1989a). The originally discovered porcine endothelial cell-derived endothelin is identical to human endothelin-1. Rat endothelin is designated endothelin-3. The human genes encoding endothelin-1 and endothelin-3 are localized to chromosome 6 (Bloch et al., 1989a) and chromosome 20 (Bloch et al., 1989b), respectively. The third endothelin gene, endothelin-2, was discovered by Southern blot analyses of rat, human, and porcine genomic DNA (Inoue et al., 1989a). Cloning and sequence analysis of mouse genomic DNA revealed a fourth endothelin-like gene, VIC or endothelin-β, that is expressed exclusively in the intestine (Saida et al., 1989). However, the same gene sequence for endothelin-β has been demonstrated in rat tissues and is currently considered an isoform of endothelin-2 rather than a fourth gene (Yanagisawa and Masaki, 1989).

VII. Structural Organization of the Endothelin Gene

The 6.8-kb preproendothelin-1 and preproendothelin-3 genes encode five exons and four introns (Bloch et al., 1989a,b; Inoue et al., 1989b). Exon 1 encodes the first 21 amino acids of preproendothelin, exon 2 encodes the mature endothelin peptide as well as the first 4 amino acids of big endothelin, exon 3 encodes the 15-amino-acid endothelin-like peptide, exon 4 encodes an intermediate portion of the prepropeptide, and exon 5 encodes the remainder of preproendothelin as well as the 3' untranslated region of preproendothelin (Bloch et al., 1989a; Inoue et al., 1989b). The promoter region of the preproendothelin-1 gene contains a TATA box and a CCAAT sequence. S-1 nuclease protection studies revealed a transcription start site 31 bp 3' of the TATA box (Inoue et al., 1989b); the CCAAT box is located 65 bp 5' to the TATA box (Bloch et al., 1989a; Inoue et al., 1989b). Several putative regulatory elements may be present in the preproendothelin 1 gene. A pair of the hexanucleotide acute phase regulatory (APR) cis-acting elements is found 5' to the coding sequence as well as in the intervening sequence of exons 1 and 2 (Inoue et al., 1989b). The APR sequences presumably mediate induction of mRNA under acute physiological stress *in vivo*. Three sequence repeats that are at least 88% homologous to the octanucleotide consensus sequence of a cis-acting AP-1/jun binding element are found in the 5' region of the gene. The AP-1/jun regulatory element is likely to be involved with phorbol ester induction of mRNA (Inoue et al., 1989b). Phorbol 12-myristate 13-acetate (PMA) and 12-O-tetradecanoyl phorbol-13-acetate (TPA) have been reported to stimulate endothelin production in various cell types (Saida et al., 1989; Ehrenreich et al., 1990,1991; Ohta et al., 1990). Further, several putative cis-acting NF-1 elements are positioned in the 5' region and in the intervening sequence of exons 4 and 5. The NF-1 DNA binding transcription factor has been found to mediate mRNA induction by transforming growth factor-β (TGF-β) (Bloch et al., 1989a; Inoue et al., 1989b). TGF-β has been demonstrated to stimulate endothelin production (Kurihara et al., 1989; Suzuki et al., 1989; Hexum et al., 1990; Ohta et al., 1990; Sunnergren et al., 1990; Casey et al., 1991). AU-repeat sequences have been demonstrated in the 3' noncoding region of the preproendothelin gene. These sequences are thought to promote specific translation-dependent destabilization of the mRNA in the cytoplasm. The presence of these sequences may be indicative of a regulatory mechanism of endothelin mRNA levels by posttranscriptional degradation (Bloch et al., 1989a; Inoue et al., 1989b). The proposed structure of the preproendothelin-1 gene is illustrated in Fig. 3.

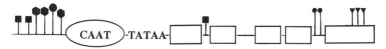

FIG. 3. Proposed structure of the preproendothelin gene. Exons for the gene are designated by the open boxes. The potential regulatory sequences are indicated by: ■ acute phase regulatory element; □, AP-1/JUN elements; ●, NF-1-binding sites; and ▼, AUUA sequences.

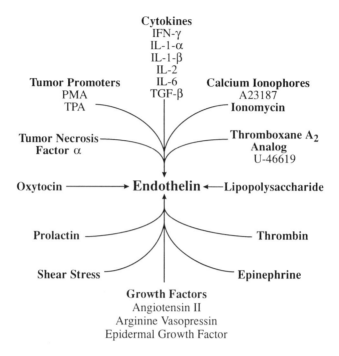

FIG. 4. Factors that stimulate endothelin production.

VIII. Factors that Stimulate Endothelin Production

Endothelin production and release has been found to be induced by mechanical stimuli and by numerous agents such as lipopolysaccharide, cytokines, growth factors, vasoactive peptides, bioactive amines, various calcium ionophores, and phorbol esters (Fig. 4). Many of these agents that promote endothelin production stimulate increases in intracellular calcium and/or an increase in protein kinase C (PKC) activity. These findings suggest that the mobilization of intracellular calcium and PKC activation may be significant events involved in activating endothelin production.

IX. Endothelin Receptors

Two distinct receptor subtypes of the endothelin (ET) family have been cloned, sequenced, and expressed functionally (Arai et al., 1990; Sakurai et al., 1990; Hosoda et al., 1991; Lin et al., 1991; Nakamuta et

TABLE I
ANALYSIS OF ENDOTHELIN RECEPTOR cDNA CLONES

cDNA library	Amino acid composition	Receptor subtype	Reference
Bovine lung	427	ET_A	Arai et al. (1990)
Rat lung	415	ET_B	Sakurai et al. (1990)
Human placenta	427	ET_A	Hosoda et al. (1991)
Rat A10 VSMC[a]	426	ET_A	Lin et al. (1991)
Human liver	442	ET_B	Nakamuta et al. (1991)
Human placenta	442	ET_B	Ogawa et al. (1991)
Human jejunum	416	ET_B	Sakamoto et al. (1991)
Human placenta	427	ET_A	Adachi et al. (1991)
Porcine cerebellum	443	ET_B	Elshourbagy et al. (1992)
Human placenta	427	ET_A	Hayzev et al. (1992)
Human lung	427	ET_A	Haendler et al. (1992)
Human lung	442	ET_B	Haendler et al. (1992)
Rat brain	442	ET_B	Cheng et al. (1993)

[a] VSMC, Vascular smooth muscle cell.

al., 1991; Ogawa et al., 1991; Sakamoto et al., 1991; Adachi et al., 1991; Elshourbagy et al., 1992; Hayzer et al., 1992; Haendler et al., 1992; Cheng et al., 1993). These receptors have been called ET_A (for aorta) and ET_B (for bronchus) (Maggi et al., 1989a). The Endothelin Receptor Nomenclature Committee of the International Union of Pharmacology recommendations designate ET_A receptors as those with affinities ET-1 = ET-2 > ET-3 and ET_B receptors as those with similar affinities for all the isopeptides. The predicted amino acid composition as well as the receptor subtype classifications of the cDNA clones are listed in Table I. Based on different affinity profiles found in various ^{125}I-labeled endothelin binding analyses, additional types of endothelin receptors have been proposed (Samson et al., 1990; Loffler and Lohrer, 1991; Masaki, 1991; Sakamoto et al., 1991; Schvartz et al., 1991; Williams et al., 1991; Sokolovsky et al., 1992).

The structure of the gene encoding a bovine ET_B receptor has seven exons with six intervening introns. The first exon encodes the first and second transmembrane-spanning domains of the receptor; each of the additional exons subsequently encodes a single transmembrane domain (Mizuno et al., 1992). A human ET_A receptor gene isolated and characterized by Hosoda et al. (1992) contains eight exons and seven introns. The 5' flanking region lacks a traditional TATA box but contains a putative SP-1 binding site upstream from the transcription start site. This gene was assigned to human chromosome 4. Northern blot analysis indicated the greatest concentration of the ET_A receptor mRNA in the aorta and in the renal mesangial cell.

Each of the cloned endothelin receptors contains seven hydrophobic sequence clusters of 22–26 residues that are intercalated between hydrophilic regions, as indicated by hydropathy analysis. The regions of hydrophobicity represent putative transmembrane-spanning segments of the receptor. This observation suggests that the endothelin receptor belongs to the superfamily of guanine nucleotide regulatory binding protein-linked receptors, which possess the characteristic seven transmembrane-spanning domains, an extracellular tail, and a cytoplasmic C-terminal tail. Further, most of the receptors contain the Asp–Arg–Tyr conserved sequence that is important for coupling of receptors to G proteins (Franke et al., 1990). This sequence motif has been demonstrated in 83% of all guanine nucleotide regulatory binding protein-linked receptors (Lin et al., 1991). The sequence analyses of all the cDNA clones indicate several potential sites for posttranslational modifications. At least one consensus site for N-linked glycosylation (Asn–X–Ser/Thr) is found in each of the receptor proteins. Further, several serine residues located in the third cytoplasmic loop and in the C-terminal tail may be phosphorylated by PKC (Kemp and Pearson, 1990).

Significant sequence disparities occur in the N-terminal region of the receptor proteins; the greatest variations exist between ET_A and ET_B receptor subtypes. In addition to the lack of sequence homology in these regions, the N terminus of the ET_A receptor contains 80 residues compared with the 101 residues of the ET_B receptor subtypes. The hydrophilic extracellular domain previously was found to play an integral role in ligand–receptor interactions (Dohlman et al., 1990). In this light, the N-terminal region may be involved in endothelin receptor–ligand interactions. The obvious dissimilarities in sequence homology may contribute to the preferred ligand specificity of the receptor molecule. One residue in particular, Lys[181], in the third transmembrane domain region of the ET_B receptor seems to be important for ligand–receptor interactions and crucial for binding of the ET_B-selective agonist STX-S6c (Zhu et al., 1992).

Differences exist between the two types of receptors in relation to the stabilities of the ligand–receptor complexes. Characterization of the ET_A and ET_B receptor–ligand complexes revealed that the ET_B receptor–ligand complex alone is resistant to acid treatment (0.2 M acetic acid and 0.5 M NaCl) (Takasuka et al., 1992).

Several specific antagonists for ET_A receptors have been isolated and characterized. BE-18257A, a cyclic pentapeptide [cyclo(D-Glu-L-Ala-allo-D-Ile-L-Ile-D-Trp)] derived from *Streptomyces misakiensis*, is a selective antagonist for ET_A receptors (Ihara *et al.*, 1991). This antagonist binds to receptors and is able to antagonize endothelin-1-induced vasoconstriction of rabbit iliac artery and pressor action in rats. In addition, the synthetic analogs of BE-18257A, BQ-123, and BQ-153 have proven to be potent antagonists of ET_A receptors since they inhibit endothelin-1-induced vasoconstriction of porcine coronary arteries (Ihara *et al.*, 1992). Spinella *et al.* (1991) developed an endothelin-1 analog in which the Cys^1–Cys^{15} disulfide linkage was replaced with an amide linkage between a diaminopropionic acid at position 1 and an alanine at position 15 [Dpr^1, Asp^{15}], this analog was found to be a specific antagonist for the ET_A receptor. This antagonist specifically inhibited endothelin-1-induced pulmonary vasoconstriction in the guinea pig lung, and competed for specific ^{125}I-labeled endothelin-1 binding sites in this tissue. A specific antagonist, [Cys^{11}–Cys^{15}]ET-1^{11-21} (IRL 1038), has been developed that inhibits ET_B receptor-mediated contraction of guinea pig ileum and tracheal smooth muscle (Urade *et al.*, 1992).

The endothelin receptor may be down-regulated or up-regulated in various pathophysiological situations. The endothelin receptor has been found to be down-regulated in streptozotocin-induced diabetes (Nayler *et al.*, 1989b) and in spontaneous hypertension (Gu *et al.*, 1990). In addition, endothelin receptor levels are up-regulated on stimulation by prolonged ischemia (Liu *et al.*, 1989), by postischemic oxygenation (Liu *et al.*, 1990), and in cyclosporin-induced toxicity (Nayler *et al.*, 1989a).

X. Localization of Endothelin

Endothelin and/or preproendothelin mRNA has been localized in a wide variety of tissues (Fig. 5). Some of the tissues in which both preproendothelin RNA and immunoreactive endothelin have been found include the vascular endothelium, heart, kidney, lung, nervous system, stomach, intestine, pancreas, spleen, eye, skin, placenta, testes, thyroid, and liver (Yanagisawa *et al.*, 1988a,b; Bloch *et al.*, (1989b; Giaid *et al.*, 1989; Haegerstrand *et al.*, 1989; Kitamura *et al.*, 1989a,b; MacCumber *et al.*, 1989; Marsden *et al.*, 1989b; Matsumoto *et al.*, 1989; Matsumura *et al.*, 1989; Saida *et al.*, 1989; Shinmi *et al.*, 1989; Yoshimi *et al.*, 1989; Sunnergren *et al.*, 1990; Yoshizawa *et al.*, 1990; Hemsen and Lundberg, 1991; Morita *et al.*, 1991; Takahashi *et*

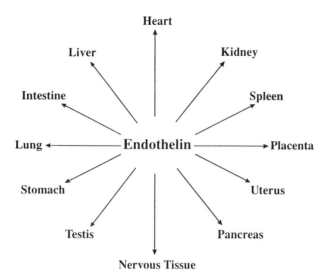

FIG. 5. Tissue distribution of endothelin.

al., 1991; Terenghi et al., 1991; van Papendorp et al., 1991; Colin et al., 1992; Naruse et al., 1992). The apparent ubiquitous localization of this peptide suggests that endothelin may play important roles in several signaling mechanisms in addition to its vasomotor regulatory capabilities.

XI. ENDOTHELIN-ACTIVATED TRANSMEMBRANE SIGNALING SYSTEMS

Endothelin activates numerous transmembrane signaling systems in various tissues and cell types. The multiple endothelin-stimulated signal transduction pathways likely contribute to the diversity of the biological responses induced by this peptide. Endothelin has been found to stimulate phospholipase C, phospholipase A_2, phospholipase D, cAMP accumulation, and Na^+/H^+ exchange (Kloog et al., 1988,1989a,b; Resink et al., 1988,1989; Ambar et al., 1989; Badr et al., 1989; Bousso-Mittler et al., 1989; Galron et al., 1989; Highsmith et al., 1989; Hirata et al., 1989b; Kai et al., 1989; Kloog and Sokolovsky, 1989; Marsden et al., 1989a; Muldoon et al., 1989; Simonson et al., 1989; Sugiura and Inagami, 1989; van Renterghem et al., 1989; Filep et al., 1990; Kramer et al., 1990; Reynolds et al., 1990; Simonson and Dunn, 1990; Abdel-Latif and Zhang, 1991; Abdel-Latif et al., 1991; Granstam et al., 1991; Konishi et al., 1991; Stojikovic et al., 1991; Vigne et al.,

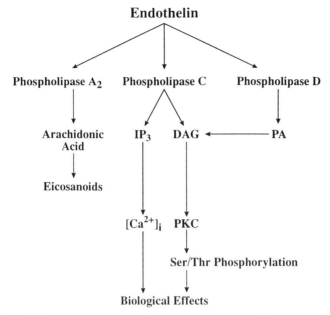

FIG. 6. Transmembrane signaling pathways activated by endothelin.

1991). Figure 6 summarizes the transmembrane signaling systems that have been demonstrated to be activated by endothelin.

Numerous reports indicate that endothelin activates phospholipase C to stimulate the phosphodiesteric hydrolysis of phosphatidylinositol-4,5-bisphosphate into phosphatidylinositol-1,4,5-trisphosphate and 1,2-diacylglycerol. The two second messengers generated participate to elevate intracellular calcium levels and to activate PKC, respectively. In vascular smooth muscle cells, the endothelin-stimulated increase in intracellular calcium possesses two components; (1) a rapid phase that is presumably a consequence of phosphatidylinositol-1,4,5-trisphosphate stimulating the release of calcium from the endoplasmic reticulum and (2) a sustained phase that is dependent on extracellular calcium availability (Miasiro et al., 1988; Marsden et al., 1989a; Danthuluri and Brock, 1990). The latter action is presumably responsible for the sustained contraction of vascular smooth muscle and is thought to be mediated by PKC (Auget et al., 1989; Ohlstein et al., 1989; Danthuluri and Brock, 1990). The increase in intracellular calcium may activate calcium-dependent protein kinases, calcium-sensitive phospholipase C, and phospholipase A_2 directly.

Endothelin has been demonstrated to activate phospholipase A_2 in various cell types. Phospholipase A_2 may be activated by direct stimulation by activation of the endothelin receptor or by an indirect pathway dependent on elevations of intracellular calcium, as mentioned previously. Arachidonic acid is liberated in this cleavage and is metabolized into a variety of bioactive mediators, including prostaglandins, leukotrienes, and thromboxanes. These eicosanoids can play integral roles in endothelin-mediated cellular responses.

Endothelin has been demonstrated to stimulate phospholipase D activity (Sunako et al., 1990). Phospholipase D hydrolyzes the polar head groups from the sn-3 position of membrane phospholipids such as phosphatidylcholine and phosphatidylethanolamine, producing a free polar head group and phosphatidic acid. Phospholipase D activation by hormonal stimulation may participate in the activation of various significant cellular signal transduction pathways. The phospatidic acid may be dephosphorylated by phosphatidic acid phosphohydrolyase to produce 1,2-diacylglycerol. The hydrolysis of phosphatidylcholine may contribute substantially to the sustained increase in 1,2-diacylglycerol levels after endothelin stimulation, since cellular phosphatidylcholine levels are considerably higher than those of phosphatidylinositol. This finding suggests that the phosphatidylcholine-derived 1,2-diacylglycerol may play a significant role in prolonged PKC stimulation.

XII. BIOLOGICAL ACTIONS OF ENDOTHELIN

Endothelin receptors have been demonstrated in numerous tissues (Fig. 7). Since specific receptors are required for biological activity of endothelin, the extensive distribution of endothelin receptors provides a foundation for the diverse multiorgan effects of endothelin. Probably the most common response of endothelin is exhibited after endothelin interaction with its receptor on the vascular smooth muscle cells to elicit its vasoconstrictive effects (de Nucci et al., 1988; Tombe et al., 1988; Marsden et al., 1989a; Power et al., 1989; Withrington et al., 1989). As mentioned previously, endothelin-induced smooth muscle contractions are prolonged and difficult to terminate. Endothelin is a more potent vasoactive agonist than phenylephrine, angiotensin II, or vasopressin (Yanagisawa et al., 1988b). In addition to its vascular effects, endothelin is responsible for a variety of diverse biological actions (Fig. 8). Endothelin has been demonstrated to stimulate contractions in nonvascular smooth muscle in tissues including vas deferens (Hiley et al., 1989), uterus (Kozuka et al., 1989), trachea (Uchida et al., 1988; Turner et al., 1989), ileum (Hiley et al., 1989), duodenum, stom-

BIOLOGICAL ACTIONS OF ENDOTHELIN 173

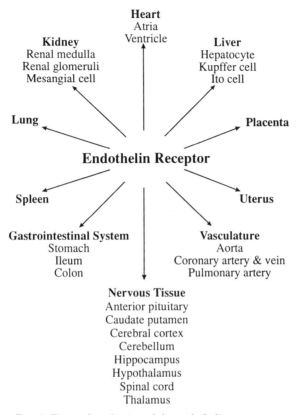

FIG. 7. Tissue distribution of the endothelin receptor.

ach (de Nucci et al., 1988), and human urinary bladder (Maggi et al., 1989b). Endothelin also has been reported to stimulate the release of EDRF from nonvascular smooth muscle in vitro (Warner et al., 1988). The release of EDRF in vivo by endothelin 1 and endothelin 3 may account for the initial depressor effect observed on administration of the endothelins (de Nucci et al., 1988; Wright and Fozard, 1988; Warner et al., 1989).

Endothelin has been shown to exhibit effects in other tissues, including the stimulation of contraction of the pulmonary bronchi (Hay et al., 1992). In addition, endothelin may influence hormonal activity, inhibiting prolactin secretion from the anterior pituitary gland (Samson et al., 1992) and stimulating the secretion of atrial natriuretic peptide (Fukuda et al., 1988,1989; Hu et al., 1988) and the production of aldosterone (Miller et al., 1989). Endothelin has mitogenic effects in a

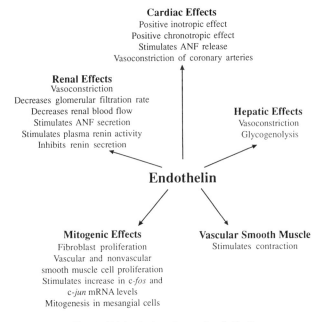

FIG. 8. Biological actions of endothelin.

variety of cell types including fibroblasts, nonvascular smooth muscle cells, and mesangial cells (Komuro et al., 1988; Brown and Littlewood, 1989; Dubin et al., 1989; Noveral et al., 1992; Ohlstein et al., 1992).

Endothelin has profound biological effects in cardiac tissue. Endothelin stimulates the contraction of isolated coronary arteries (Yanagisawa et al., 1988a,b) and has positive inotropic and chronotropic effects in rat and guinea pig isolated atria and ventricles (Ishikawa et al., 1988a,b; Yanagisawa et al., 1988b; Ambar et al., 1989; Moravec et al., 1989). Endothelin-stimulated positive inotropism was demonstrated in human atria and ventricles; the increase in contractility was more profound in the atrium than in the ventricle (Davenport et al., 1989) and phosphatidylinositol hydrolysis was increased on administration of endothelin in the rat heart (Kloog et al., 1988; Ambar et al., 1989b).

Endothelin receptors have been localized in the renal artery and vein, vascular bundle, glomerulus, renal papilla, and cortical and medullary regions of the kidney (Koseki et al., 1989; Davenport et al., 1989; Orita et al., 1989; Nunez et al., 1990). Endothelin induces vasoconstriction in the kidney, which consequently increases renal vascular resistance and reduces the renal blood flow and glomerular flow rate (GFR) (Firth et al., 1988; Goetz et al., 1988; Badr et al., 1989; Cairnes et al., 1989; Lopez-Farre et al., 1989; Miller et al., 1989). The decrease in the GFR has been proposed to occur through a mechanism in which endothelin-stimulated mesangial cells contract, induce glomerular capillary vasoconstriction, and consequently lower the ultrafiltration coefficient (Badr et al., 1989; King et al., 1989; Orita et al., 1989). In conjunction with the endothelin-induced decrease in renal hemodynamics, a decrease in urine flow and sodium excretion occurs in addition to an attenuation in fractional sodium and lithium excretion from the proximal tubule (Miller et al., 1989). These findings indicate that endothelin promotes the reabsorption of fluid and sodium in the proximal tubule. Atrial natriuretic peptide and aldosterone levels increase on endothelin exposure (Miller et al., 1989). Endothelin-1 has been found to attenuate renin release from isolated glomeruli, juxtaglomerular cells, and kidney slices *in vitro* (Rakugi et al., 1988; Takagi et al., 1988; Matsumura et al., 1989). The *in vitro* effect on renin release may be overshadowed by the indirect *in vivo* effects of endothelin, decreasing renal blood flow and the GFR and increasing blood pressure, peripheral resistance, and plasma renin activity.

XIII. Pathophysiology of Endothelin

Numerous reports have been published regarding the possible role of endothelin in a variety of pathophysiological conditions. Figure 9 illustrates the numerous situations in which tissue or plasma endothelin levels correlate with various pathophysiological scenarios. Increased circulating levels of endothelin have been associated with various pathophysiological conditions in the heart including angina pectoralis, congestive heart failure, and myocardial infarction (Miyauchi et al., 1989; Cavero et al., 1990; Schrader et al., 1990; Lam et al., 1991; Stewart et al., 1991a; Toyo-oko et al., 1991; Tsuji et al., 1991; Cody et al., 1992). Increased plasma endothelin levels also have been detected in renal-associated illnesses such as cyclosporin-induced nephrotoxicity, glomerular inflammation, and acute and chronic renal failure (Cairns et al., 1988; Firth et al., 1988; Kon et al., 1990; Schrader et al., 1990; Tomita et al., 1990; Awazu et al., 1991; Simonson and Dunn, 1991; Deray et al., 1992; Fogo et al., 1992; Perico et al., 1992). Various pathophysiological pulmonary conditions have been correlated with endothelin levels including asthma, pulmonary tumors, and pulmonary hypertension (Giaid et al., 1990; Mattoli, et al., 1991; Stewart et al., 1991b; Cody et al., 1992). Also, endothelin has been reported to corre-

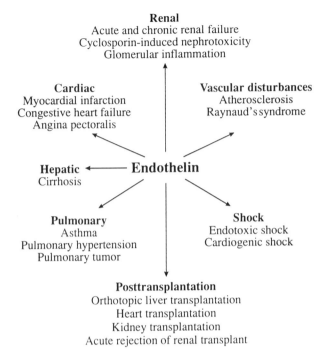

FIG. 9. Correlation of endothelin with pathophysiological conditions.

late with various postoperative pathophysiological conditions including the period following heart, kidney, and orthotopic liver transplantation (Schrader et al., 1990; Lerman et al., 1991,1992; Textor et al., 1992). Increased endothelin has been found in numerous diseases such as acute ischemic cerebral stroke, atherosclerosis, essential hypertension, Raynaud's phenomenon, pre-eclampsia, ulcerative colitis, uremia, and arthritis (Luscher et al., 1989; Saito et al., 1989; Shichiri et al., 1989; Kohno et al., 1990,1991; Taylor et al., 1990; Biondi et al., 1991; Yamane et al., 1991; Rachmilewitz et al., 1992; Tsunoda et al., 1992; Wharton et al., 1992; Ziv et al., 1992). Endothelin may participate in the pathology of various autoimmune diseases such as scleroderma and systemic lupus erythematosis (Julkunen et al., 1991; Kahaleh, 1991). Increased levels of endothelin also have been demonstrated in endotoxic shock (Morel et al., 1989; Nakamura et al., 1991; Pittet et al., 1991; Vemulapalli et al., 1991; Weitzberg et al., 1991; Uchihara et al., 1992; Voerman et al., 1992).

XIV. Endothelin and the Liver

As indicated in Section IV, endothelin is an extremely potent vasoactive peptide that acts on numerous tissues to stimulate a wide variety of biological effects. Most notably, endothelin may play a significant role in maintaining vasomotor tone in mammalian vascular beds in concert with other vasoactive agents. When administered intravenously, ^{125}I-labeled endothelin is distributed primarily in the lung, kidney, and liver (Anggard et al., 1989; Hoyer et al., 1989; Shiba et al., 1989; Sirvio et al., 1990; Furuya et al., 1991). Localization of injected endothelin in the liver suggests that the liver has the capacity to bind, exhibit biological responses to, and/or clear this peptide mediator from the circulation. Also, the literature has noted that several pathophysiological situations can exhibit elevated tissue or plasma levels of endothelin, including hepatic cirrhosis (Schrader et al., 1990; Gulberg et al., 1992; Uchihara et al., 1992; Uemasu et al., 1992; Veglio et al., 1992) and kidney and liver transplantation sequelae (Schrader et al., 1990; Lerman et al., 1991; Yamakado et al., 1991; Textor et al., 1992). Further, circulating levels of immunoreactive endothelin have been demonstrated to increase on administration of endotoxin and have been found to correlate with the severity of the endotoxemia (Morel et al., 1989; Sugiura and Inagami, 1989; Nakamura et al., 1991; Pittet et al., 1991; Vemulapalli et al., 1991; Uchihara et al., 1992; Voerman et al., 1992). This latter observation is important because liver cells, especially the hepatic Kupffer cells, are the major cell types in the body involved in endotoxin clearance and degradation.

For these reasons, this laboratory became interested in the signaling mechanisms in which endothelin may participate in the mammalian liver. Since the liver is a highly vascularized organ that is subject to neural, mechanical, and hormonal regulation, endothelin may participate in the regulation of hepatic circulation and metabolism. As shown in Fig. 10, when introduced into the perfused liver, endothelin stimulates an increase in glucose production, transient alterations in oxygen hepatic consumption, and a profound sustained increase in hepatic portal pressure in a dose-dependent manner (Gandhi et al., 1990; Serradeil-Le Gal et al., 1991; Roden et al., 1992). The vasoconstrictive response is resistant to washout by perfusion buffer following the endothelin infusion and is absolutely dependent on the presence of calcium in the extracellular medium. The endothelin-induced responses in the perfused rat liver are not subject to regulation by α- or β-adrenergic agonists, glucagon, or platelet activating factor. Further, pros-

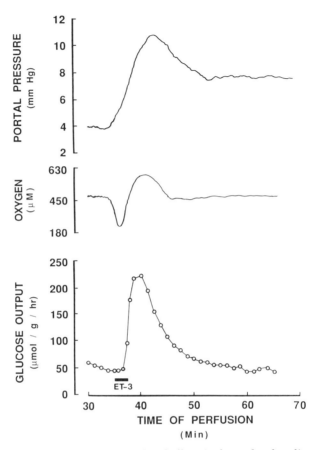

FIG. 10. Characteristic endothelin-3-induced effects in the perfused rat liver. Rat liver was perfused for 35 min prior to infusion with 3 nM endothelin-3 for 2 min. Hepatic portal pressure, oxygen consumption, and glucose production were measured. Representative tracings characterize the hemodynamic and metabolic effects of endothelin in the liver. Reproduced with permission from Gandhi et al., (1990).© The American Society for Biochemistry & Molecular Biology.

taglandins, leukotrienes, and thromboxanes have no effect on the endothelin-induced response in the perfused rat liver. Also, the glycogenolytic response in the perfused rat liver is not mediated via the secondary production of cyclooxygenase-derive eicosanoids. Coinfusion of indomethacin or ibuprofen, effective inhibitors of the cyclooxygenase pathway, causes no alteration in the extent of the duration of the glycogenolytic response of the liver when stimulated with endothelin.

Sequential infusions of endothelin into the perfused rat liver result in an attenuation of the endothelin-induced glucose production, suggesting that endothelin-mediated glycogenolytic activation is a receptor-mediated event and that this process or response is subject to some form of regulation or desensitization mechanism. Several lines of evidence demonstrate that the endothelin-induced glycogenolytic response occurs through the direct activation of endothelin receptors on the parenchymal cell. Interestingly, endothelin has been demonstrated to stimulate the elevation of phosphorylase a and intracellular calcium in isolated hepatocytes (Serradeil-Le Gal et al., 1991). Also of importance in this issue is the finding that specific high-affinity receptors for endothelin are found on rat hepatocytes in primary culture (Gandhi et al., 1992a). Cross-linking experiments suggest the presence of endothelin-1-specific ET_A and endothelin nonselective ET_B receptors on the rat hepatocyte. Endothelin-1 and-3 stimulate phosphoinositide metabolism via a pertussis toxin-sensitive guanine nucleotide regulatory protein. The hepatocyte endothelin receptors(s) are downregulated after prolonged exposure to homologous ligand.

The hepatocyte and the Kupffer cell are capable of internalizing and catabolizing radiolabeled endothelin (Gandhi et al., 1993). Whereas Kupffer cells are able to convert endothelin to a single peptide metabolite, the hepatocyte is capable of complete degradation of endothelin, including deiodination of the iodinated tyrosine residue of ^{125}I-radiolabeled endothelin. These observations indicate that the liver has the ability to be a primary location of clearance and degradation of systemic endothelin that may be generated at some remote site in the vasculature.

Endothelin-induced increases in hepatic portal pressure are complex and are likely to involve multiple cell types and mechanisms. The hepatic cell type responsible for the endothelin-induced portal pressure increases may be the nonparenchymal Ito cell, which possesses subepithelial myofibroblast characteristics. The Ito cell has been reported to facilitate contraction of the sinusoids via contraction of the Ito cell actin-rich cytoplasmic extensions (Wake, 1980), which is believed to be stimulated by the mobilization of intracellular calcium. Endothelin has been shown to induce phosphoinositide metabolism and elevation of intracellular calcium in hepatocytes, Ito cells and Kupffer cells. Specific endothelin receptors have been demonstrated on the Ito cell (Furuya et al., 1992). These findings suggest that endothelin plays a significant physiological role in the regulation of microvascular hepatic blood flow via the Ito cell. Kupffer cells do not seem to play a role in the endothelin-3-stimulated elevations in portal pressure since the endothelin-induced glycogenolytic response is not altered in rat livers exposed to gadolinium chloride, a rare earth metal that depresses reticuloendothelial cell function.

Endothelin may play a significant role in the hyperglycemic burst that occurs in the early stages of sepsis. The level of circulating endothelin was found to increase in endotoxic shock; the majority of this endothelin is derived from vascular endothelial cells. Further, hepatic endothelial cells have been shown to produce endothelin when exposed to Kupffer cells that have been treated with lipopolysaccharide (Ehrenreich et al., 1990). Therefore, endothelin may exhibit both paracrine and autocrine responses in the liver during normal and pathophysiological scenarios.

The transmembrane signal transduction pathways stimulated by endothelin have been investigated in cultured rat Kupffer cells (Gandhi et al., 1992b). Endothelin-3 does not appear to activate phospholipase A_2 in the Kupffer cell. However, this peptide stimulates the exclusive production of prostaglandin E_2, indicating that the production of this eicosanoid is not regulated at the level of phospholipase A_2 but at another as yet undefined level. Possible points of regulation include the various enzymes of the cyclooxygenase pathway such as prostglandin endoperoxidase synthase, prostaglandin H synthase, or endoperoxidase isomerase E. The intracellular concentration of arachidonic acid is extremely low with the majority of the arachidonic acid esterified to the sn-2 position of cellular diacylphosphoglycerols (Irvine, 1982). Free arachidonic acid may be made available to the cell through the concerted actions of diacylglycerol lipase and phospholipase C, since the calcium-requiring lipase cleaves arachidonic acid from the sn-2 position of membrane diacylglycerol.

The effects of endothelin on phospholipase C activation also have been investigated in isolated Kupffer cells in primary culture (Gandhi et al., 1992b). Endothelin-3 stimulates phosphodiesteric hydrolysis of membrane polyphosphoinositides, leading to inositol phosphate production typical of a classical calcium signaling mechanism. Endothelin-3-stimulated activation of phospholipase C is not subject to heterologous ligand regulation by other mediators. This finding agrees with studies in the perfused rat liver in which endothelin-3 effects on glucose production, hepatic oxygen consumption, and hemodynamic actions were not affected by other agonists. However, endothelin-stimulated phosphodiesteric hydrolysis of phosphoinositides is attenuated on pretreatment with endothelin, indicating a homologous ligand-induced regulation of the endothelin response in the Kupffer cell.

The Kupffer cell possesses high-affinity receptors for endothelin-3

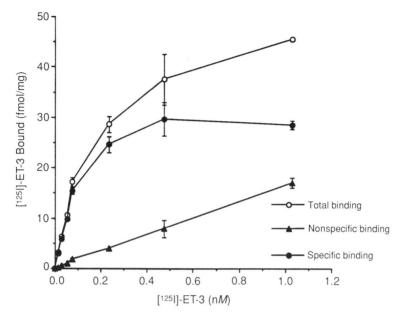

FIG. 11. Saturation analysis of [^{125}I]-endothelin-3 binding to rat cultured Kupffer cells at 37°C: 2×10^6 Kupffer cells in primary culture were incubated in Hank's buffer containing 0.1% BSA at the indicated concentration [^{125}I]-endothelin-3 (ET-3) for 2 hr at 37°C. The binding reaction was terminated by washing the cells thoroughly with ice-cold Hank's buffer with 0.1% BSA. The cells were solubilized in 0.5 M NaOH and radioactivity was determined in a Beckman Gamma 8000 scintillation counter. Total binding (○) and nonspecific binding (▲) of [^{125}I]-endothelin-3 to the Kupffer cell were determined in the absence and presence of 1000-fold excess unlabeled endothelin-3, respectively. Specific binding (●) was determined by subtracting the nonspecific binding parameter from total binding. The values represent means ± SD of duplicate determinations from a representative of two experiments.

with a dissociation constant of 0.07–0.21 nM and a B_{max} of approximately 5800 receptors per Kupffer cell, as determined by ^{125}I-labeled endothelin-3 radioligand binding analyses (Fig. 11); (K. Stephenson and M. S. Olson, unpublished results). Competition experiments in which various unlabeled endothelin isopeptides were incubated concomitantly with ^{125}I-labeled endothelin-3 indicate that the Kupffer cell endothelin receptor does not discriminate among endothelin isopeptides. Such nonselectivity is characteristic of an ET_B type endothelin receptor. The receptor–ligand complex is extremely stable since a significant portion of associated ^{125}I-labeled endothelin-3 cannot be removed by a 1000-fold excess of unlabeled peptide.

The Kupffer cell endothelin receptor is subject to regulation by a cAMP-dependent protein kinase and PKC. Prolonged treatment with cAMP analogs and forskolin induces a significant increase in ^{125}I-labeled endothelin-3 binding sites on the Kupffer cell in primary culture, as a result of an increase in the number of cell-surface endothelin receptors rather than an increase in the receptor binding affinity. Further, short-term pretreatment with PMA, an agent that activates PKC, decreases ^{125}I-labeled endothelin-3 binding on the Kupffer cell. On the other hand, prolonged pretreatment with the phorbol ester increases ^{125}I-labeled endothelin-3 binding. In both cases, activation of PKC alters the receptor density on the cell surface and not the receptor affinity.

The results of endothelin receptor regulation studies with cAMP and PKC activators suggest that endothelin responses in the Kupffer cell may not participate in the acute hepatic metabolic responses to endothelin stimulation, but may play a significant role in pathophysiological responses of the hepatic reticuloendothelial cells, certainly on a more prolonged time scale. This hypothesis is supported by the lack of Kupffer cell involvement in the initial hemodynamic and glycogenolytic responses induced by endothelin in the isolated perfused rat liver. The increase in Kupffer cell endothelin receptors after prolonged periods of time may contribute to the pathophysiological response of the liver to various conditions such as ischemia–reperfusion injury and sepsis.

Endothelin may be involved in generating potent lipid mediators (Mustafa and Olson, unpublished results) that participate in an inflammatory response to facilitate the infiltration of neutrophils. Endothelin stimulation of cultured Kupffer cells induces the synthesis of platelet activating factor. This response requires extracellular calcium, involves a guanine nucleotide regulatory binding protein for coupling, and results in the activation of both phospholipase A_2 and PKC.

Up-regulation of the endothelin receptor on the Kupffer cell may prove beneficial to both the liver and the organism. Increased levels of prostaglandin E_2 produced by endothelin-stimulated Kupffer cells may provide cytoprotective effects for the traumatized liver, as well as aid in hepatic regeneration, since this eicosanoid acts as a mitogen in the liver. Further, since the Kupffer cell has been found to metabolize endothelin, increased numbers of endothelin receptors on the cell surface may facilitate the removal of endothelin from the circulation during pathophysiological scenarios that result in an increased generation of endothelin, eg, sepsis, liver cirrhosis, and orthotopic liver transplantation sequelae.

XV. Conclusions

Endothelin is a very potent vasoactive peptide mediator that is generated by endothelial cells exposed to a variety of injurious or stimulatory conditions. Most organs and tissues possess receptors for endothelin. Several transmembrane signaling pathways are activated after interaction of endothelin with its receptor(s). Elevated tissue or plasma levels of endothelin have been associated with numerous pathophysiological situations. However, no single disease entity appears to be associated exclusively with endothelin synthesis or response(s). Well-documented responses of endothelin are known in both vascular cells and parenchymal cells. Of particular interest to this laboratory are endothelin synthesis and responses in the liver. Our contention is that endothelin is an important signaling molecule in cell–cell interactions between sinusoidal cells and the hepatocytes, since the liver is required to respond to systemic trauma episodes such as sepsis or specifically to hepatic trauma. Hepatic endothelial cells are likely to generate this potent peptide mediator whereas Kupffer cells, Ito cells, and hepatocytes possess receptors with which to respond to endothelin.

During the past 5 years, an impressive and imposing body of literature has developed describing this interesting signaling molecule. The next 5 years of developments in the physiological and pathophysiological significance of this family of peptides and their receptors will be even more dramatic.

Acknowledgments

Katherine Stephenson is supported by a fellowship grant from Berlex Laboratories, Inc. Merle S. Olson is supported by NIH Grant DK-19473 and by the Robert A. Welch Foundation (AQ-728).

REFERENCES

Abdel-Latif, A. A., and Zhang, Y. (1991). Species differences in effects of endothelin-1 on myoinositol trisphosphate accumulation, cyclic AMP formation and contraction of isolated iris sphincter of rabbit and other mammalian species. *Invest. Opthal. Vis. Sci.* **32,** 2432–2438.

Abdel-Latif, A. A., Zhang, Y., and Yousufzai, S. Y. K. (1991). Endothelin-1 stimulates the release of arachidonic acid and protaglandins in rabbit iris sphincter smooth muscle activation of phospholipase A_2. *Curr. Eye Res.* **10,** 259–265.

Adachi, M., Yang, Y. Y., Furuichi, Y., and Miyamoto, C. (1991). Cloning and characterization of cDNA encoding human A-type endothelium receptor. *Biochem. Biophys. Res. Commun.* **180,** 1265–1272.

Agricola, K. M., Rubanyi, G., Paul, R. J., and Highsmith, R. F. (1984). Characterization of a potent coronary artery vasoconstrictor produced by endothelial cells in culture. *Fed. Proc.* **43,** 899.

Ambar, I., Kloog, Y., Kochva, E., Wollberg, Z., Bdolah, A., Oron, U., and Sokolovsky, M. (1989a). Characterization and localization of a novel neuroreceptor for the peptide sarafotoxin. *Biochem. Biophys. Res. Commun.* **157**, 1104–1110.
Ambar, I., Kloog, Y., Schvartz, I., Hazum, E., and Sokolozsky, M. (1989b). Competitive interaction between endothelin and sarafotoxin: Binding and phosphoinositide hydrolysis in rat atria and brain. *Biochem. Biophys. Res. Commun.* **155**, 167–172.
Anggard, E., Galton, S., Raae, G., Thomas, R., McLoughlin, L., de Nucci, G., and Vane, J. R. (1989). The fate of radioiodonated endothelin-1 and endothelin-3 in the rat. *J. Cardiovasc. Pharmacol.* **13**(Suppl. 5), S46–S49.
Arai, H., Hori, S., Aramori, I., Ohkubo, H., and Nakanishi, S. (1990). Cloning and expression of a cDNA encoding an endothelin receptor. *Nature (London)* **348**, 730–732.
Auget, M. S., Delaflotte, P., Chabrier, E., and Branquet, P. (1989). Comparative effects of endothelin and phorbol 12,13-dibutyrate in rat aorta. *Life Sci.* **45**, 2051–2059.
Awazu, M., Sugiura, M., Inagami, T., Ichikawa, T., and Kon, V. (1991). Cyclosporin promotes glomerular binding *in vivo. J. Am. Soc. Nephrol.* **1**, 1253–1258.
Badr, K. F., Murray, J. J., Breyer, M. D., Takahashi, K., Inagami, T., and Harris, R. C. (1989). Mesangial cell, glomerular and renal vascular resistance response to endothelin in the rat kidney. Elucidation of signal transduction pathways. *J. Clin. Invest.* **83**, 336–342.
Bdolah, A., Wollberg, G., Fleminger, G., and Kochiva, E. (1989). SRTX-d, a new native peptide of the endothelin/sarafotoxin family. *FEBS Lett.* **256**, 1–3.
Biondi, M. L., Marasini, B., Bassani, C., and Agostoni, A. (1991). Increased plasma endothelin levels in patients with Raynaud's phenomenon. *N. Engl. J. Med.* **329**, 1139–40.
Bloch, D., Friedrich, S., Lee, M., Eddy, R., Shows, T., and Quertermous, T. (1989a). Structural organization and chromosomal assignment of the gene encoding endothelin. *J. Biol. Chem.* **264**, 10851–10857.
Bloch, D., Eddy, R., Shows, T., and Quertermous, T. (1989b). cDNA cloning and chromosomal assignment of the gene encoding endothelin-3. *J. Biol. Chem.* **264**, 18156–18161.
Bousso-Mittler, D., Kloog, Y., Wolberg, Z., Bdolah, A., Kochiva, E., and Sokolovsky, M. (1989). Functional endothelin/sarafotoxin receptors in the rat uterus. *Biochem. Biophys. Res. Commun.* **162**, 952–957.
Brown, K. D., and Littlewood, C. J. (1989). Endothelin stimulates DNA synthesis in Swiss 3T3 cells: synergy with polypeptide growth factors. *Biochem. J.* **263**, 977–980.
Brown, S. C., Donlan, M. E., and Jeffs, P. W. (1990). Structural studies of endothelin by CD and NMR. *In* "Peptides: Chemistry, Structure and Biology." (J. E. Rivier and G. R. Marshall, eds.), p. 595 ESCOM, Leiden, The Netherlands.
Cade, C., Lumma, W. C., Jr., Mohan, R., Rubanyi, G. M., and Parker-Botelho, L. H. (1990). Lack of biological activity of preproendothelin [110–130] in several endothelin assays. *Life Sci.* **47**, 2097–2103.
Cairnes, H. S., Rogerson, M., Fairbanks, L. D., Westwick, J., and Neild, G. H. (1988). Endothelin and cyclosporin nephrotoxicity. *Lancet* **2**, 1496–1497.
Cairnes, H. S., Rogerson, M. P. E., Fairbanks, L. D., Neild, G. H., and Westwick, J. (1989). Endothelin induces an increase in renal vascular resistance and a fall in glomerular filtration rate in the rabbit isolated perfused kidney. *Br. J. Pharmacol.* **98**, 155–160.
Casey, N. L., Word, R. A., and MacDonald, P. C. (1991). Endothelin-1 gene expression and regulation of endothelin mRNA and protein biosynthesis in avascular human amnion. Potential source of amniotic fluid endothelin. *J. Biol. Chem.* **266**, 5762–5768.

Cavero, P. G., Miller, W. L., Heublin, D. M., Margulies, K. B., and Burnett, J. C., Jr. (1990). Endothelin in experimental congestive heart failure in the anesthetized dog. *Am. J. Physiol.* **259,**F312–F317.
Cheng, H. F., Su, Y. M., Yeh, J. R., and Chang, K. J. (1993). Alternative transcript of the nonselective-type endothelin receptor from rat brain. *Mol. Pharmacol.* **44,** 533–538.
Cody, R. J., Haas, G. J., Binkley, P. F., Capers, Q., and Kelly, R. (1992). Plasma endothelin correlates with the extent of pulmonary hypertension in patients with congestive heart failure. *Circulation* **85,** 504–509.
Colin, I., Berbinschi, A., Denef, J. F., and Ketelslegers, J. M. (1992). Detection and identification of endothelin-1 immunoreactivity in rat and porcine thyroid follicular cells. *Endocrinol.* **130,** 544–546.
Danthuluri, N. R., and Brock, T. A. (1990). Endothelin receptor-coupling mechanisms in vascular smooth muscle: role for protein kinase C. *J. Pharmacol. Exp. Therapeut.* **254,** 393–399.
Davenport, A. P., Kaumann, A. J., Hall, J. A., Nunez, D. J., and Brown, M. J. (1989). [^{125}I] Endothelin binding in mammalian tissues: relation to human atrial inotropic effects and coronary contraction. *Br. J. Pharmacol.* **96,** 102 p.
DeMey, J. G., and Vanhoutte, P. M. (1982). Heterologous behavior of the canine arterial and venous wall: Importance of the endothelium. *Circ. Res.* **51,** 439–447.
DeMey, J. G., and Vanhoutte, P. M. (1983). Anoxia and endothelium-dependent reactivity in canine femoral artery. *J. Physiol. (London)* **335,** 65–74.
De Nucci, G., Thomas, G. R., D'Orleans-Juste, P., Antunes, E., Walden, C. Warner, T. D., and Vane, J. R. (1988). Pressor effects of circulating endothelin are limited by its removal in the pulmonary circulation and by the release of prostacyclin and endothelium-derived relaxing factor. *Proc. Natl. Acad. Sci. USA,* **85,** 9797–9800.
Deray, G., Carayon, A., Maistre, G., Benhmida, M., Masson, F., Barthelemy, C., Petitclere, T., and Jacobs, C. (1992). Endothelin in chronic renal failure. *Nephrol. Dial. Transplant.* **7,** 300–305.
Doherty, A. M., Cody, W. L., Leitz, N. L., DePue, P. L., Taylor, M. D., Repundalo, S. T., Hingorani, G. P., Major, T. C., Pansk, R. L., and Taylor, D. G. (1991). Structure–activity studies of the C-terminal region of the endothelins and the sarafotoxins. *J. Cardiovasc. Pharmacol.* **17**(Suppl. 7), S59–S61.
Dohlman, J. G., DeLoof, H., and Segrest, J. P. (1990). Charge distributions and amphipathicity of receptor-binding-alpha-helices. *Mol. Immunol.,* **27,** 1009–20.
Dubin, D., Pratt, R. E., Cooke, J. P., and Dzau, V. J. (1989). Endothelin, a potent vasoconstrictor, is a vascular smooth muscle mitogen. *J. Vasc. Med. Biol.* **1**(No. 3), 150–154.
Ehrenreich, H., Anderson, R. W., Fox, C. H., Rieckmann, P., Hoffman, G. S., Travis, W. D., Coligan, J. E., Kehrl, J. H., and Fauci, A. S. (1990). Endothelins, peptides with potent vasoactive properties, are produced by human macrophages. *J. Exp. Med.* **172,** 1742–1748.
Ehrenreich, H., Anderson, R. W., Ogino, Y., Reickmann, P. R., Costa, T., Wood, G. P., Coligan, J. E., Kehrl, J. H., and Fauci, A. S. (1991). Selective autoregulation of endothelins in primary astrocyte cultures: Endothelin receptor-mediated potentiation of endothelin-1 secretion. *New Biol.* **3,** 135–141.
Elshourbagy, N. A., Ice, J. A., Korman, D. R., Nuthalaganti, P., Sylvester, D. R., Dilella, A. G., Sutiphong, J. A., and Kumar, C. S. (1992). Molecular cloning and characterization of the major endothelin receptor subtype in porcine cerebellum. *Mol. Pharmacol.* **41,** 465–473.
Endo, S., Inooka, H., Ishibashi, Y., Kitada, C., Mitzuta, E., and Fujino, M. (1989). Solu-

tion conformation of endothelin determined by nuclear magnetic resonance and distance geometry. *FEBS Lett.* **257,** 149.

Filip, J. G., Battistini, B., and Sirois, P. (1990). Endothelin induces thromboxane release and contraction of isolated guinea pig airways. *Life Sci.* **47,** 1845–1850.

Firth, J. D., Raine, A. E., Ratcliffe, P. J., and Ledingham, J. G. (1988). Endothelin: an important factor in acute renal failure. *Lancet* **19,** 1179–1181.

Fogo, A., Hellings, S. E., Inagami, T., and Kon, V. (1992). Endothelin receptor antagonism is protective in *in vivo* acute cyclosporin toxicity. *Kidney Int.* **42,** 770–774.

Franke, R. R., Konig, B., Sakmar, T. P., Khorana, H. G., and Hoffman, K. P. (1990). Rhodopsin mutants that bind but fail to activate transducin. *Science* **250,** 123–125.

Fukuda, Y., Hirata, Y., Yoshimi, H., Kojima, T., Kobayashi, Y., Yanagisawa, M., and Masaki, T. (1988). Endothelin is a potent secretagogue for atrial natiuretic peptide in cultured rat atrial myocytes. *Biochem. Biophys. Res. Commun.* **155,** 167–172.

Fukuda, Y., Hirata, Y., Taketani, S. Kojima, T., Oikawa, S., Nakazato, H., and Kobayashi, Y. (1989). Endothelin stimulates accumulations of cellular atrial natiuretic peptide and its messenger RNA in rat cardiocytes. *Biochem. Biophy. Res. Commun.* **164,** 1431–1436.

Furchgott, R. F. (1984). The role of the endothelium in the responses of vascular smooth muscles to drugs. *Annu. Rev. Pharmacol. Toxicol.* **24,** 175–197.

Furchgott, R. F., and Zawadski, J. V. (1980). The obligatory role of endothelial cells in the relaxation of arterial smooth muscle by acetylcholine. *Nature (London)* **288,** 373–376.

Furuya, S., Naruse, S., Nakayama, T., Furuya, K., and Nokihara, K. (1991). Localization of [^{125}I] endothelin-1 in rat tissues observed by electron-microscopic radioautography. *J. Cardiovasc. Pharmacol.* **17**(Suppl. 7), S452–S454.

Furuya, S., Naruse, S., Nakayama, T., and Nokihara, K. (1992). Binding of [^{125}I] endothelin-1 to fat storing cells in the liver by electron microscopic radioautography. *Anat. Embryol.* **185,** 97–100.

Galron, R., Kloog, R., Bdolah, A., and Sokolovsky, M. (1989). Functional endothelin/sarafotoxin receptors in rat heart myocytes: Structure–activity relationships and receptor subtypes. *Biochem. Biophys. Res. Commun.* **163,** 936–943.

Gandhi, C. R., Stephenson, K., and Olson, M. S. (1990). Endothelin, a potent peptide agonist in the liver. *J. Biol. Chem.* **265,** 17432–17435.

Gandhi, C. R., Behal, R. H., Harvey, S. A. K., Nouchi, T. A., and Olson, M. S. (1992a). Hepatic effects of endothelin. Receptor characterization and endothelin-induced signal transduction in hepatocytes. *Biochem. J.* **287,** 897–904.

Gandhi, C. R., Stephenson, K., and Olson, M. S. (1992b). A comparative study of endothelin- and platelet activating factor-mediated signal transduction and prostaglandin synthesis in rat Kupffer cells. *Biochem. J.* **281,** 485–492.

Gandhi, C. R., Harvey, S. A. K., and Olson, M. S. (1993). Hepatic effects of endothelin: Metabolism of endothelin by liver-derived cells. *Arch. Biochem. Biophys.* **305,** 38–46.

Giaid, A., Gibson, S., Ibrahim, N., Legon, Bloom, S., Yanagisawa, M., Masaki, T., Varndell, I., and Polak, J. (1989). Endothelin-1, an endothelium-derived peptide, is expressed in neurons of the human spinal cord and dorsal root ganglia. *Proc. Natl. Acad. Sci. USA* **86,** 7634–7638.

Giaid, A., Hamid, Q. A., Springall, D. R., Yanagisawa, M., Shinmi, O., Sawamura, T., Masaki, T., Kimura, S., Corrin, B., and Polak, J. M. (1990). Detection of endothelin immunoreactivity and mRNA in pulmonary tumors. *J. Pathol.* **162,** 15–22.

Gillespie, M. N., Owasoyo, J. O., McMurtry, I. F., and O'Brien, R. F. (1986). Sustained

coronary vasoconstriction provoked by a peptidergic substance released from endothelial cells in culture. *J. Pharmacol. Exp. Therapeut.* **236**, 339–343.
Goetz, K. L., Wang, B. C., Madured, J. B., Zhu, J. L., and Leadley, R. J., Jr. (1988). Cardiovascular, renal and endocrine responses to intravenous endothelin in conscious dogs. *Am. J. Physiol.* **255**, R1064–R1068.
Granstam, E., Wang, L., and Bill, A. (1991). Effects of endothelins (ET-1, ET-2 and ET-3) in the rabbit eye: role of prostaglandins. *Eur. J. Pharmacol.* **194**, 217–223.
Gu, X. H., Casley, D. J., Cincotta, M., and Nayler, W. (1990). [^{125}I]Endothelin-1 binding to brain and cardiac membranes from nomotensive and spontaneously hypertensive rats. *Eur. J. Pharmacol.* **177**, 205–209.
Gulberg, V., Gerbes, A. L., Vollmar, A. M., and Paumgartner, G. (1992). Endothelin-3 like immunoreactivity in plasma of patients with cirrhosis of the liver. *Life Sci.* **51**, 1165–1169.
Haegerstrand, A., Hemsen, A., Gillis, C., Larsson, O., and Lundberg, J. (1989). Endothelin: Presence in human umbilical vessels, high levels in fetal blood and potent constrictor effect. *Acta. Physiol. Scand.* **137**, 541–542.
Haendler, B., Hechler, U., and Schleuning, W. D. (1992). Molecular cloning of human endothelin (ET) receptors ET_A and ET_B. *J. Cardiovasc. Pharmacol.* **20**(Suppl. 12), S1–S4.
Harder, D. R. (1987). Pressure-induced myogenic activation of cat cerebral arteries is dependent on intact endothelium. *Circ. Res.* **60**, 102–107.
Hay, D. W. (1992). Pharmacological evidence for distinct endothelin receptors in guinea pig broncus and aorta. *Br. J. Pharmacol.* **106**, 759–761.
Hayzer, D. J., Rose, P. M., Lynch, J. S., Webb, M. L., Kienzle, B. K., Liu, E. C., Bogosian, E. A., Brinson, E., and Runge, M. S. (1992). Cloning and expression of a human endothelium receptor: subtype A. *Am. J. Med. Sci.* **304**, 231–238.
Hemsen, A., and Lundberg, J. M. (1991). Presence of endothelin-1 and endothelin-3 in peripheral tissues and central nervous system. *Regul. Peptides* **36**, 71–83.
Hexum, T., Hoeger, C., Rivier, J., Baird, A., and Brown, M. (1990). Characterization of endothelin secretion by vascular endothelial cells. *Biochem. Biophys. Res. Commun.* **167**, 294–300.
Hickey, K. A., Rubanyi, G., Paul, R. J., and Highsmith, R. F. (1985). Characterization of a coronary vasoconstrictor produced by cultured endothelial cells. *Am. J. Physiol.* **248**, C550–C556.
Highsmith, R. F., Pang, D. C., and Rappoport, R. M. (1989). Endothelial cell-derived vasoconstrictors: Mechanisms of action in vascular smooth muscle. *J. Cardiovasc. Pharmacol.* **13**(Suppl. 5), S36–S44.
Hiley, C. R. (1989). Functional studies on endothelin catch up with molecular biology. [Synopsis of work by T. Kimura *et al.*, and of work of others as presented at The First William Harvey Workshop on Endothelin, London, December 5–6, 1988]. *Trends Pharmacol. Sci.* **10**, 47.
Hiley, C. R., Pelton, J. T., and Miller, R. C. (1989). Effects of endothelin on field stimulated rat vas deferens and guinea pig ileum. *Br. J. Pharmacol.* **96**, 104P.
Hirata, Y., Yoshimi, H., Emori, T., Marumo, F., Watanabe, T. X., Kumagaye, S., Nakajima, K., Kimura, T., and Sakakibara, S. (1989a). Interaction of synthetic sarafotoxin with rat vascular endothelin receptors. *Biochem. Biophys. Res. Commun.* **162**, 441–447.
Hirata, Y., Fukuda, Y., Yoshimi, H., Emori, T., Shichiri, M., and Marumo, F. (1989b). Specific receptor for endothelin in cultured rat cardiocytes. *Biochem. Biophys. Res. Commun.* **160**, 1438–1444.

Holden, W. E., and McCall, E. (1983). Hypoxic vasoconstriction of porcine pulmonary artery strips *in vitro* requires an intact endothelium. *Am. Rev. Resp. Dis.* **127**, 301.

Holden, W. E., and McCall, E. (1984). Hypoxia-induced contractions of porcine pulmonary artery strips depend on intact endothelium. *Expl. Lung Res.* **7**, 101–112.

Hosoda, K., Nakao, K., Arai, H., Suga, S.-I., Ogawa, Y., Mukoyama, M., Shirakami, G., Saito, Y., Nakanishi, S., and Imura, H. (1991). Cloning and expression of human endothelin-1 receptor cDNA. *FEBS Lett.* **287**, 23–26.

Hosoda, K., Nakao, K., Tamura, N., Arai, H., Ogawa, Y., Suga, S., Nakanishi, S., and Imura, H. (1992). Organization, structure, chromosomal assignment and expression of the gene encoding the human endothelin-A receptor. *J. Biol. Chem.* **267**, 18797–18804.

Hoyer, D., Walker, C., and Palacios, J. M. (1989). [^{125}I]endothelin-1 binding sites: Autoradiographic studies in the brain and periphery of various species including humans. *J. Cardiovasc. Pharmacol.* **13**(Suppl. 5), S162–S165.

Hu, J. R., Berniger, U. G., and Lang, R. E. (1988) Endothelin stimulates atrial natriuretic peptide (ANP) release from rat atria. *Eur. J. Pharmacol.* **158**, 177–178.

Ihara, M., Fukuroda, T., Sacki, T., Nishikibe, M., Kojiri, K., Suda, H., and Yano, M. (1991). An endothelin receptor (ET_A) antagonist isolated from *Streptomyces misakiensis*. *Biochem. Biophys. Res. Commun.* **178**, 132–137.

Ihara, M., Noguchi, K., Saeki, T., Fukuroda, T., Tsuchida, S., Kimura, S., Fukami, T., Ishikawa, K., Nishikabe, M., and Yano, M. (1992). Biological profiles of highly potent novel endothelin antagonists selective for the ET_A receptor. *Life Sci.* **50**, 247–255.

Ikegawa, R., Matsumura, Y., Tsukahara, Y., Takaoka, M., and Morimoto, S. (1990a). Phosphoramidon, a metalloproteinase inhibitor, supresses the secretion of endothelin-1 from cultured endothelial cells by inhibiting a big endothelin-1 converting enzyme. *Biochem. Biophys. Res. Commun.* **171**, 669–675.

Ikegawa, R., Matsumura, Y., Takaoka, M., and Morimoto, S. (1990b). Evidence for a pepstatin-sensitive conversion of porcine big endothelin-1 by the endothelial cell extract. *Biochem. Biophys. Res. Commun.* **167**, 860–866.

Immer, H., Eberle, I., Fischer, W., and Moser, E. (1988). Solution synthesis of endothelin. 20th European Peptide Symposium, Abstract No. 98.

Inoue, A., Yanagisawa, M., Kimura, S., Kasuya, Y., Miyauchi, T., Goto, K., and Masaki, T. (1989a). The human endothelin family: Three structurally and pharmacologically distinct isopeptides predicted by three separate genes. *Proc. Natl. Acad. Sci. USA* **86**, 2863–2867.

Inoue, A., Yanagisawa, M., Takuwa, Y., Mitsui, Y., Kobayashi, M., and Masaki, T. (1989b). The human preproendothelin-1 gene. *J. Biol. Chem.* **264**, 14954–14959.

Irvine, R. F. (1982). How is the level of free arachidonic acid controlled in mammalian cells? *Biochem. J.* **204**, 3–16.

Ishikawa, T., Yanagisawa, M., Kimura, S., Goto, K., and Masaki, T. (1988a). Positive inotropic action of novel vasoconstrictor peptide endothelin on guinea pig atria. *Am. J. Physiol.* **255** (Heart Circ. Physiol. 24), H970–H973.

Ishikawa, T., Yanagisawa, M., Kimura, S., Goto, K., and Masaki, T. (1989b). Positive chronotropic effects of endothelin, a novel endothelium-derived vasoconstrictor peptide. *Pflügers Arch.* **413**, 108–110.

Itoh, Y., Yanigisawa, M., Ohkubo, S., Kimura, S., Kosaka, T., Inoue, A., Ishida, N., Mitsui, Y., Onda, H., Fujino, M., and Masaki, T. (1988). Cloning and sequence analysis of cDNA encoding the precursor of a human endothelium-derived vasoconstrictor peptide, endothelin: Identity of human and porcine endothelin. *FEBS Lett.* **231**, 440–444.

Julkunen, H., Saijonmaa, O., Gronhagen-Riska, C., Teppo, A. M., and Fyhrquist, F. (1991). Raised plasma concentrations of endothelin-1 in systemic lupus erythematosis. *Ann. Rheum. Dis.* **50**, 526–527.

Kahalen, M. B. (1991). Endothelin, an endothelial-dependent vasoconstrictor in scleroderma. Enhanced production and fibrotic action. *Arthritis Rheum.* **34**, 978–983.

Kai, H., Kanaide, H., and Nakamura, M. (1989). Endothelin-sensitive intracellular Ca^{2+} store overlaps with caffeine-sensitive one in rat aortic smooth muscle cells in primary culture. *Biochem. Biophys. Res. Commun.* **158**, 235–243.

Katusic, Z. S., Shepherd, J. T., and Vanhoutte, P. M. (1987). Endothelium-dependent contraction to stretch in canine basilar arteries. *Am. J. Physiol.* **252**, H671–H673.

Kemp, B. E., and Pearson, R. B. (1990). Protein kinase recognition and sequence motifs. *Trends Biochem. Sci.* **15**, 342–346.

Kimura, S., Kasuya, Y., Sawamura, T., Shinmi, O., Sugita, Y., Yanagisawa, M., Goto, K., and Masaki, T. (1988). Structure-activity relationships of endothelin: Importance of the C-terminal moiety. *Biochem. Biophys. Res. Commun.* **156:** 1182–1186.

King, A. J., Brenner, B. M., and Anderson, S. (1989). Endothelin: a potent renal and systemic vasoconstrictor peptide. *Am. J. Physiol.* **256**(Renal Fluid Electrolyte Physiol. 25), F1051–F1058.

Kitamura, K., Tanaka, T., Kato, J., Eto, T., and Tanaka, K. (1989a). Regional distribution of immunoreactive endothelin inporcine tissue: Abundance in inner medulla of kidney. *Biochem. Biophys. Res. Commun.* **161**, 348–352.

Kitamura, K., Tanaka, T., Kato, J., Ogawa, T., Eto, T., and Tanaka, K. (1989b). Immunoreactive endothelin in rat kidney inner medulla: Marked decrease in spontaneously hypertensive rats. *Biochem. Biophys. Res. Commun.* **162**, 38–44.

Kloog, Y., and Sokolovsky, M. (1989). Similarities in mode and sites of action of sarafotoxins and endothelins. *Trends Pharmacol. Sci.* **10**, 212–214.

Kloog, Y., Ambar, I., Sokolovsky, M., and Wollberg, Z. (1988). Sarafotoxin, a novel vasoconstrictor peptide: Phosphoinositide hydrolysis in rat heart and brain. *Science* **242**, 268–270.

Kloog, Y., Bousso-Mittler, D., Bdolah, A., and Sokolovsky, M. (1989a). Three apparent receptor subtypes for the endothelin/sarafotoxin family. *FEBS Lett.* **253**, 199–202.

Kloog, Y., Ambar, I., Kochva, E., Wollberg, Z., Bdolah, A., and Sokolovsky, M. (1989b). Sarafotoxin receptors mediate phosphoinositide hydrolysis in various rat brain regions. *FEBS Lett.* **242**, 387–390.

Kobayashi, Y. (1990). Solution conformation of endothelin. *In* "Peptides: Chemistry, Structure and Biology" (J. E. Rivier and G. R. Marshall, eds.), p. 552, ESCOM, Leiden, The Netherlands.

Kohno, M., Yasunari, K., Murakawa, K., Yokakawa, K., Horio, T., Fukui, T., and Takeda, T. (1990). Plasma immunoreactive endothelin in essential hypertension. *Am. J. Med.* **88**, 614–618.

Kohno, M., Murakawa, K., Horio, T., Yokokawa, K., Yasunari, K., Fukui, T., Takeda, T. (1991). Plasma immunoreactive endothelin-1 in experimental malignant hypertension. *Hypertension* **18**, 93–100.

Komuro, I., Kurihara, H., Sugiyama, T., Takaku, F., and Yazaki, Y. (1988). Endothelin stimulates c-*fos* and c-*myc* expression and proliferation of vascular smooth muscle cells. *FEBS Lett.* **238**, 249–252.

Kon, V., Sugiura, M., Inagami, T., Hoover, R. L., Frogo, A., Harvie, B. R., and Ichikawa, I. (1990). Cyclosporin causes endothelin-dependent acute renal failure. *Kidney Int.* **37**, 1487–1491.

Konishi, F., Knodo, T., and Inagami, T. (1991). Phospholipase D in cultured rat vascular

smooth muscle cells and its activation by phorbol ester. *Biochem. Biophys. Res. Commun.* **179**, 1070–1076.
Koseki, C., Imai, M., Hirata, Y., Yanagisawa, M., Masaki, T. (1989). Autoradiographic distribution in rat tissues of binding sites for endothelin: a neuropeptide? *Am. J. Physiol.* **256**, R858–R866.
Kozuka, M., Ito, T., Hirose, S., Yakahashi, K., and Hagiwara, H. (1989). Endothelin induces two types of contractions of rat uterus: Phasic contractions by way of voltage channels and developing contractions through a second type of calcium channels. *Biochem. Biophys. Res. Commun.* **159**, 317–323.
Kramer, B. K., Smith, T. W., and Kelly, R. A. (1990). Endothelin and increased contractility in adult rat ventricular myocytes: Role of intracellular alkalosis induced by activation of the protein kinase C-dependent Na^+/H^+ exchanger. *Circ. Res.* **68**, 269–279.
Kumagaye, S., Kuroda, H., Nakajima, K., Watanabe, T. X., Kimura, T., Masaki, T., and Sakakibara, S. (1988). Synthesis and secondary structure determination of porcine endothelin: An endothelium-derived vasoconstricting peptide. *Int. J. Peptide Protein Res.* **32**, 519–526.
Kurihara, H., Yoshizumi, T., Sugiyama, T., Takatu, F., Yanagisawa, M., Masaki, T., Hamaoki, M., Kato, H., and Yazaki, Y. (1989). Transforming growth factor-β stimulates the expression of endothelin mRNA by vascular endothelial cells. *Biochem. Biophys. Res. Commun.* **159**, 1435–1440.
Lam, H. C., Takahashi, K., Ghatei, M. A., Warrens, A. N., Rees, A. J., and Bloom, S. R. (1991). Immunoreactive endothelin in human plasma, urine, milk, and saliva. *J. Cardiovasc. Pharmacol.* **17**(Suppl. 7), S390–S393.
Lee, S. Y., Lee, C. Y., Chen, Y. M., and Kochva, E. (1986). Coronary vasospasm as the primary cause of death due to the venom of the burowing asp *Atractaspis engaddensis*. *Toxicon* **24**, 285–291.
Lerman, A., Click, R. L., Narr, B. J., Weisner, R. H., Krom, R. A. F., Textor, S. C., and Burnett, J. C., Jr. (1991). Elevation of plasma endothelin associated with systemic hypertension in humans following orthotopic liver transplantation. *Transplantation*, **51**, 646–650.
Lerman, A., Kubo, S. H., Tschumperlin, L. K., and Burnett, J. C., Jr. (1992). Plasma endothelin concentrations on humans with end-stage heart failure and after heart transplantation. *J. Am. Coll. Cardiol.* **20**, 849–853.
Lin, H. Y., Kaji, E. H., Winkel, G. K., Ives, H. E., and Lodish, H. E. (1991). Cloning and functional expression of a vascular smooth muscle endothelin-1 receptor. *Proc. Natl. Acad. Sci. USA* **88**, 3185–3189.
Liu, J. J., Caseley, D. J., and Nayler, W. G. (1989). Ischaemia causes externalization of endothelin-1 binding sites in rat cardiac membranes. *Biochem. Biophys. Res. Commun.* **164**, 1220–1225.
Liu, W., Shiue, G. H., and Tam, J. P. (1990). A novel strategy for the deprotection of S-acetamido-methyl containing peptides: An approach to the efficient synthesis of endothelin. *In* "Peptides: Chemistry, Structure and Biology." (J. E. Rivier and G. R. Marshall, eds.), p. 271. ESCOM, Leiden, The Netherlands.
Loffler, B.-M., and Lohrer, W. (1991). Different endothelin receptor affinities in dog tissues. *J. Receptor Res.* **11**, 293–298.
Lopez-Farre, A., Montanes, I., Millas, I., and Lopez-Novoa, J. M. (1989). Effect of endothelin on renal function in rats. *Eur. J. Pharmacol.* **163**, 187–189.
Luscher, T. F., Yang, Z., Diederich, D., and Buhler, F. R. (1989). Endothelium-derived vasoactive substances: Potential role in hypertension, atherosclerosis and vascular occlusion. *J. Cardiovasc. Pharmacol.* **14**(Suppl. 6), S63–S69.

MacCumber, M., Ross, C., Glaser, B., and Snyder, S. (1989). Endothelin: visualization of mRNAs by *in situ* hybridization provides evidence for local action. *Proc. Natl. Acad. Sci. USA* **86**, 7285–7289.

Maggi, C. A., Guilani, S., Patacchini, R., Santicioli, Rovero, P., Giachetti, A., and Meli, A. (1989a). The C-terminal hexapeptide, endothelin-(16–21), discriminates between different endothelin receptors. *Eur. J. Pharmacol.* **166**, 121–122.

Maggi, C. A., Guilani, S., Patacchini, R., Santicioli, P., Turini, D., Barbanti, G., and Meli, A. (1989b). Potent contractile activity of endothelin on the human isolated urinary bladder. *Br. J. Pharmacol.* **96**, 755–757.

Marsden, P. A., Danthuluri, N. R., Brenner, B. M., Ballerman, B. J., and Brock, T. A. (1989a). Endothelin action on vascular smooth muscle involves inositol trisphosphate and calcium mobilization. *Biochem. Biophys. Res. Commun.* **158**, 86–93.

Marsden, P. A., Dorfman, D. M., Brenner, B. M., Orkin, B. J., and Ballermann, B. J. (1989b). Endothelin: Gene expression, release and action in cultured cells of the renal glomerulus. *Am. J. Hypertens.* **2**, 49A.

Masaki, T. (1991). Tissue specificity of the endothelin-induced response. *J. Cardiovasc. Pharmacol.* **17**,(Suppl. 7), S1–S4.

Matsumoto, H., Suzuki, N., Onda, H., and Fujino, M. (1989). Abundance of endothelin-3 in rat intestine, pituitary gland and brain. *Biochem. Biophys. Res. Commun.* **164**, 74–80.

Matsumura, Y., Nakase, K., Ikegawa, R., Hayashi, K., Ohyama, T., and Morimoto, S. (1989). The endothelium-derived vasoconstrictor peptide endothelin inhibits renen release *in vitro*. *Life Sci.* **44**, 149–157.

Mattoli, S., Soloperto, M., Marini, M., and Fasoli, A. (1991). Levels of endothelin in the bronchoalveolar lavage fluid of patients with symptomatic asthma and reversible airflow obstruction. *J. Allergy Clin. Immunol.* **88**(3 Pt. 1), 376–384.

McMahon, E. G., and Palomo, M. A. (1990). Phosphoramidon blocks the pessor activity of big endothelin-1 *in vivo*. *J. Vasc. Med. Biol.* **2**, 174.

Miasiro, N., Yamamoto, H., Kanaide, H., and Nakamura, M. (1988). Does endothelin mobilize calcium from intracellular store sites in rat aortic vascular smooth muscle cells in primary culture? *Biochem. Biophys. Res. Commun.* **156**, 312–317.

Miller, W. L., Redfield, M. M., and Burnett, J. C. (1989). Integrated cardiac, renal, and endocrine actions of endothelin. *J. Clin. Invest.* **83**, 317–320.

Miyauchi, T., Yanagisawa, M., Tomizawa, T., Sugishita, Y., Suzuki, N., Fujino, M. Ajisaka, R., Goto, K., and Masaki, T. (1988). Increased plasma levels of endothelin-1 and big endothelin-1 in acute myocardial infarction. *Lancet* **2**(8653), 53–54.

Mizuno, T., Saito, Y., Itakura, M., Ito, F., Ito, T., Moriyama, E., Hagiwara, H., and Hirose, S. (1992). Structure of the bovine ETB endothelin receptor gene. *Biochem. J.* **287**, 305–309.

Moncada, S., Gryglewski, R., Bunting, S., and Vane, J. R. (1980). An enzyme isolated from arteries transforms prostaglandin endoperoxidase to an unstable substance that inhibits platelet aggregation. *Nature (London)* **263**, 663–665.

Moncada, S., Radomski, M. W., and Palmer, R. M. J. (1988). Endothelium-derived relaxation factor: Identification as nitric oxide and role in the control of vascular tone and platelet function. *Biochem. Pharmacol.* **37**, 2495–2501.

Moravec, C. S., Reynolds, E. E., Stewart, R. W., and Bond, M. (1989). Endothelin is a positive inotropic agent in the human and rat heart *in vitro*. *Biochem. Biophys. Res. Commun.* **159**, 14–18.

Morel, D. R., Lacroix, S., Hemsen, A., Steinig, D. A., Pittet, J. F., and Lundberg, J. M. (1989). Increased plasma and pulmonary lymph levels of endothelin during endotoxic shock. *Eur. J. Pharmacol.* **167**, 427–428.

Morita, S., Kitamura, K., Yamamoto, Y., Eto, T., Osada, Y., Sumiyoshi, A., Koono, M., and Tanaka, K. (199). Immunoreactive endothelin in human kidney. *Ann. Clin. Biochem.* **28,** 267–271.

Muldoon, L. L., Rodland, K. D., Forsythe, M. L., and Magun, B. E. (1989). Stimulation of phosphatidylinositol hydorlyusis, diacylglycerol release and gene expression in response to endothelin, a potent new agonist for fibroblasts and smooth muscle. *J. Biol. Chem.* **264,** 8529–8536.

Nakajima, K., Kubo, S., Kumagaye, S.-I., Nishio, H., Tsunemi, M., Inui, T., Kuroda, H., Chino, N., Watanabe, T. X., Kimura, T., and Sadadibara, S. (1989a). Structure–activity relationships of endothelin: Importance of charged groups. *Biochem. Biophys. Res. Commun.* **163,** 424–429.

Nakajima, K., Kumagaye, S., Nishio, H., Kuroda, H., Watanabe, T., Kobayashi, Y., Tamaoki, H., Kimura, T., and Sakakibara, S. (1989b). Synthesis of endothelin-1 analogues, endothelin-3 and sarafotoxin S6b: Structure activity relationships. *J. Cardiovasc. Pharmacol.* **13**(*Suppl. 5*), S8–S12.

Nakaki, T., Nakayama, M., Yamamoto, S., and Kato, R. (1989). Endothelin-mediated stimulation of DNA synthesis in vascular smooth muscle cells. *Biochem. Biophys. Res. Commun.* **158,** 880–883.

Nakamura, T., Kasai, K., Sekiguchi, Y., Banba, N., Takahashi, K., Fmoto, T., Hattori, Y., and Shimoda, S. (1991). Elevation of plasma endothelin concentrations during endotoxic shock in dogs. *Eur. J. Pharmacol.* **205,** 277–282.

Nakamuta, M., Takayanagi, R., Sakai, Y., Sakamoto, S., Hagiwara, H., Mizuno, R., Saito, Y., Hiurose, S., Yamamoto, M., and Nawata, H. (1991). Cloning and sequence analysis of a cDNA encoding human non-selective type of endothelin receptor. *Biochem. Biophys. Res. Commun.* **177,** 34–39.

Naruse, M., Naruse, K., Nishikawa, T., Yoshihara, I., Ohsumi, K., Suzuki, N., Demura, R., and Demura, H. (1992). Endothelin-3 immunoreactivity in gonadotrophs of the human anterior pituitary. *J. Clin. Endocrinol. Metab.* **74,** 968–972.

Nayler, W., Gu, X., Casley, S., Panagiotopoulow, S., Liu, J., and Mottram, P. (1989a). Cyclosporine increases endothelin-1 binding site density in cardiac cell membranes. *Biochem. Biophys. Res. Commun.* **163,** 1270–1274

Nayler, W., Liu, J., Panagiotopoulos, S., and Casey, D. J. (1989b). Streptozotocin-induced diabetes reduces the density of endothelin-1 binding sites in rat cardiac membranes. *Br. J. Pharmacol.* **97,** 993–995.

Nishikori, K., Akiyama, H., Inagaki, Y., Ohta, H., Kashiwabara, T., Iwamatsu, A., Nomizu, M., and Morita, A. (1991). Receptor binding affinity and biological activity of C-terminal elongated forms of endothelin-1. *Neurochem. Int.* **18,** 535–539.

Nomizu, M., Inagake, Y., Iwamatsu, A., Kashiwabara, T., Ohta, H., Morita, A., Nishikori, K., Otaka, A., Fujii, N., and Yajima, H. (1990). Application of two-step hard acid deprotection/cleavage procedures to the solid-phase synthesis of the putative precursor of human endothelin. *In* "Peptides: Chemistry, Structure, and Biology" (J. E. Rivier and G. R. Marshall, eds.) p. 276. ESCOM, Leiden, The Netherlands.

Noveral, J. P., Rosenberg, S. M., Anbar, R. A. Pawlowski, N. A., and Grunstein, M. M. (1992). Role of endothelin-1 in regulating proliferation of cultured rabbit airway smooth muscle cells. *Am. J. Physiol.* **263**(3 Pt. 1), L317–L324.

Nunez, D. J. R., Brown, M. J., Davenport, A. P., Neylon, C. B., Schofield, J. P., and Wyse, R. K. (1990). Endothelin-1 mRNA is widely expressed in porcine and human tissues. *J. Clin. Invest.* **85,** 1537–1541.

O'Brien, R. F., and McMurtry, I. F. (1984). Endothelial cell (EC) supernatents contract bovine pulmonary artery (PA) rings. *Am. Rev. Resp. Dis.* **129,** A337.

O'Brien, R. F., Robbins, R. J., and McMurtry, I. F. (1987). Endothelial cells in culture produce a vasoconstrictor substance. *J. Cell Physiol.* **132**, 263–270.

Ogawa, Y., Nakao, K., Arai, H., Nakagawa, O., Hosoda, K., Suga, S., Nakanishi, S., and Imura, H. (1991). Molecular cloning of a non-isopeptide-selective human endothelin receptor. *Biochem. Biophys. Res. Commun.* **178**, 248–255.

Ohlstein, E. H., Horohonich, S., and Hay, D. W. P. (1989). Cellular mechanisms of endothelin in rabbit aorta. *J. Pharmacol. Exp. Therapeut.* **250**, 548–555.

Ohlstein, E. H., Arleth, A., Bryan, H., Elliot, J. D., and Sung, C. P. (1992). The selective endothelin ET_A receptor antagonist BQ 123 antagonizes endothelin-1 mediated mitogenesis. *Eur. J. Pharmacol.* **225**, 347–350.

Ohta, K., Hirata, Y., Imai, T., Kanno, K., Emori, T., Shichiri, M., and Marumo, F. (1990). Cytokine-induced release of endothelin-1 from porcine renal epithelial cell line. *Biochem. Biophys. Res. Commun.* **169**, 578–584.

Okada, K., Miyazaki, Y., Takada, J., Matsuyama, K., Yamaki, T., and Yano, M. (1990). Conversion of big endothelin-1 by membrane-bound metalloendopeptidase in cultured bovine endothelial cells. *Biochem. Biophys. Res. Commun.* **171**, 1192–1198.

Orita, Y., Fugiwara, Y., Ochi, S., Takama, T., Fukunaga, M., and Yokoyama, K. (1989). Endothelin-receptors in rat renal glomeruli. *J. Cardiovasc. Pharmacol.* **13**(Suppl. 5), S159–S161.

Palmer, R. M. J., Ferrige, A. G., and Moncada, S. (1987). Nitric oxide release accounts for the biological activity of endothelium-derived relaxing factor. *Nature (London)* **327**, 524–526.

Palmer, R. M. J., Ashton, D. S., and Moncada, S. (1988). Vascular endothelial cells synthesize nitric oxide from L-arginine. *Nature (London)* **333**, 664–666.

Perico, N., Ruggenenti, P., Gaspari, F., Mosconi, L., Benigni, A., Amuchastegui, C. S., Gasparini, F., and Remuzzi, G. (1992). Daily renal hypoperfusion induced by cyclosporin with renal transplantation. *Transplantation* **54**, 56–60.

Perlman, D., and Halvorson, H. O. (1983). A putative signal peptidase recognition site and sequence in eukaryotic and prokaryotic signal peptides. *J. Mol. Biol.* **167**, 391–409.

Pittet, J. F., Morel, D. R., Hemsen, A., Gunning, K., Lacroix, J. S., Suter, P. M., and Lundberg, J. M. (1991). Elevated plasma endothelin-1 concentrations are associated with the severity of illness in patients with sepsis. *Ann Surg.* **213**, 261–264.

Power, R. F., Wharton, J., Salas, S., Knase, S., Ghatei, N., Bloom, S. R., and Polak, J. M. (1989). Autoradiographic localization of endothelin binding sites in human and porcine coronary arteries. *Eur. J. Pharmacol.* **160**, 199–200.

Rachmilewitz, D., Eliakim, R., Ackerman, Z., and Karmeli, F. (1992). Colonic endothelin-1 immunoreactivity in active ulcerative colitis. *Lancet* **339**, 1062.

Rakugi, H., Nakamura, M., Saito, H., Higaki, J., and Ogihara, T. (1988). Endothelin inhibits renin release from isolated rat glomeruli. *Biochem. Biophys. Res. Commun.* **155**, 1244–1247.

Resink, T. J., Scott-Burden, T., and Buhler, F. R. (1988). Endothelin stimulates phospholipase C in cultured vascular smooth muscle cells. *Biochem. Biophys. Res. Commun.* **157**, 1360–1368.

Resink, T. J., Scott-Burden, T., and Buhler, F. R. (1989). Activation of phospholipase A_2 by endothelin in cultured vascular smooth muscle cells. *Biochem. Biophys. Res. Commun.* **158**, 279–286.

Reynolds, E. E., Mok, S. S., and Kurokawa, S. (1989). Phorbol ester dissociates endothelin-stimulated phosphoinositide hydrolysis and arachidonic acid release in vascular smooth muscle cells. *Biochem. Biophys. Res. Commun.* **160**, 868–873.

Roden, M., Vierhapper, H., Liener, K., and Waldhause, W. (1992). Endothelin-1-stimulated glucose production *in vitro* in the isolated perfused rat liver. *Metabolism*, **41**, 290–295.
Rubanyi, G. M., and Parker-Botelho, L. H. (1991). Endothelins. *FASEB J.* **5**, 2713–2720.
Rubanyi, G. M., and Vanhoutte, P. M. (1984). Hypoxia releases a vasoconstrictor substance from the coronary arterial endothelium. *Circulation* **70**, 122.
Rubanyi, G. M., and Vanhoutte, P. M. (1985). Hypoxia releases a vasoconstrictor substance from the canine vascular endothelium, *J. Physiol. (London)* **364**, 45–56.
Saida, K., Mitsui, Y., and Ishida, N. (1989). A novel peptide, vasoactive intestinal constrictor, of a new (endothelin) peptide family. *J. Biol. Chem.* **264**, 14613–14616.
Saito, Y., Nakao, K., Mukoyama, M., and Imura, H. (1989). Increased plasma endothelin level in patients with essential hypertension. *N. Engl. J. Med.* **322**, 305.
Sakamoto, A., Yanagisawa, M., Sakurai, T., Takuwa, Y., Yanagisawa, H., and Masaki, T. (1991). Cloning and functional expression of human cDNA for the ET_B endothelin receptor. *Biochem. Biophys. Res. Commun.* **178**, 656–663.
Sakurai, T., Yanagisawa, M., Takuwa, Y., Miyaazaki, H., Kimura, S., Goto, K., and Masaki, T. (1990). Cloning of a cDNA encoding a non-isopeptide-selective subtype of the endothelin receptor. *Nature (London)* **348**, 732–735.
Samson, W. K. (1992). The endothelin-A receptor subtype transduces the effects of the endothelins in the anterior pituitary gland. *Biochem. Biophys. Res. Commun.* **187**, 590–595.
Samson, W. K., Skala, K. D., Alexander, B. D., and Huang, F. L. S. (1990). Pituitary site of action of endothelin: Selective inhibition of prolactin release *in vitro. Biochem. Biophys. Res. Commun.* **169**, 737–743.
Saudek, V. Hoflack, J., and Pelton, J. T. (1989). ^{1}H-NMR study of endothelin, sequence-specific assignment of the spectrum and a solution structure. *FEBS Lett.* **257**, 145.
Schrader, J., Tebbe, U., Borrics, M., Ruschitzka, F., Schoel, G., Kandt, M., Warneke, G., Zuchner, C., Weber, M. H., Neu, U., Rath, W., and Henning, H. V. (1990). Plasma-endothelin bei normalperson und patienten mit nephrologisch-rheumatologischen und kardiovaskularen erkrankungen. *Klin. Wochenschr.* **68**, 774–779.
Schvartz, I., Ittoop, O., and Hazum, E. (1991). Direct evidence for multiple endothelin receptors. *Biochemistry*, **30**, 4325–4327.
Serradiel-Le Gal, C. Jouneaux, C., Sanchez-Bueno, A., Raufaste, D., Rache, B., Preaux, A. M., Maffrand, J. P. Cobbold, P. H., Hanoune, J., and Lotersztajn, S. (1991). Endothelin action in rat liver: receptors, free Ca^{2+} oscillations and activation of glycogenolysis. *J. Clin. Invest.* **87**, 133–138.
Shiba, R., Yanagisawa, M., Miyauchi, T., Ishii, Y., Kimura, S., Uchiyama, Y., Masaki, T., and Goto, K. (1989). Elimination of intravenously injected endothelin-1 from the circulation of the rat. *J. Cardiovasc. Pharmacol.* **13**(Suppl. 5), S98–S101.
Shichiri, M., Hirata, Y., Ando, K., Emori, T., Ohta, K., Kimoto, S., Inoue, A., and Marumo, F. (1989). Plasma endothelin levels in patients with hypertension and end-stage renal failure. *Circulation* **80**(Suppl 2) II, 125.
Shinmi, O., Kimura, S., Sawamura, T., Sugita, Y., Yoshizawa, T., Uchiyama, Y., Yanagisawa, M., Goto, K., Masaki, T., and Kanazawa, I. (1989). Endothelin-3 is a novel neuropeptide: Isolation and sequence determination of endothelin-1 and endothelin-3 in porcine brain. *Biochem. Biophys. Res. Commun.* **164**, 587–593.
Simonson, M. S., and Dunn, M. J. (1990). Endothelin-1 stimulate contraction of rat glomerular mesangial cells and potentiates β-adrenergic mediated cyclic adenosine monophosphate accumulation. *J. Clin. Invest.* **85**, 790–797.

Simonson, M. S., and Dunn, M. J. (1991). Endothelin peptides: a possible role in glomerular inflammation. *Lab. Invest.* **64,** 1–4.

Simonson, M. S., Wann, S., Men, P., Dubyak, G. R., Kester, M., Nakazato, Y., Sedor, J. R., and Dunn, M. J. (1989). Endothelin stimulates phospholipase C, Na^+/H^+ exchange, c-*fos* expression and mitogenesis in rat mesangial cells. *J. Clin. Invest.* **83,** 708–712.

Sirvio, M. L., Metsarinne, K., Saijoonmaa, O., and Fyhrquist, F. (1990). Tissue distribution and half-life of ^{125}I-endothelin in the rat: importance of pulmonary clearance. *Biochem. Biophys. Res. Commun.* **167,** 1191–1195.

Sokolovsky, M., Ambar, I., and Galron, R. (1992). A novel subtype of endothelin receptors. *J. Biol. Chem.* **267,** 20551–20554.

Spinella, M. J., Krystek, S. R., Peapus, D. H., Wallace, B. A., Brunner, C., and Anderson, T. T. (1989). A proposed structural model of endothelin. *Peptide Res.* **2,** 286.

Spinella, M. J., Palik, A. B., Everitt, J., and Andersen, T. T. (1991). Design and synthesis of a specific endothelin-1 antagonist. Effects on pulmonary vasoconstriction. *Proc. Natl. Acad. Sci. USA* **88,** 7443–7446.

Stewart, D. J., Kubac, G., Costello, K. B., and Cernacek, P. (1991a). Increased plasma endothelin-1 in the early hours of acute myocardial infarction. *J. Am. Coll. Cardiol.* **18,** 38–43.

Stewart, D. J., Levy, R. D., Cernacek, P., and Langleben, D. (1991b). Increased plasma endothelin-1 in pulmonary hypertension: marker or mediator of disease? *Intern. Med.* **114,** 464–469.

Stojikovic, S. S., Iida, T., Merelli, F., and Catt, K. J. (1991). Calcium signaling and secretory responses in endothelin-stimulated anterior pituitary cells. *Mol. Pharmacol.* **39,** 762–770.

Sugiura, M., and Inagami, T. (1989). Endothelin action: Inhibition by a protein kinase C inhibitor and involvement of phosphoinositides. *Biochem. Biophys. Res. Commun.* **158,** 170–176.

Sunako, M., Kawahara, Y., Kirata, K., Tsuda, T., Yokoyama, M., Fukuzaki, H., and Takai, Y. (1990). Mass analysis of 1,2-diacylglycerol in cultured rabbit vascular smooth muscle cells: Comparison of stimulation by angiotensin II and endothelin. *Hypertension* **15,** 84–88.

Sunnergren, K. P., Word, R. A., Sambrook, J. F., MacDonald, P. C., and Casey, M. L. (1990). Expression and regulation of endothelin procursor mRNA in avascular human amnion. *Mol. Cell. Endocrinol.* **68,** R7–R14.

Suzuki, N., Matsumoto, H., Kitada, C., Kimura, S., and Fujino, M. (1989). Production of endothelin-1 and big endothelin-1 by tumor cells with epithelial-like morphology. *J. Biochem.* **106,** 736–741.

Takagi, M., Matsuoka, H., Atarashi, K., and Yagi, S. (1988). Endothelin: a new inhibitor of renin release. *Biochem. Biophys. Res. Commun.* **157,** 1164–1168.

Takahashi, K., Ghatei, M. A., Jones, P. M., Murphy, J. K., Lam, H. C., O'Halloran, D. J., and Bloom, S. R. (1991). Endothelin in human brain and pituitary gland: Presence of immunoreactive endothelin, endothelin messenger ribonucleic acid and endothelin receptors. *J. Clin. Endocrinol. Metab.* **72,** 693–699.

Takasaki, C., Yanagisawa, M., Kimura, S., Goto, K., and Masaki, T. (1988a). Similarity of endothelin to snake venom toxin. *Nature (London)* **335,** 303.

Takasaki, C., Tamiya, N., Bdolah, A., Wollberg, Z., and Kochva, E. (1988b). Sarafotoxins S6: Several isotoxins from Atractaspis engaddensis (burrowing asp) venom that affect the heart. *Toxicol.* **26,** 543–548.

Takasuka, T., Adachi, M., Miyamoto, C., Furuichi, Y., and Watanabe, T. (1992). Characterization of endothelin receptors ET_A and ET_B expressed in COS cells. *J. Biochem.* **112,** 396–400.

Taylor, R. N., Varma, M., Teng, N. N., and Roberts, J. M. (1990). Women with preeclampsia have higher endothelin levels than women with normal pregnancies. *J. Clin. Endocrin. Metab.* **71,** 1675–1677.

Terenghi, G., Bull, H. A., Bunker, C. B., Springall, D. R., Shao, Y., Wharton, J., Dowd, P. M., and Polak, J. M. (1991). Endothelin-1 in human skin: immunohistochemical, receptor binding, and functional studies. *J. Cardiovasc. Pharmacol.* **17**(Suppl. 7), 467–470.

Textor, S. C., Wilson, D. J., Lerman, A., Romero, J. C., Burnett, J. C., Jr., Weisner, R., Dickson, E. R., and Krom, R. A. (1992). Renal hemodynamics, urinary eicosanoids and endothelin after liver transplantation. *Transplantation* **54,** 74–80.

Tombe, Y., Miyauchi, T., Saito, A., Yanagisawa, M., Kimura, S., Goto, K., and Masaki, T. (1988). Effects of endothelin on the renal artery from spontaneously hypertensive and Wistar Kyoto rats. *Eur. J. Pharmacol.* **152,** 373–376.

Tomita, K., Ujiie, K., Nakanishi, T., Tomura, S., Matsuda, O., Ando, K., Shichiri, M., Hirata, Y., and Marumo, F. (1990). Plasma endothelin levels in patients with acute renal failure. *N. Engl. J. Med.* **321,** 1127.

Topiol, S. (1987). The deletion model for the origin of receptors. *Trends Biochem. Sci.* **12,** 419.

Toyo-oko, T., Aizawa, T., Suzuki, N., Hirata, Y., Miyauchi, T., Shin, W. S., Yanagisawa, M., Masaki, T., and Sugimoto, T. (1991). Increased plasma level of endothelin-1 and coronary spasm induction in patients with vasospastic angina pectoralis. *Circulation,* **83,** 476–483.

Tsuji, S., Sawamura, A., Watanabe, H., Takihara, K., Park, S. E., and Azuma, J. (1991). Plasma endothelin levels during myocardial ischemia and reperfusion. *Life Sci.* **48,** 1745–1749.

Tsunoda, K., Abe, K., Yoshinaga, K., Furuhashi, N., Kimura, H., Tsujiei, M., and Yajima, A. (1992). Maternal and umbilical venous levels of endothelin in women with preeclampsia. *J. Hum. Hypertens.* **6,** 61–64.

Turner, N. C., Power, F. R., Polak, J. M., Bloom, S. R., and Dollery, C. T. (1989). Contraction of rat tracheal smooth muscle by endothelin. *Br. J. Pharmacol.* **96,** 103P.

Uchida, Y., Ninomiya, H., Saotome, M., Nomura, A., Ohtsuka, M., Yanagisawa, M., Goto, K., Masaki, T., and Hasegawa, S. (1988). Endothelin, a novel vasoconstrictor peptide, as a potent vasoconstrictor. *Eur. J. Pharmacol,* **154,** 227–228.

Uchihara, M., Izumi, N., Sato, C., and Marumo, F. (1992). Clinical significance of elevated plasma endothelin concentration in patients with cirrhosis. *Hepatology,* **16,** 95–99.

Uemasu, J., Matsumoto, H., and Kawasaki, H. (1992). Increased plasma endothelin levels with liver cirrhosis. *Nephron* **60,** 380.

Urade, Y., Fujitani, Y., Oda, K., Watanabe, T., Umemura, I., Takai, M., Okada, T., Sakata, K., and Karaki, H. (1992). An endothelin B receptor-selective antagonist: IRL 1038, [Cys11–Cys15]-endothelin(11–21). *FEBS Lett.* **311,** 12-16.

Vanhoutte, P. M., Rubanyi, G. M., Miller, V. M., and Houston, D. S. A. (1986). Modulation of vascular smooth muscle contraction by the endothelium. *Annu. Rev. Physiol.* **48,** 307–320.

Van Papendorp, C. L., Cameron, I. T., Davenport, A. P., King, A., Barker, P. J., Huskisson, N. S., Gilmor, R. A., Brown, M. J., and Smith, S. K. (1991). Localization and endogenous concentration of endothelin-like immunoreactivity in human placenta. *J. Endocrinol.* **131**, 507–511.

Van Renterghem, C., Vigne, P., Barhanin, J., Schmid-Alliana, A., Frelin, C., and Lazdunski, M. (1989). Molecular mechanism of action of the vasoconstrictor peptide endothelin. *Biochem. Biophys. Res. Commun.* **157**, 977–985.

Veglio, F., Pinna, G., Melchio, R., Rabbia, F., Panarelli, M., Gagliardi, B., and Chiandussi, L. (1992). Plasma endothelin levels in cirrhotic patients. *J. Hepatol.* **15**, 85–87.

Vemulapalli, S., Chiu, P. J., Rivelli, M., Foster, C. V., and Syvertz, E. J. (1991). Modulation of circulating levels of endothelin in hypertension and endotoxemia in rats. *J. Cardiovasc. Pharmacol.* **18**, 895–903.

Vigne, P., Ladoux, A., and Frelin, D. (1991). Endothelins activate Na+/H+ exchange in brain capillary endothelial cells via a high affinity endothelin-3 receptor that is not coupled to phospholipase. C. *J. Biol. Chem.* **266**, 5925–5928.

Voerman, H. J., Stehouwer, C. D., van Kamp, G. J., Strack van Schijndel, R. J., Groeneveld, A. B., and Thijs, L. J. (1992). Plasma endothelin levels are increased during septic shock. *Crit. Care Med.* **20**, 1097–1101.

Wake, K. (1980). Perisinusoidal stellate cells (Fat-storing cells, interstitial cells, lipocytes), their related structure in and around liver sinusoids, and vitamin A storage cells in extrahepatic organs. *Int. Rev. Cytol.* **66**, 303–353.

Warner, T., de Nucci, G., and Vane, J. R. (1988). Release of EDRF by endothelin in the rat isolated perfused mesentery. *Br. J. Pharmacol.* **95**, 723P.

Warner, T., de Nucci, G., and Vane, J. R. (1989). Rat endothelin is a vasodilator in the isolated perfused mesentery of the rat. *Eur. J. Pharmacol.* **159**, 325–326.

Weitzberg, E., Lundberg, J. M., and Rudehill, A. (1991). Elevated plasma levels of endothelin in patients with sepsis syndrome. *Circ. Shock.* **33**, 222–227.

Wharton, J., Rutherford, R. A., Walsh, D. A., Mapp, P. I., Knock, G. A., Blake, D. R., and Polak, J. M. (1992). Autoradiographic localization and analysis of endothelin-1 binding sites in human synovial tissue. *Arthritis Rheum.* **35**, 894–899.

Williams, D. L., Jr., Jones, K. L., Pettibone, D. J., Lis, E. V., and Clineschmidt, B. V. (1991). Sarafotoxin S6c: An agonist which distinguishes between endothelin receptor subtypes. *Biochem. Biophys. Res. Commun.* **175**, 556–561.

Withrington, P. J., de Nucci, G., and Vane, J. R. (1989). Endothelin-1 causes vasoconstriction and vasodilation in the blood perfused liver of the dog. *J. Cardiovasc. Pharmacol.* **13**(Suppl 5), S209–S210.

Wright, C. E., and Fozard, J. R. (1988). Regional vasodilation is a prominent feature of the haemodynamic sponse to endothelin in anaesthetized, spontaneously hypertensive rats. *Eur. J. Pharmacol.* **155**, 210–203.

Yamakado, M., Hirata, Y., Matsuoka, H., and Sugimoto, T. (1991). Pathophysiological role of endothelin in renal transplant. *J. Cardiovasc. Pharmacol.* **17**(Suppl. 7), S477–S479.

Yamane, K., Kashiwagi, H., Suzuki, N., Miyauchi, T., Yanagisawa, M., Goto, K., and Masaki, T. (1991). Elevated plasma levels of endothelin-1 in systemic sclerosis. *Arthritis Rheum.* **34**, 342–344.

Yanagisawa, M., and Masaki, T. (1989). Endothelin, a novel endothelium-derived peptide. *Biochem. Pharmacol.* **38**, 1877–1883.

Yanagisawa, M., Kurihara, H., Kimura, S., Tomobe, Y., Kobarashi, M., Mitsui, Y., Goto, K., and Masaki, T. (1988a). A novel potent vasoconstrictor peptide produced by vascular endothelial cells. *Nature (London)* **332**, 411–415.

Yanagisawa, M., Akhiro, I., Ishikawa, T., Kasuya, Y., Kimura, S., Kumagaye, S. H., Nakijima, K., Watanabe, T., Sakakibara, S., Goto, D., and Masaki, T. (1988b). Primary structure, synthesis and biological activity of rat endothelin and endothelium-derived vasoconstrictor peptide. *Proc. Natl. Acad. Sci. USA* **85**, 6964–6967.

Yoshimi, H., Hirata, Y., Fukuda, T., Kawano, Y., Emori, T., Kuramochi, M., Omae, T., and Marumo, F. (1989). Regional distribution of immunoreactive endothelin in rats. *Peptides* **10**, 805–808.

Yoshizawa, T., Shinmi, O., Giad, A., Yanagisawa, M., Gibson, S., Kimura, S., Uchigama, Y., Polak, J. M., Masaki, T., and Kanazawa, I. (1990). Endothelin: A novel peptide in the posterior pituitary system, *Science* **247**, 462–464.

Zhu, G., Wu, L. H., Mauzy, C., Egloff, A. M., Mirzadegan, T., and Chung, F. Z. (1992). Replacement of lysine-181 by aspartic acid in the third transmembrane spanning region of the endothelin type B receptor reduces its affinity to the endothelin peptides and sarafotoxin without affecting G-protein coupling. *J. Cell. Biochem.* **50**, 159–164.

Ziv, I., Fleminger, G., Djaldetti, R., Achiron, A., Melmed, E., and Solovsky, M. (1992). Elevated plasma endothelin-1 levels in patients with acute ischemic cerebral stroke. *Stroke* **23**, 1014–1016.

Zoja, C., Orisio, S., Perico, N., Benigni, A., Morigi, M., Benatte, L., Rambaldi, A., and Remuzzi, G. (1991). Constitutive expression of endothelin gene in cultured human mesangial cells and its modulation by transforming growth factor-β, thrombin and a thromboxane A_2 analogue. *Lab. Invest.* **64**, 16–20.

Cyclic ADP–Ribose: Metabolism and Calcium Mobilizing Function

HON CHEUNG LEE,* ANTONY GALIONE,[†] AND TIMOTHY F. WALSETH[‡]

Departments of *Physiology and ‡Pharmacology
University of Minnesota
Minneapolis, Minnesota
†Department of Pharmacology
Oxford University
Oxford OX1 3QT, England

I. Introduction
II. Cyclic ADP–Ribose
   A. Discovery
   B. Structure
   C. Endogenous Levels in Tissues
III. Enzymes Involved in the Metabolism of Cyclic ADP–Ribose
   A. ADP–Ribosyl Cyclase
   B. Cyclic ADP–Ribose Hydrolase
   C. Lymphocyte CD38
   D. Relationship with NAD^+ Glycohydrolase
   E. Regulation by Cyclic GMP
IV. Cyclic ADP–Ribose-Dependent Ca^{2+} Release
   A. Sea Urchin Egg as a Model System
   B. Ca^{2+} Stores
   C. Relationship with Ca^{2+}-Induced Ca^{2+} Release
   D. Mammalian Systems
V. Cyclic ADP–Ribose Receptor
   A. Specific Binding to Sea Urchin Egg Microsomes
   B. Cyclic ADP–Ribose Antagonists
   C. Photoaffinity Labeling of cADPR Binding Sites
VI. Physiological Roles of Cyclic ADP–Ribose
   A. Fertilization
   B. Insulin Secretion
VII. Conclusion
   References

I. INTRODUCTION

Many extracellular stimuli including hormones, neurotransmitters, and growth factors interact with cell-surface receptors and result in an increase in intracellular free Ca^{2+} (Berridge, 1993). The Ca^{2+} increase serves as a universal cell regulator and constitutes an important intracellular signal that can trigger a wide range of cellular responses, depending on the phenotype of the cell (Campbell, 1983). Cells maintain their resting intracellular free Ca^{2+} at submicromolar concentrations against an enormous concentration gradient across the plasma membrane (Carafoli, 1987), and have developed elaborate and extensive mechanisms for maintaining this delicate balance. Specialized transport proteins, such as Ca^{2+}–ATPases and Na^+/Ca^{2+} exchangers, are present in the plasma membrane and can translocate Ca^{2+} across the membrane against its electrochemical gradient. Similarly, unique Ca^{2+}–ATPases that are sensitive to the inhibitor thapsigargin are present in internal membranes. These proteins sequester Ca^{2+} from the cytoplasm into organellar stores, mainly the endoplasmic reticulum. During Ca^{2+} signaling, these Ca^{2+} stores are often transiently liberated through the opening of specific Ca^{2+} channels resulting in a Ca^{2+} transient (Gill, 1989). Because of the spatial separation between the plasma and endoplasmic reticular membranes, diffusable molecules are required to convey the hormonal message at the cell surface to the Ca^{2+} channels of the internal membranes.

The two classes of Ca^{2+} channel most widely found in the internal membranes are the inositol trisphosphate (IP_3) and ryanodine receptors (Ferris and Snyder, 1992; Sorrentino and Volpe, 1993). Activation of many cell surface receptors leads to the production of the second messenger IP_3, which diffuses to the endoplasmic reticulum, binds to the IP_3 receptor, and activates the Ca^{2+} channel (Berridge, 1993). The regulation of ryanodine receptors has been less clear. However, evidence suggests that it too may be controlled by an intracellular messenger, cyclic ADP–ribose (cADPR; Galione *et al.*, 1991; Lee, 1993a; Meszaros *et al.*, 1993).

Cyclic ADP-ribose is a novel endogenous metabolite of NAD^+ that is active in mobilizing intracellular Ca^{2+} in mammalian as well as invertebrate cell preparations in the nanomolar concentration range. Where its mechanism of action has been studied, cADPR appears to release Ca^{2+} from a ryanodine-sensitive but IP_3-insensitive store (Galione, 1993; Lee, 1993b). Indirect evidence suggests that cADPR levels may be sensitive to extracellular stimuli (Takasawa *et al.*, 1993; Galione *et al.*, 1993b), indicating that cADPR is not only a regulator of intracellular free Ca^{2+} but also joins IP_3 as a Ca^{2+}-mobilizing second messenger.

In this chapter, we describe the discovery of cADPR as a novel endogenous Ca^{2+}-mobilizing agent, and how it has fulfilled most of the criteria necessary for it to be considered a second messenger. The enzymatic pathways for the synthesis and degradation of the metabolite are summarized. Current knowledge of the properties of its intracellular receptor and the mechanism of its Ca^{2+}-mobilizing activity are discussed. Finally, we document the physiological roles of cADPR in two specific cellular systems: the sea urchin egg, an invertebrate cell, and the pancreatic β cell, a mammalian system.

II. Cyclic ADP–Ribose

A. Discovery

Cyclic ADP–ribose was discovered in 1987 during investigations of the mechanism of Ca^{2+} mobilization in sea urchin eggs (Clapper et al., 1987). The approach taken was to develop an in vitro system that would allow free access to internal Ca^{2+} stores. The first attempt was to permeabilize the eggs with high voltage discharge, which produced only transient permeabilization (Clapper and Lee, 1985). Although various substances such as IP_3 could be introduced into the permeabilized cells, the preparation was not found to be suitable as a routine assay. The alternative approach of homogenization was adopted. Responsive homogenates can be prepared routinely from Lytechinus pictus eggs with a regular homogenizer equipped with a glass pestle (Clapper and Lee, 1985), whereas a gentler technique of N_2 decavitation was needed for Strongylocentrotus purpuratus eggs (Lee, 1991). In the presence of ATP, the endogenous Ca^{2+} pump in the homogenates sequesters the contaminating Ca^{2+} and lowers the ambient concentration of Ca^{2+} to the submicromolar range, allowing the convenient use of fluorescence Ca^{2+} indicators for monitoring the movement of Ca^{2+}. The preparation is highly stable and very responsive to IP_3, which elicits a large and immediate Ca^{2+} release (Clapper and Lee, 1985). A large quantity of egg homogenates can be prepared routinely and stored frozen without loss of responsiveness for long periods of time. The preparation is, therefore, ideally suited for use as a bioassay for Ca^{2+} release activators.

NAD^+ was tested with this assay system in addition to other possible candidates for Ca^{2+} release activators. Shortly after fertilization of sea urchin eggs, a large and rapid conversion of a third of the cellular NAD^+ to NADP and then to NADPH is known to occur (Epel, 1980). The time this occurs corresponds to the time of mobilization of intracellular Ca^{2+} stores. Addition of NAD^+ to egg homogenates elicits as much Ca^{2+} release as that induced by IP_3, although the kinetics of the two release processes are very different (Clapper et al., 1987). In contrast to the immediate release triggered by IP_3, the NAD^+-dependent release shows a prominent initial delay of 1–4 min. The release is also stereospecific, requiring β-NAD+; the α form has no effect. The time delay and the stereospecificity suggest that enzymatic conversion is involved, a hypothesis that is substantiated since pre-incubation of NAD+ with egg extracts produces an active metabolite that can release Ca^{2+} without a delay (Clapper et al., 1987). A high-performance liquid chromatography (HPLC) procedure was developed to purify the metabolite which, at the time, was called E-NAD+, an abbreviation for "enzyme-activated NAD+" (Clapper et al., 1987). As described in Section II,B, structural determination shows that E-NAD+ is actually cyclized ADP–ribose and was thus named cADPR (Lee e al., 1989). The structure of cADPR is shown in Fig. 1.

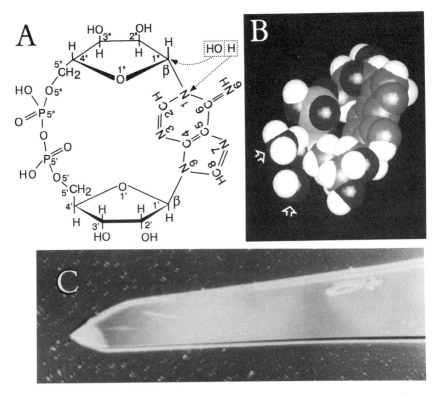

FIG. 1. Structure of cyclic ADP–ribose. (A) Structure of cyclic ADP–ribose. Hydrolysis of the molecule at positions indicated by the arrows produces ADP–ribose. (B) CPK view of cyclic ADP–ribose. The two arrowheads indicate two of the three water molecules co-crystallized with cyclic ADP–ribose. (C) Crystal of cyclic ADP–ribose. The width of the crystal is about 0.1 mm.

Although cADPR releases about the same amount of Ca^{2+} as IP_3, this activator is much more potent; the half-maximal effective concentration is ~17 nM, 5–7 times lower than that of IP_3 when assayed with the same homogenate preparation (Dargie et al., 1990). Cyclic ADP–ribose is not simply a nonspecific Ca^{2+} ionophore, since it selectively releases Ca^{2+} from microsomes but not from mitochondria. Further, a high dose of cADPR can desensitize the microsomes to subsequent addition of cADPR but not to IP_3 (Dargie et al., 1990). This activator also does not function as an inhibitor of the Ca^{2+} pump, since the Ca^{2+} leak elicited by blocking the pump with thapsigargin (Lee, 1993a) or by removal of ATP (Dargie et al., 1990) is much slower than the release induced by cADPR. As described in Section V, cADPR appears to operate through a specific receptor system.

Cyclic ADP–ribose not only is active in releasing Ca^{2+} in vitro. Microinjection of the metabolite into live eggs also induces transient elevation of internal Ca^{2+} and triggers a massive exocytic reaction, the cortical reaction, leading to the formation of the fertilization envelope (Dargie et al., 1990). The half-maximal effective dose for activating the egg is about 60 nM and the activation occurs in the absence of external Ca^{2+}, indicating that cADPR can indeed mobilize internal Ca^{2+} stores in live cells. In addition to inducing Ca^{2+} release and the cortical reaction, cADPR can also parthenogenically activate the eggs to undergo multiple cycles of nuclear membrane breakdown and reformation, as well as DNA synthesis (Dargie et al., 1990). Results obtained in vitro and in vivo show clearly that cADPR is one of the most effective metabolites for mobilizing internal Ca^{2+} stores in sea urchin eggs.

The approach that led to the discovery of cADPR is quite different from the approaches used to discover other second messengers such as cyclic AMP and IP_3. In the case of cAMP, the stimulus and the response were first identified and used as an assay for the second messenger. In the case of IP_3, the stimuli and the messenger were discovered first, whereas the Ca^{2+}-mobilizing activity of IP_3 was the last component that completed the pathway. The approach used to identify cADPR was totally different; first, a bioassay was developed to identify possible Ca^{2+} release activators, which led to the discovery of cADPR. The unknown component in this pathway is the stimulus that elevates the cellular levels of cADPR. As discussed in later sections, three important advances are bringing this issue close to resolution. First, in pancreatic islets, cADPR levels appear to increase in response to glucose concentrations that are sufficient to stimulate insulin secretion (Takasawa et al., 1993). Second, inhibition of the cADPR pathway is now shown to be sufficient to suppress fertilization-induced Ca^{2+}

mobilization, provided the IP_3 pathway is also blocked (Galione et al., 1993a; Lee et al., 1993a). Third, the synthesizing enzyme of cADPR is now found to be regulated by a cGMP-dependent process (Galione et al., 1993b).

B. STRUCTURE

A large-scale procedure using three HPLC steps was developed to purify sufficient cADPR for structural determination (Lee et al., 1989). Beginning with crude egg extracts incubated with NAD^+, the overall purification was about 284-fold with a 25% yield. The final product was judged pure because it eluted as a single peak from five different HPLC columns. To determine the structure, four different approaches were used (Lee et al., 1989). The first was radioactive labeling. Radioactive NAD^+ with labels at various positions of the molecule was used as a precursor. If the radioactive label was on the adenine ring or on the adenylate phosphate, it was conserved after conversion to cADPR. However, label was lost if it was on the nicotinamide group. These results indicate that the nicotinamide group of NAD^+ is modified during the enzymatic conversion.

The second approach was proton nuclear magnetic resonance (NMR), which verified that the nicotinamide group was completely removed from NAD^+ (Lee et al., 1989). In addition, the characteristics of the NMR spectrum of cADPR suggested that both anomeric carbons of the two ribosyl units bonded to nitrogen. Two-dimensional correlated Overhauser spectroscopy (COSY) NMR indicated that the two ribose units were also intact.

The third approach was mass spectrometry, which allows precise determination of the mass of the molecular ion to an accuracy >2 ppm (Lee et al., 1989). The mass of cADPR was found to be exactly one water molecule less than that of ADP-ribose. Phosphate determination showed that cADPR contained 2 mol phosphate. The molecular formula of cADPR was determined by examining all possible combinations of phosphates, carbon, hydrogen, nitrogen, and oxygen that would give a mass within 3 ppm of the measured mass. The knowledge of the precise mass, and that the molecule contains two phosphates, allowed unique specification of the molecular formula of cADPR as $C_{15}H_{20}N_5O_{13}P_2$.

The only structure that is consistent with all these results is a cyclic compound formed by removal of the nicotinamide group of NAD^+ and linking of the anomeric carbon of the second ribose to a nitrogen of the adenine group. In other words, the metabolite is a cyclized ADP–ribose as shown in Fig. 1A. This cyclic structure has precisely the atomic composition, down to the number of hydrogen atoms, specified by the molecular formula.

The final approach to verify the structure was analyzing the hydrolytic products of cADPR. Although the molecule is quite stable, it does spontaneously hydrolyze with a half-time of about 10 days at room temperature under acidic conditions. Even at 37°C, the half-time for spontaneous hydrolysis is about 24 hr (Lee and Aarhus, 1993). When the hydrolytic products were analyzed, only one was found that was unequivocally shown to be ADP–ribose, a compound larger than its precursor cADPR by one water molecule (Lee et al., 1989). This result is precisely the one that would be predicted from the cyclic structure shown in Fig. 1. As depicted in the figure, hydrolysis of the linkage between the second ribose and the amino group of the adenine would result in addition of an –OH group to the ribose and an –H to the amino group. The product would be ADP-ribose, as was shown to be true experimentally.

Evidence described earlier definitely shows that cADPR is a cyclic compound with a linkage between the adenine group and the second ribose. Two possible sites of cyclization at the adenine ring are consistent with all the evidence, either at the N6-amino group or at the N1 position of the adenine ring. The N6 site was proposed because of the simplicity of the model (Lee et al., 1989). Results based on comparison of the ultraviolet (UV) absorption spectra of model compounds with that of cADPR suggests that N1 linkage may be more probable (Kim et al., 1993b). The model compounds used were N1- or N6-substituted adenosine derivatives. The possible effect of the cyclic structure on the UV spectrum of cADPR was not considered.

In order to resolve this issue, we have used X-ray crystallography to directly determine the site of cyclization (Lee et al., 1994). A crystal of cADPR is shown in Fig. 1C. In its pure form, cADPR can be readily crystallized into large plate-like crystals. The one shown in Fig. 1C has the width of about 0.1 mm. The length of some of these crystals are as long as 1–2 mm. The chemical structure of cADPR based X-ray crystallography data is shown in Fig. 1A. The site of cyclization is at the N1 instead of N6. The two glycosidic linkages at N1 and N9 are both in the β-configuration. A CPK view of the molecule is shown in Fig. 1B. Two of the three water molecules co-crystallized with cADPR are indicated by arrow heads.

C. ENDOGENOUS LEVELS IN TISSUES

Using the egg homogenate as a bioassay for cADPR, a procedure was developed to measure endogenous cADPR levels in tissues (Walseth et al., 1991). The procedure involves freeze-clamping tissues rapidly at liquid nitrogen temperature to halt metabolism. The powdered tissues are then extracted with perchloric acid while they are still frozen. Radioactive cADPR can be synthesized from radioactive NAD^+ using ADP–ribosyl cyclase (described in Section III,A) and can be used as a tracer. The endogenous cADPR in the acid extracts can be purified by HPLC. The amounts in the fractions are determined by the ability to release Ca^{2+} from egg homogenates. By comparison with the Ca^{2+}-releasing activity induced by known amounts of authentic cADPR, the endogenous cADPR in the fractions can be quantified. To verify that the Ca^{2+}-releasing activity in the purified fractions is indeed due to endogenous cADPR, the homogenates can be desensitized first with high doses of cADPR; then they should be unresponsive to the active fractions. Using this procedure, the endogenous levels of cADPR in rat tissues were found to range from 1 pmol cADPR/mg protein in the heart to ~3 pmol cADPR/mg protein in the liver (Walseth et al., 1991).

Using a similar approach, the endogenous cADPR levels in pancreatic islets were found to increase 3- to 4-fold after treatments with high concentrations of glucose that resulted in stimulation of insulin secretion (Takasawa et al., 1993). More interestingly, drugs that can produce diabetes in animals, such as streptozotocin, can abolish the glucose-induced increase in endogenous cADPR. These results not only demonstrate that cADPR is a naturally occurring metabolite present in a variety of tissues, but also suggest that the endogenous cADPR levels are responsive to stimuli, consistent with a second messenger role for cADPR.

Currently, the only available assay with sufficient specificity for cADPR is the bioassay using egg homogenates. This assay is capable of detecting 4–5 pmol cADPR (Walseth et al., 1991). Several findings show that the sensitivity of the egg homogenates to cADPR can be increased about 10-fold by pretreatment with caffeine (Lee, 1993a). This increase would bring the sensitivity of the bioassay to a range similar to that of the radioreceptor assay for IP_3 (Bredt et al., 1989). The development of a specific antagonist for cADPR-dependent Ca^{2+} release (Walseth and Lee, 1993) would also simplify the demonstration of specificity of the assay. Nevertheless, the bioassay is still somewhat cumbersome and prone to interference present in crude cell extracts, making it necessary for a prepurification by HPLC. The development of a radioimmunoassay with high sensitivity and specificity, similar to the one available for cAMP, would facilitate the identification of stimuli that alter cADPR levels in various systems.

III. Enzymes Involved in the Metabolism of Cyclic ADP–Ribose

The metabolic pathways for cADPR are shown in Fig. 2. The synthetic pathway for cADPR is mediated by ADP–ribosyl cyclase, which catalyzes the cyclization of NAD^+. The degradation pathway consists

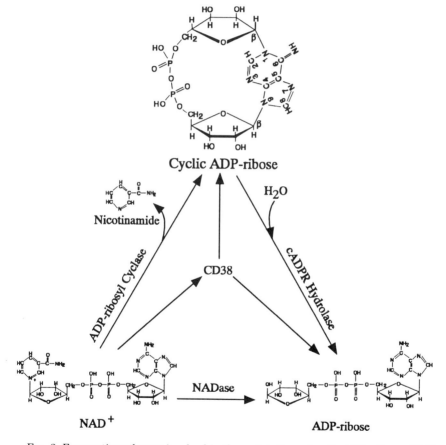

FIG. 2. Enzymatic pathways involved in the metabolism of cyclic ADP–ribose. CD38 is a lymphocyte antigen that is also a bifunctional enzyme, catalyzing both the synthesis and the hydrolysis of cyclic ADP–ribose.

of hydrolysis of cADPR to ADP–ribose and is catalyzed by cADPR hydrolase. In addition, a lymphocyte protein called CD38 and an enzyme purified from spleen have been shown to be able to catalyze the synthesis and the hydrolysis of cADPR (Howard *et al.*, 1993; Kim *et al.*, 1993a).

A. ADP–RIBOSYL CYCLASE

As described in Section II,A, ADP–ribosyl cyclase was discovered as an activity in sea urchin egg homogenates that activates NAD^+ to release Ca^{2+}. This discovery led to the identification of cADPR (Clapper *et al.*, 1987). ADP–ribosyl cyclase has since been detected in a variety of mammalian (including human), avian, amphibian, and invertebrate tissues (Rusinko and Lee, 1989; Lee *et al.*, 1993b; Lee and Arhus, 1993). Table I compares the cyclase activity in various tissue extracts. The activity in sea urchin egg extracts is actually the lowest, about one million-fold lower than in *Aplysia* ovotestis extracts and 200-fold lower than in mammalian brain extracts. Data presented in Table I are meant to be used for qualitative comparison only, because different temperatures were used in assaying the activity in vertebrate and invertebrate tissues, and because cADPR hydrolase was present in these extracts. The enzymatic activity of ADP–ribosyl cyclase does not require any exogenous cofactors such as magnesium, in contrast with other NAD^+-utilizing enzymes such as dehydrogenases.

TABLE I
WIDESPREAD OCCURRENCE IN ANIMAL TISSUES OF ADP–RIBOSYL CYCLASE AND cADPR HYDROLASE[a]

Tissue	ADP–ribosyl cyclase (nmol/mg/hr)	cADPR hydrolase (nmol/mg/hr)	Reference
Human RBC ghosts	8.4	5.4	Lee *et al.* (1993b)
Dog brain	62.7 ± 6.2	165.4 ± 29.4	Lee and Aarhus (1993)
Chick embryonic brain	20.6 ± 2.8	21.2 ± 5.8	Lee and Aarhus (1993)
Salamander brain	4.5 ± 0.1	4.6 ± 0.5	Lee and Aarhus (1993)
Sea urchin egg	0.3 ± 0.05	3.8 ± 1.6	Lee and Aarhus (1993)
Aplysia ovotestis	1.3×10^6	—	Lee and Aarhus (1991)

[a] Experimental details are described in the references. Red blood cell is abbreviated RBC. cADPR hydrolase activity in *Aplysia* ovotestis extracts was not determined.

Soluble and membrane-bound forms of the ADP–ribosyl cyclase have been reported. The soluble form is present in dog testis, sea urchin egg ovaries and testis (H. C. Lee, unpublished results), and, most abundantly, in the ovotestis of *Aplysia californica* (Table I). The *Aplysia* cyclase has been purified and sequenced (Glick et al., 1991; Hellmich and Strumwasser, 1991; Lee and Aarhus, 1991). In sea urchin eggs and most mammalian tissues, the enzyme is membrane bound (Lee, 1991; Lee and Aarhus, 1993). Two membrane-bound cyclases have been purified, a 40-kDa protein called CD38 from lymphocyte and a 39-kDa protein from spleen. Both are bifunctional enzymes catalyzing the synthesis and degradation of cADPR (Section III,C).

The *Aplysia* enzyme is a protein of 29 kDa that was originally identified as an NADase because it was thought to degrade NAD^+ to ADP–ribose (Hellmich and Strumwasser, 1991). The enzyme assays used, however, did not distinguish between ADP–ribose and cADPR. Using HPLC and NMR techniques, researchers have now demonstrated that this enzyme exclusively makes cADPR from NAD^+, so it has been renamed ADP–ribosyl cyclase to distinguish it from conventional NADases that produce ADP–ribose (Lee and Aarhus, 1991). As shown in Fig. 2, the reaction catalyzed by this enzyme produces one molecule of cADPR and releases one molecule of nicotinamide from each molecule of NAD^+ (Lee and Aarhus, 1991). High resolution cation-exchange chromatography resolves the homogeneous enzyme into multiple active forms differing in charge (Hellmich and Strumwasser, 1991; Lee and Aarhus, 1991). The specific activity of the purified cyclase is about 1.2 mmol/min/mg (Lee and Aarhus, 1991), corresponding to a turnover rate of about $580\ sec^{-1}$, ranking it among the most catalytically active enzymes. The *Aplysia* ADP–ribosyl cyclase has proven to be very useful in largescale production of cADPR, as well as in synthesis of various analogs of cADPR (Section V,B).

The amino acid sequence of the *Aplysia* cyclase has been determined (Glick et al., 1991). Comparison of the amino acid sequence of ADP–ribosyl cyclase with sequences in GenBank (release 69) revealed that 86 of the 285 amino acids in the ADP–ribosyl cyclase are identical to human lymphocyte antigen CD38 (Jackson and Bell, 1990). An additional 110 amino acids are conservative substitutions, giving an overall homology of the two molecules of 69% (States et al., 1992). More intriguing is the finding that the 10 cysteine residues in the ADP–ribosyl cyclase sequence are all perfectly aligned with those in the CD38 sequence, suggesting that the tertiary structures of the two proteins, with all the disulfide linkages formed, may be very similar (Howard et al., 1993). The murine homolog of CD38 has been cloned and sequenced (Harada *et al.*, 1993). Figure 3 shows the sequences of the three proteins. The similarities and differences in enzymatic activities of CD38 and ADP–ribosyl cyclase are discussed in Section III,C. Also indicated in Fig. 3, a stretch of 19 amino acids is highly conserved among all three sequences. The conserved region in CD38 and the *Aplysia* cyclase may represent the site for NAD$^+$ binding or for catalysis.

Sequence comparison also shows several notable differences between *Aplysia* cyclase and CD38. As indicated in the sequence in Fig. 3, the N-terminal amino acid in the mature *Aplysia* cyclase purified

```
                 Cytoplasmic Domain    Transmembrane Domain
HuCD38      MANCEFSPVSGDKPCCRLSRRAQLCLGVSILVLILVVLAVVV...PRWRQTWSGP
MoCD38      MANYEFSQVSGDRPGCRLSRKAQIGLGVGLLVLIALVVGIVVILLRPRSLLVWTGE
ApCyclase                           DLEMSPVAIIACVCLAVTLT......SISPSEAIV
                                                                        ▲
            Extracellular Domain>                                     ◄►
HuCD38      GTTKRFPETVLARCV.....KYTEI.HPEMRHVDCQSVWDAFKGAFISKHFCNITE
MoCD38      PTTKHFSDIFLGRCL.....IYTQILRPEMRDQNCQEILSTFKGAFVSKNPCNITR
ApCyclase   PTT.RELENVFLGRCKDYEITRYLDIL.PRVR.SDCSALWKDFFKAFSFKNPCDLDL

                                                          ├─Conserved Domain─
HuCD38      EDYQPLMKLGTQTVPCNKILLWSRIKDLAHQFTQVQRDMFTLEDTLLGYLADDLTW
MoCD38      EDYAPLVKLVTQTIPCNKTLFWSKSKHLAHQYTWIQGKMFTLEDTLLGYVADDLRW
ApCyclase   GSYKDFFTSAQQQLPKNKVMFWSGVYDEAHDYANTGRKYITLEDTLPGYMLNSLVW

              ┐                                                    ◄►
HuCD38      CGEFNTSKINYQSCPDWRKDCSNNPVSVFWKTVSRRFAEAACDVVIVMLNGSRSKI
MoCD38      CGDPSTSDMNYVSCPHWSENCPNNPITMFWKVISQKFAEDACGVVQVMLNGSLREP
ApCyclase   CGQRANPGFNEKVCRDF.KTCPVQARESFWGMASSSYAHSAEGEVTYMVDGSNPKV

                ◄►
HuCD38      ..FDKNSTFGSVEVHNLQPEKVQTLEAWVIHGGREDSRDLCQDPTIKELESIISKR
MoCD38      ..FYKNSTFGSLEVFSLDPNKVHKLQAWVMHDIEGASSNACSSSSLNELKMIVQKR
ApCyclase   PAYRPDSFFGKYELPNLT.NKVTRVKVIVLHRLGEKIIEKCGAGSLLDLEKLVKAK

HuCD38      NIQFSCKNIYRPDKFLQCVKNPEDSSCTSEI
MoCD38      NMIFACVDNYRPARFLQCVKNPEHPSCRLNT
ApCyclase   HFAFDCVENPRAVLFLLCSDNPNARECRLAK
```

FIG. 3. Sequence homology between ADP–ribosyl cyclase and CD38. *Aplysia* ADP–ribosyl cyclase (ApCyclase), lymphocyte antigen CD38 from mouse (MoCD38), and lymphocyte antigen CD38 from human (HuCD38). Black boxes indicate conserved cysteine. Hatched boxes indicate identical amino acids. Open boxes indicate conservative substitutions. The double-headed arrows indicate potential glycosylation sites in HuCD38 and MoCD38. The triangle indicates the N-terminal amino acid of ApCyclase. References for sequence data are Jackson and Bell (1990) for HuCD38, Glick *et al.* (1991) for ApCyclase, and Harada *et al.* (1993) for MoCD38.

FIG. 4. Crystal of *Aplysia* ADP–ribosyl cyclase. The dimension of the crystal is about 0.2 mm × 0.2 mm × 0.1 mm. The crystal was prepared by David Levitt, Department of Physiology, University of Minnesota, Minneapolis.

from ovotestis is isoleucine (Glick *et al.*, 1991). Thus, the hydrophobic sequence corresponding to the transmembrane domain of CD38 may be removed, probably after translation. This result is consistent with the idea that the cyclase is a soluble protein. The *Aplysia* cyclase sequence also lacks the glycosylation sites that are present in the CD38 sequence.

Taking advantage of the unusual abundance of ADP–ribosyl cyclase in *Aplysia* ovotestis, a highly efficient method using cation-exchange chromatography has been developed for large-scale purification of the cyclase (Lee and Aarhus, 1991). Levitt has succeeded in crystallizing the *Aplysia* ADP–ribosyl cyclase. Figure 4 shows a representative crystal of the enzyme. X-Ray diffraction data to the resolution of about 3 Å have been obtained by Levitt (University of Minnesota, Minneapolis; unpublished data). Solution of the three-dimensional crystal structure of the cyclase will certainly provide important information about the reaction mechanism as well as the regulation of this enzyme.

B. Cyclic ADP–Ribose Hydrolase

If cADPR is indeed a second messenger, a pathway must exist for degrading it so its action can be terminated. The presence of an enzyme that can degrade cADPR was noted during measurements of ADP–ribosyl cyclase activity in various tissue extracts (Rusinko and Lee, 1989). Researchers found that incubation of tissue extracts with NAD^+ resulted in production of cADPR and progressive increase in Ca^{2+} release activity. However, after reaching a peak, the Ca^{2+} release activity declined, indicating that cADPR produced was destroyed by a degradation enzyme in the incubation mixture. This degradative activity was found to be especially high in brain tissue extracts. Using HPLC and NMR techniques, the degradation product was identified as ADP–ribose (Lee and Aarhus, 1993). Since a unique N-glycosidic linkage between the adenine group and the terminal ribosyl moiety of cADPR is hydrolyzed (see Fig. 2), the enzyme was named cyclic ADP–ribose hydrolase. The hydrolase activity, like ADP–ribosyl cyclase, is widely distributed. The occurrence of the two activities, in fact, appears to be correlated, that is, when the cyclase activity is high, as in brain extracts, the hydrolase activity is also high (Table I). The only exception is in the case of *Aplysia* ovotestis, where the cyclase activity is so high that the hydrolase activity is insignificant in comparison. The widespread occurrence of these enzymes indicates that cellular levels of cADPR are likely to be tightly regulated and that cADPR metabolism may have important biological consequences.

Of all the enzymes summarized in Table I, only the hydrolase from the dog brain extracts has been partially characterized (Lee and Aarhus, 1993). The enzyme activity is associated with membranes and, similar to the cyclase, also does not appear to require any exogenous cofactors. This enzyme displays a broad pH maximum near neutral. Kinetic analysis shows that the enzyme has a K_m of 160 μM for cADPR and a V_{max} of 1.2 μmol/mg/hr. Currently, nothing is known about the regulation of cADPR hydrolase activity.

C. Lymphocyte CD38

As discussed in Section III,A, *Aplysia* cyclase is homologous to CD38, a lymphocyte surface antigen. The question of whether CD38 possesses the cyclase activity has been addressed. A soluble form of CD38 was engineered by constructing and expressing a cDNA with the hydrophobic sequence removed from the N-terminal of CD38 (Howard et al., 1993). In addition, a flag sequence was introduced to aid in subsequent purification. The resulting recombinant CD38 was a soluble protein of 40 kDa and was purified to homogeneity, first by immunoaffinity columns constructed with antibodies to the flag sequence and then by HPLC. Incubation of the purified CD38 with NAD^+ resulted in progressive production of cADPR. The unexpected finding was that the purified CD38 exhibited, in addition to the ADP–ribosyl cyclase activity, the ability to hydrolyze cADPR (Howard et al., 1993). CD38 thus appears to be a bifunctional enzyme, apparently designed to metabolize cADPR. Perhaps ectodistribution of CD38-like enzymes on cell surfaces is a general property of cells of hemopoietic origin. Indeed, in red blood cells, the cyclase and the hydrolase activity have also been observed to be on the extracellular surface, suggesting that a CD38-like enzyme may also be responsible for the activities (Lee et al., 1993b).

Another 39-kDa protein has been purified 16,000-fold from the spleen (Kim et al., 1993a). Similar to CD38, this protein is also a bifunctional enzyme catalyzing the synthesis and the hydrolysis of cADPR. The enzymatic reaction was shown to be fully reversible in the sense that, if cADPR and nicotinamide were given as substrates, the enzymes could produce NAD^+ as product. The cellular origin of this enzyme has not been determined. Since the majority of splenic cells are lymphocytes and hemopoietic cells, this bifunctional enzyme may, in fact, be CD38. The similarity of the molecule weight and catalytic properties are very suggestive.

Bifunctional enzymes are not very common. The most extensively studied mammalian bifunctional enzyme is the 6-phosphofructo-2-kinase/fructose-2,6-bisphosphatase (FK/FP) in liver. Similar to the bifunctional enzymes involved in cADPR metabolism, this enzyme also catalyzes the hydrolysis and the synthesis of a biologically important molecule, fructose-2,6-bisphosphate, the steady-state concentrations of which determine the rate of gluconeogenesis (Pilkis, 1990). The hydrolytic and synthetic activities of FK/FP are, in turn, regulated by a cAMP-dependent mechanism. Phosphorylation of FK/FP by cAMP-dependent kinase results in activation of the bisphosphatase activity and inhibition of the kinase activity. This event presumably would result in a decrease in cellular steady-state concentration of fructose-2,6-bisphosphate, leading to stimulation of gluconeogenesis (Pilkis and Granner, 1992). A similar mechanism could be involved in regulating cADPR levels in cells. The fact that the cyclase and hydrolase activities of cADPR always seem to be present together in a variety of tissues (Table I) is consistent with both activities being manifestations of a single CD38-like enzyme. That both enzyme activities are associated with membranes is also consistent with the membrane-bound nature of CD38. Investigators have observed that the production of cADPR in sea urchin egg homogenates can be stimulated by a cGMP-dependent mechanism (Section III,E). That cGMP-dependent phosphorylation of a CD38-like bifunctional enzyme in the egg homogenates could be responsible for the observation is intriguing. Indeed, examination of the sequences in Fig. 2 shows that the *Aplysia* cyclase and CD38 from mouse and human have numerous consensus phosphorylation sites for cGMP-dependent kinase on their sequences: R/KXS/T, R/KXXS/T, R/KR/KSX/T, R/KXXXS/T, and S/TXR/K (Pearson and Kemp, 1991).

Indeed, the possibility that a single CD38-like enzyme is responsible for the cyclase and the cADPR hydrolase activities observed in tissues is attractive since it could provide a unifying way of thinking about the multiplicity of enzymes involved in the metabolism of cADPR. Researchers postulate that the enzyme responsible for regulating the intracellular concentration of cADPR is a membrane-bound bifunctional enzyme such as CD38. The cyclase and hydrolase activity of this bifunctional enzyme can be modulated by mechanisms such as protein phosphorylation, which would directly lead to alteration of the cellular cADPR levels. The expression of this enzyme at the surface of B lymphocytes could be the result of specific modifications during its synthesis, such as glycosylation or linkage with associated proteins that promote surface expression. Both these mechanisms have been shown to be operative in directing surface expression of various antigens and receptors in lymphocytes (Matsuuchi et al., 1992; Tiff et al., 1992). A corollary of this scenario is that the *Aplysia* cyclase could be an aberrant form of the CD38-like enzyme, with modifications that inactivate its hydrolase activity. This may very well be the case since the *Aplysia* cyclase has been shown to be present mainly in granules of *Aplysia* eggs (Hellmich and Strumwasser, 1991). Therefore, the protein may be modified for the purpose of storage. In any case, the *Aplysia* cyclase provides a good reference for delineating the structural basis for the functional differences between CD38 and *Aplysia* cyclase. With the knowledge of the sequences of both enzymes and the availability of molecular biological techniques, this problem may not be too difficult to solve.

The physiological role of expressing a bifunctional enzyme metabolizing cADPR on the surface of lymphocytes has not been elucidated.

Ecto-expression of CD38 in B lymphocyte is known to be stage specific, appearing only in the early and late stages of differentiation (Jackson and Bell, 1990; Malavasi *et al.*, 1992). That the ecto-expression of CD38 may be related to cellular differentiation is supported by findings in HL-60 cells, a promyelocytic cell line derived from a leukemic patient. These cells can be triggered by various substances such as dibutyryl-cAMP, retinoic acid, and phorbol ester to terminally differentiate into granulocytes, monocytes, or macrophages (Breitman *et al.*, 1980; Collins, 1987). Of all these substances, only retinoic acid specifically induces surface expression of CD38 in close correlation with cellular differentiation to granulocytes (Kontani *et al.*, 1993). In addition, antibodies against CD38 can activate proliferation of B lymphocytes in the presence of other co-stimulants such as interleukin 4 (Malavasi *et al.*, 1992; Howard *et al.*, 1993). Investigators have postulated that CD38 is a surface receptor and that ligand binding may be responsible for cellular activation. Despite extensive research, no such ligand has ever been identified. The discovery of the enzymatic activities of CD38 suggests that this molecule may not be a surface receptor, but that its biological activity may reside in its ability to metabolize cADPR.

CD38 could exert its enzymatic effects either internally or externally. In the former case, one could envision that antibody binding could result in internalization of surface CD38, which could then exert its enzymatic effects intracellularly. The observation of cocapping of CD38 with the B cell antigen receptor IgM after B cell activation is consistent with such a notion (Malavasi *et al.*, 1992).

Alternatively, CD38 may exert its enzymatic effect extracellularly. Because of exocytosis or cell death, NAD^+ could be released and be available as a substrate for the production of cADPR by CD38. A necessary corollary of this scenario is that cADPR must have as yet unidentified extracellular functions, perhaps through binding to a surface receptor. This possibility is consistent with the finding that cADPR applied externally can potentiate B lymphocyte proliferation in the presence of other co-stimulants, whereas ADP–ribose produces inhibition (Howard *et al.*, 1993). Thus, the two metabolic products of CD38, when applied externally, can modulate lymphocyte differentiation in an opposite manner, further indicating that the biological function of CD38 is likely to reside in its enzymatic activities. Indeed, cADPR and ADP–ribose produced by CD38 on the surface of B lymphocytes may affect behavior of other hemopoietic cells in a paracrine fashion. This type of interaction would be especially important in lymphoid tissues such as the spleen, where B lymphocytes undergo terminal differentiation. The finding of a CD38-like bifunctional enzyme in the spleen is consistent with this notion (Kim et al., 1993a). This scenario for cADPR is analogous to that for ATP, which is normally used as an energy source intracellularly, but can be released under some conditions to exert its extracellular functions through the purinergic receptors present on the surface of cells (Cusack and Hourani, 1990), as was shown to occur in mast cells. Degranulation of mast cells releases ATP, which then acts on surface purinergic receptors and results in spreading of Ca^{2+} signals from cell to cell (Osipchuck and Cahalan, 1992). The similarity can be extended since the extracellular surface of many cells appears to have substantial ecto-ATPase and ecto-kinase activities (Dicker et al., 1980; Lin, 1989). Further, one report indicates that ATP can actually function as a fast excitatory neurotransmitter in brain (Edwards et al., 1992). Similarly, cAMP, a known intracellular second messenger, also functions extracellularly as a chemoattractant for *Dictyostelium* (Caterina and Devreotes, 1991). The extracellular and intracellular functions of these metabolites are, therefore, quite different. By analogy, the putative extracellular function of cADPR may or may not be related to its intracellular Ca^{2+}-mobilizing activity. In any case, the discovery of the enzymatic activities of CD38 is likely to have important biological implications.

D. Relationship with NAD^+ Glycohydrolase

NAD^+ glycohydrolase (EC 3.2.2.5), more commonly referred to as NADase, is classified as an enzyme that hydrolyzes the glycosyl linkage between nicotinamide and the terminal ribosyl unit of NAD^+ to produce ADP–ribose. The discoveries of cADPR and the enzymes involved in its metabolism raise an important issue concerning the classification of NADase. Clearly ADP–ribosyl cyclase and NADase are two different enzymes because they catalyze two different reactions. Both use NAD^+ as substrate, but the products are very different. ADP–ribosyl cyclase produces cADPR, whereas NADase produces ADP–ribose. cADPR hydrolase and NADase are also different because the substrates are different although the product, ADP–ribose, is the same. Indeed, a commercial NADase (Sigma, N-9629) can effectively hydrolyze NAD^+ but does not have any detectable cADPR hydrolase activity, even when assayed at 6-fold higher concentrations of the enzyme (H. C. Lee, unpublished observation). The situation with CD38-like enzymes is less clear. CD38-like enzymes are bifunctional, catalyzing the synthesis and the degradation of cADPR. The overall reaction catalyzed by these enzymes is, therefore, identical to that of an NADase. However, clearly the two reactions are different mechanistically. The bifunctional reaction involves the formation of a cADPR intermediate, whereas the NADase reaction, at least in the original classification, does not involve cADPR at all.

NADases have been known for more than 50 years. The molecular masses of various NADases range from ~25 kDa to ~90 kDa and are found in prokaryotes and eukaryotes from plants to human (see Price and Pekala, 1987, for a review). Soluble and membrane-bound NADases have been described; the latter are found in plasma membrane and in microsomes. The widely varying physical properties of NADases suggest that they do not constitute a homogeneous class of enzymes. Indeed, some NADases can catalyze a transglycosidation reaction and/or ADP–ribosylation reaction whereas others cannot (Price and Pekala, 1987). The discoveries of cADPR and the enzymes involved in its metabolism add to this complexity.

The confusion between NADase and the enzymes involved in cADPR metabolism is 2-fold. First is the mistaken identity of the reaction products. Previously used assay systems for NADases that are based on measuring generation of nicotinamide or disappearance of NAD^+ cannot distinguish between an NADase and an ADP-ribosyl cyclase. Further, ADP–ribose and cADPR can be difficult to separate on some of the chromatographic systems used to study NADase in the past, especially without knowing the existence of cADPR. Second, the combined presence of ADP–ribosyl cyclase and cADPR hydrolase in tissues can result in an overall conversion of NAD^+ to ADP–ribose and nicotinamide, a reaction identical to that catalyzed by NADase. Therefore, distinguishing between the pathways is very difficult if the enzyme preparations used are not homogeneous.

The confusion described so far is mainly technical and can be readily resolved using the current knowledge of the existence of cADPR. The confusion raised by the discovery of CD38-like enzymes is much more complicated. The substrates and the products of CD38-like enzymes and NADase are the same. The two reactions can be distinguished only if the intermediate, cADPR, and/or the hydrolase activity (determined with cADPR as substrate) of the bifunctional reaction is specifically measured. Depending on how closely matched the cyclase and the hydrolase activities of the bifunctional reaction are, the steady-state concentration of cADPR can be quite low and difficult to measure, as was shown to be the case for the lymphocyte CD38 (Howard et al., 1993) and for the ecto-activities on red cells (Lee et al., 1993b). Two examples suggest that the confusion between CD38-like enzymes and NADases may be quite pervasive.

First, researchers long have known that retinoic acid-induced differentiation of HL-60 cells results in the ecto-expression of an NADase in the membranes of the differentiated cells (Hemmi and Breitman, 1982). Kontani et al. (1993) have now identified the NADase expressed as CD38. Second, red blood cells contain an NADase on the exterior surface of the cell (Pekala and Anderson, 1978). Lee et al. (1993b) have now shown that the exterior surface of intact red cells can catalyze the synthesis and hydrolysis of cADPR. This situation may be yet another example of misclassification of a CD38-like enzyme as an NADase.

Although the reaction products of ADP–ribosyl cyclase and NADase are totally different, mechanistically the two enzymes may be closely related. Researchers generally accept that the NADase reaction proceeds via a covalent intermediate of enzyme–ADP–ribose (reviewed by Price and Pekala, 1987). Nucleophilic attack of the enzyme intermediate by water results in hydrolysis of the covalent linkage and the release of the product, ADP–ribose. Formation of cADPR by the ADP–ribosyl cyclase can also proceed via a similar mechanism. The main difference would be the accessibility of water to the active site. One can envision that, if the conformation of the ADP–ribosyl cyclase is such that the active site is secluded and inaccessible to water, the intramolecular attack from the adeninyl nitrogen would, instead, result in the formation and release of cADPR. Indeed, the formation of enzyme-bound cADPR may be an intermediate step of the reaction catalyzed by the CD38-like enzymes, as has been proposed for the splenic bifunctional enzyme (Kim et al., 1993a). Because of these similarities, all cADPR-metabolizing enzymes could be classified as NADases. The main disadvantage of doing so is that the biologically most important aspect of the reaction, namely, the metabolism of cADPR, is obscured by the emphasis on the degradative nature of the NADase. Further, this classification would also perpetuate the unnecessary confusion of these two classes of enzyme. Grouping all the enzymes involved in the metabolism of cADPR into a separate class would be more appropriate. Indeed, as long as NADase has been known, its existence and function have always been an enigma. No convincing evidence suggests its participation in any cellular functions (Pekala and Anderson, 1978). If some, if not all, of the NADases are in fact ADP–ribosyl cyclases or CD38-like bifunctional enzymes, these ubiquitous enzymes do indeed have a very important cellular function.

E. REGULATION BY CYCLIC GMP

One essential criterion needed to be fulfilled before unambiguously assigning cADPR as a second messenger, that is, the intracellular concentration of cADPR had to be demonstrated to be under the control of extracellular stimuli acting at the cell surface. By analogy to the phosphoinositide pathway, for cADPR to function as a second messenger, its synthetic enzyme ADP–ribosyl cyclase would be expected to, like phosphoinositidase C, be regulated either directly or indirectly by cell-surface receptor activation (Berridge, 1993). An unexpected insight into this problem came from experiments on sea urchin eggs that were designed to examine signal transduction mechanisms at fertilization. Whereas numerous agents including IP_3 and cADPR can mobilize intracellular free Ca^{2+} on microinjection into the sea urchin egg, the time course differs markedly from that seen during fertilization (Whitaker and Swann, 1993). However, one intriguing candidate as a sperm-induced second messenger is cyclic GMP (cGMP), since its pattern of Ca^{2+} release most closely resembles that seen at fertilization (Whalley et al., 1992).

Cyclic GMP is synthesized from GTP in reactions catalyzed by two isoforms of guanylate cyclase (Schmidt et al., 1993). The first isoform is particulate and constitutes a family of cell-surface receptors possessing cytoplasmic domains that have intrinsic guanylate cyclase activity while the extracellular domains are capable of binding a variety of peptide hormones, including atrial natriuretic factor and a chemoattractant peptide secreted from eggs. The binding of these peptides to the receptor domain results in enhancement of the guanylate cyclase activity (Schulz et al., 1989). The second form of guanylyl cyclase is soluble; its enzymatic activity is stimulated by the gaseous transmitters nitric oxide and carbon monoxide (Lowenstein and Snyder, 1992).

The effects of cGMP on Ca^{2+} release in sea urchin eggs have been examined at the level of the intact cell and of cell homogenates. When microinjected into sea urchin eggs, cGMP causes a Ca^{2+} transient similar to that seen at fertilization (Whalley et al., 1992). First, a latency of ~20 sec occurs before the Ca^{2+} transient is initiated; second, the time course of the Ca^{2+} transient is more protracted than that induced by IP_3 (Swann and Whitaker, 1986) or cADPR (Dargie et al., 1990) which, similar to that induced by the fertilizing sperm, can last for as long as 10 min. The action of cGMP involves mobilization from intracellular stores, since extracellular Ca^{2+} is not required for the response, and appears to be indirect, since it failed to release Ca^{2+} in digitonin-permeabilized eggs (Whalley et al., 1992). Further, the effect in intact eggs was not blocked by heparin, indicating that its mode of action was distinct from the IP_3 receptor. However, Ca^{2+} mobilization by extracellular application of the membrane permeant analog dibutyryl-cGMP in intact eggs was blocked by prior microinjection of the ryanodine receptor antagonist ruthenium red, indicating that the ryanodine receptor is an essential element involved in the response to cGMP (Galione et al., 1993b). In sea urchin egg homogenates, cGMP was shown to release Ca^{2+} only in the presence of low concentrations (10 μM) of β-NAD^+, the precursor of cADPR synthesis catalyzed by ADP–ribosyl cyclase. A hallmark of cGMP-induced Ca^{2+} release in egg homogenates was the long latency before release occurred which, for 60 μM cGMP in the presence of β-NAD^+ (10 μM), was as long as 30 sec (Galione et al., 1993b). This result is consistent with an indirect action of cGMP on the Ca^{2+}-release mechanism. The involvement of the IP_3 receptor and IP_3 in mediating this effect was ruled out by the finding that this effect of cGMP was insensitive to inhibition by heparin and could be observed in homogenates desensitized by pretreatment with high concentrations of IP_3. Prior release of Ca^{2+} by the ryanodine receptor agonists, caffeine, and ryanodine rendered the homogenates unresponsive to cGMP, providing further support for the conclusion that cGMP-stimulated Ca^{2+} release was via a ryanodine receptor, as observed in intact eggs. That cADPR might be mediating the cGMP effect was shown by the fact that cGMP also failed to mobilize Ca^{2+} from homogenates desensitized by a high concentration of cADPR. The requirement of NAD^+ for the action of cGMP suggested that cGMP could act by promoting the conversion of NAD^+ to cADPR. Analysis of [^{14}C]NAD^+ metabolism by thin layer chromatography (TLC) in egg homogenates demonstrated that cGMP stimulated the conversion of NAD^+ to a metabolite that comigrated with ADP–ribose, the hydrolytic product of cADPR (Galione et al., 1993b). This result, in conjunction with the fact that cGMP had no effect on the rate of [^3H]cADPR hydrolysis to ADP–ribose, supported the idea that cGMP activates ADP–ribosyl cyclase, perhaps through cGMP-dependent protein phosphorylation, resulting in enhanced conversion of NAD^+ to cADPR. Whether cGMP-dependent protein kinase (Hoffman et al., 1992) directly phosphorylates ADP–ribosyl cyclase has yet to be demonstrated.

Since the activation of a number of cell-surface receptors leads to activation of guanylate cyclase, either directly or via the production of intermediate messengers such as nitric oxide (Schmidt et al., 1993), this pathway may be a way in which cell-surface events are linked to cADPR production. This pathway could also lead to regenerative Ca^{2+} signals such as propagating waves and oscillations, since many possible sites for interaction between cGMP and Ca^{2+} are available (Schmidt et al., 1993). In particular, nitric oxide synthesis by nitric oxide synthases are Ca^{2+}–calmodulin dependent (Bredt and Snyder, 1990). The scheme would involve initiation by nitric oxide which, through cGMP, activates cADPR production and results in Ca^{2+} mobilization. The rise in Ca^{2+} can lead to further nitric oxide production through the calmodulin pathway and, hence, to further Ca^{2+} mobilization. The negative feedback loop of the scheme is provided by inhibited guanylate cyclase as a result of increases in intracellular Ca^{2+} (Schmidt et al., 1993), making the pathway more likely to generate intercellular Ca^{2+} waves (Sanderson et al., 1990). Alternatively, the presence of positive and negative feedback interactions between cGMP and Ca^{2+} might lead to oscillations in intracellular Ca^{2+}. Indeed, nitric oxide-induced mobilization of Ca^{2+} from ryanodine-sensitive stores has been observed in interstitial cells of the mammalian colon and has been proposed as a mechanism for the amplification of nitric oxide production in the mammalian gut (Publicover et al., 1993). Whether cADPR is an intermediate in this reaction remains to be shown.

IV. Cyclic ADP-Ribose-Dependent Ca^{2+} Release

A. Sea Urchin Egg as a Model System

The sea urchin egg has long been a favorite model system for studying regulation of cellular functions by ionic mechanisms. Unfertilized eggs released into seawater are in a dormant state characterized by a depressed metabolism and a slightly acidic internal pH. In this state, the eggs have only a limited lifespan of 1–2 days. After fertilization, the eggs undergo a dramatic transformation and convert to a reductive state conducive to synthetic activities. This change is followed by resumption of protein and DNA synthesis, and the developmental program is set into motion. The fertilized eggs eventually develop into adults. In this sense, they will achieve immortality. All these changes that eventually immortalize a dying cell are, in fact, triggered by two ionic changes: alkalinization of internal pH and transient elevation of internal Ca^{2+} (Epel, 1980). The pH change is a result of activation of a Na^+/H^+ exchanger in the plasma membrane. The increase in internal Ca^{2+} is caused by mobilization of intracellular stores since external Ca^{2+} is not required. The immediate consequence of the Ca^{2+} elevation is the induction of a massive exocytosis, the cortical reaction, resulting in the formation of the fertilization envelope. This proteinaceous membrane represents a mechanical block to polyspermy.

Studies show that the IP_3 and the cADPR receptor are involved in mediating the Ca^{2+} mobilization associated with fertilization (Galione et al., 1993a; Lee et al., 1993a). Since the process is initiated by an external stimulus, the sperm, interacting with a surface receptor, it is analogous to other ligand–receptor mediated events. Conceptually different are the five other Ca^{2+} transients that occur afterward. These Ca^{2+} changes do not require external stimuli and are temporally correlated with various developmental events such as pronucleus migration, the streak stage, nuclear membrane breakdown, metaphase, and cleavage (Poenie et al., 1985). The source of Ca^{2+} for these later transients also appears to be the internal stores, since the early development of the embryo, similar to the fertilization process, can proceed normally for at least the first seven cleavages in the absence of external Ca^{2+} (Schmit et al., 1982). The second messengers responsible for mediating these later Ca^{2+} changes have not been identified. Once the egg is fertilized, it will develop in total isolation. Therefore, all the developmental events, in conjunction with the multiple Ca^{2+} transients associated with them, are likely to be controlled internally by mechanisms that are turned on at fertilization. The initial and the later Ca^{2+} transients are fundamentally different, so different second messenger systems are likely to be involved. Clearly the Ca^{2+}-regulation systems in the egg are complex, and multiple pathways are present. The egg is thus a good model system for elucidating various mechanisms for Ca^{2+} mobilization.

In addition to having multiple Ca^{2+}-regulation systems, the sea urchin egg also offers several technical advantages as a model system. Large quantities of homogeneous cells can be obtained readily for biochemical analyses. Various preparations have been developed for investigation at different levels and aspects of Ca^{2+} regulation. At the whole-cell level, the eggs are large and sturdy enough for various micromanipulations, such as microinjection. Indeed, the first demonstration of the functional presence of the IP_3-dependent system in sea urchin eggs was done by microinjecting IP_3 (Whitaker and Irvine, 1984). Microinjection of various specific antagonists has been employed to obtain the first direct evidence for a physiological role for cADPR in the Ca^{2+}-signaling process in the eggs (Lee et al., 1993a).

A permeabilized cell system using electroporation has been developed to allow access to the cell interior without microinjection (Clapper and Lee, 1985; Swezey and Epel, 1989). The procedure can be applied to a population of cells and, depending on whether Ca^{2+} is present or not, the resealing of the permeabilized eggs can be controlled (Swezey and Epel, 1989). Using this method, various metabolites including IP_3 can be delivered into the eggs.

At the subcellular level, the cortical and homogenate preparations are the most useful for investigating mechanisms of Ca^{2+} mobilization. The cortical preparation was originally developed for studying the dependency of the exocytosis reaction of cortical granules on Ca^{2+} (Vacquier, 1975). Researchers showed that calmodulin is needed to confer Ca^{2+} sensitivity on the exocytosis reaction. The presence of an inhibitory antibody to calmodulin raised the Ca^{2+} concentration required for triggering exocytosis from 5 μM to more than 1 mM (Steinhardt and Alderton, 1982). The cortical preparation has been adapted for investigation of Ca^{2+} uptake and release by cortical endoplasmic reticulum (Terasaki and Sardet, 1991).

The egg homogenate preparation was developed specifically for studying the mechanisms of Ca^{2+} mobilization (Clapper and Lee, 1985; Clapper et al., 1987). The preparation can be used without further fractionation, providing a good representation of most, if not all, of the Ca^{2+}-release mechanisms in the cell. Detailed characterization of this preparation led to the discovery of cADPR. In addition to the IP_3- and cADPR-dependent Ca^{2+}-release systems, a third independent system is sensitive to NADP (Clapper et al., 1987). This Ca^{2+}-release system has been not yet been fully characterized. Clearly the homogenate preparation has the simplicity and accessibility of a cell-free system, yet it contains a wealth of possibilities ready to be explored.

To date, information obtained from the homogenate preparation has been applicable not only on the whole-egg level but also to other cell systems. Thus, cADPR was discovered in the *in vitro* preparation and has been shown to be active in intact eggs (Dargie et al., 1990; Rusinko and Lee, 1989) as well as in various mammalian systems (Section IV,D). Conversely, the Ca^{2+}-mobilizing effect of cGMP was first demonstrated in intact eggs (Whalley et al., 1992) and later verified in the homogenate system (Galione et al., 1993b). More importantly, the simplicity of the *in vitro* system has allowed the elucidation of the relationship between the cGMP effect and the cADPR-dependent Ca^{2+}-release mechanism (Galione et al., 1993b). The ability to address the same question at various levels of complexity from organelle to intact cell has been an invaluable asset of the sea urchin egg as a model system, and promises to be an important avenue that will advance our understanding of the mechanisms of Ca^{2+} mobilization.

Another important advantage of the homogenate preparation is that it allows identification and purification of various components of a complex pathway such as Ca^{2+} mobilization. The plasma membrane and the associated cortical granules can be easily removed as large complexes by low speed centrifugation, since the gentle homogenization largely preserves their association. Additional fractionation of egg homogenates can be conveniently accomplished by Percoll density centrifugation, which cleanly separates the Ca^{2+} stores from mitochondria and yolk granules (Clapper and Lee, 1985; Clapper et al., 1987; Lee, 1991). All the IP_3- and cADPR-sensitive Ca^{2+}-release activities, as well as the cADPR binding activity, are associated with the Ca^{2+} stores, which co-purified with glucose-6-phosphatase activity, a marker for endoplasmic reticulum. The separation of the Ca^{2+} stores on the Percoll gradient was found to be ATP dependent. In the absence of ATP, some of the microsomes lose the accumulated Ca^{2+} and migrate as a diffuse band. With ATP, the Ca^{2+} is retained and a narrow microsomal band results (Clapper and Lee, 1985). This property can potentially be exploited to separate different Ca^{2+} stores further. For example, the IP_3-sensitive stores could be discharged first, so they would migrate differently on the Percoll gradient than the IP_3-insensitive stores. In any case, the egg homogenate system should prove to be invaluable for purification and reconstitution of the Ca^{2+}-release pathways.

B. Ca^{2+} STORES

1. *IP_3- and Ryanodine-Sensitive Ca^{2+} Stores*

The seminal observation that IP_3 acts by mobilizing internal Ca^{2+} has focused much attention on intracellular Ca^{2+} stores. Cell fractionation studies measuring binding or functional parameters of exogenously applied pharmacological probes have revealed that Ca^{2+}-release mechanisms are predominantly microsomal. Some question has remained about whether the main location of these channels is on the general endoplasmic reticulum or in a specialized region called the calciosome (Volpe et al., 1988). One approach for investigating the subcellular localization of various Ca^{2+}-release channels employs single-cell Ca^{2+} imaging to examine spatial aspects of Ca^{2+} signals activated by pharmacological agonists of these channels (Burgoyne et al., 1989). Another successful approach is electron microscopy in conjunction with specific immunocytochemical probes of the channels (Sharp et al., 1993). The results indicate that the IP_3- and ryanodine-sensitive stores overlap to varying degrees, depending on cell types. No overriding principle governing their subcellular distribution seems to exist (Meldolesi et al., 1990). At extremes, certain cells have only a single type of Ca^{2+}-release channel. For example, *Xenopus* and hamster eggs and oocytes seem to contain only the IP_3-sensitive channels (Miyazaki, et al., 1992; Galione et al., 1993a; Kume et al., 1993), whereas cardiac and skeletal muscle mainly express the ryanodine receptors

(Fleisher and Inui, 1989). Most cells, however, including sea urchin eggs (Galione *et al.*, 1993a), neurons (Sharp *et al.*, 1993), and smooth muscle (Chen and van Breeman, 1992), contain both types of channels. In cells expressing dual Ca^{2+}-release channels, some appear to contain a single store with two release mechanisms (Galione *et al.*, 1993a), whereas others appear to have different mechanisms distributed on separate Ca^{2+} pools (Burgoyne *et al.*, 1989). In hippocampal neurons, ryanodine receptors are found in axons and dendritic spines whereas IP_3 receptors are found mainly in cell bodies; both channels colocalize to dendritic shafts (Sharp *et al.*, 1993).

2. *Cyclic ADP–Ribose-Sensitive Ca^{2+} Stores*

The nature of the cADPR-sensitive Ca^{2+} pools has been investigated mainly in sea urchin eggs. Fractionation of egg homogenates by Percoll density centrifugation showed that cADPR-sensitive Ca^{2+} stores were well separated from mitochondria, but comigrated with a marker enzyme of the endoplasmic reticulum, glucose-6-phosphatase (Clapper *et al.*, 1987; Lee, 1991). The nonmitochondrial nature is consistent with the fact that cADPR-sensitive Ca^{2+} release is not affected by mitochondrial inhibitors (Clapper *et al.*, 1987; Dargie *et al.*, 1990). Several lines of evidence suggest that the IP_3- and cADPR-sensitive pools substantially overlap (Dargie *et al.*, 1990). First, IP_3- and cADPR-sensitive microsomes comigrate during fractionation using Percoll density centrifugation. Second, IP_3- and cADPR-induced Ca^{2+} releases are nonadditive. Finally, an inverse relationship between cADPR- and IP_3-induced Ca^{2+} release was detected. Sequential additions of the two agents to egg homogenates showed that, if one agent had released more Ca^{2+}, less was available for the other to release. In all cases, the total amount of releasable Ca^{2+} was similar. These results are consistent with two overlapping Ca^{2+} pools.

An interesting property observed in egg homogenates is that high concentrations of cADPR can desensitize the microsomes so subsequent additions of cADPR would not produce more Ca^{2+} release (Clapper *et al.*, 1987). The desensitization is not the result of elevation of Ca^{2+} concentration since the effect persists even after the released Ca^{2+} is resequestered. The desensitized microsomes, however, are fully responsive to other Ca^{2+}-release agents such as IP_3. This desensitizing effect is not unique to cADPR, since similar results can be obtained with IP_3 (Clapper and Lee, 1985). Further, researchers have shown that the sensitivity of the refractory microsomes to IP_3 could be restored only if the vesicles were washed to remove IP_3. A simple interpretation of these results would be that cADPR and IP_3 are releasing

Ca^{2+} from two separate Ca^{2+} pools. This interpretation is, however, in contradiction with other results described earlier showing that the two pools overlap. However, unlike in many mammalian systems (Ferris and Snyder, 1992), the Ca^{2+}-release channels in sea urchin egg microsomes could potentially be desensitized by their own ligands in a Ca^{2+}-independent manner. Irrespective of the exact mechanism, this apparent desensitization has proved to be a very useful property of the microsomes, and has been used to demonstrate the presence of multiple Ca^{2+}-release mechanisms in egg microsomes (Clapper et al., 1987). The apparent desensitization has since been used to show the specificity of the bioassay for cADPR (Walseth et al., 1991), to identify a ryanodine receptor-like channel as mediator of the cADPR-dependent Ca^{2+} release (Galione et al., 1991), and to demonstrate the production of a cADPR-like metabolite in pancreatic β cells following glucose treatments (Takasawa et al., 1993).

The question of Ca^{2+} pools has also been addressed at the intact egg level. Microinjection of IP_3, the nonhydrolyzable analog of IP_3, totally discharged the Ca^{2+} stores in the eggs; subsequently, not even Ca^{2+} ionophore could release more Ca^{2+} (Whalley et al., 1992). This result indicates only a single Ca^{2+} store is present in the eggs and is consistent with the results obtained from egg homogenates described earlier. In a separate study, the Ca^{2+} release at fertilization was found to be unaffected by heparin or prior activation of the IP_3 pathway with GTP-γ-S. Reseachers concluded that multiple stores of Ca^{2+} are released at fertilization (Rakow and Shen, 1990; Crossley et al., 1991). Strictly speaking, these results only suggest the presence of multiple Ca^{2+}-release mechanisms in the eggs, and not necessarily multiple stores.

Indeed, researchers now know that IP_3- and cADPR-sensitive Ca^{2+}-release mechanisms are involved in mediating the Ca^{2+} release at fertilization (Lee et al., 1993a). The mobilization of Ca^{2+} from a cADPR-sensitive store is supported by data showing that, after the fertilization-induced Ca^{2+} transient, cADPR-sensitive Ca^{2+} release appears refractory (Shen and Buck, 1993). Further, results using Ca^{2+} imaging showed a redundancy of Ca^{2+}-release mechanisms involved in the Ca^{2+} wave at fertilization; IP_3 or ryanodine was independently able to initiate and propagate the wave. Although the amplitude of the wave is smaller when relying on a single mechanism, the time course and spatial aspect look remarkably similar, suggesting that the IP_3 and ryanodine receptors employed in Ca^{2+} signaling at fertilization have a similar and homogeneous distribution in the egg (Galione et al., 1993b). IP_3 and cADPR injections into sea urchin eggs have been reported to induce similar Ca^{2+} waves (Shen and Buck, 1993). However, a different spatiotemporal pattern of Ca^{2+} release was observed for cADPR- and ryanodine-induced responses. The difference may reflect activation of different ryanodine receptors or differences in the mechanism of activation of the same channels by these two agents since, although cADPR and ryanodine appear to release Ca^{2+} by activating the same channels, the time course of release induced by these agents is very different (Galione et al., 1991).

All these results indicate the definite existence of multiple Ca^{2+}-release mechanisms in the eggs. Whether these mechanisms are distributed on different Ca^{2+} stores is not totally clear. Fluorescence microscopy using a lipophilic dye to stain the endoplasmic reticulum of the eggs suggests that the entire reticulum in the eggs is an interconnected compartment (Terasaki and Jaffe, 1991), consistent with the idea of a single Ca^{2+} store. Further, researchers found that the endoplasmic reticulum became fragmented and then reformed during the first 1–3 min after fertilization (Jaffe and Terasaki, 1993), temporally correlated with the Ca^{2+} changes at fertilization. This extensive reorganization of the internal membrane network and, therefore, the Ca^{2+} pools could be caused by the Ca^{2+} changes. The possibility that the Ca^{2+} stores are reshuffled following egg activation should be taken into account in interpreting some of the effects on Ca^{2+} stores in intact eggs.

Immunocytochemical staining has been used to localize the distribution of ryanodine receptor-like proteins in sea urchin eggs. Biochemical studies using an antibody raised against skeletal muscle ryanodine receptors reveal cross-reactivity with a protein of 380 kDa in the eggs (McPherson et al., 1992), smaller than the major known subtypes of ryanodine receptors (McPherson and Campbell, 1993). This protein seems to be localized to the cortical regions of the egg. However, its relationship to the cADPR receptor remains in doubt since ryanodine-mediated Ca^{2+} waves can propagate right through the egg cytoplasm and are not restricted to the cortex (Galione et al., 1993b). Whether other subtypes of ryanodine receptors are more uniformly distributed in this cell remains to be shown.

C. Relationship with Ca^{2+}-Induced Ca^{2+} Release

The concept of Ca^{2+}-induced Ca^{2+} release (CICR) originated from studies on excitation–contraction coupling in muscles. Investigators found that a small rapid increase in extravesicular Ca^{2+} could cause a much larger release of stored Ca^{2+} from skinned muscle fibers or sarcoplasmic reticulum vesicles (Fleischer and Inui, 1989). CICR is believed to be mediated by the ryanodine receptors/Ca^{2+} channels, the major Ca^{2+}-release channels of muscle sarcoplasmic reticulum (McPherson and Campbell, 1993; Sorrentino and Volpe, 1993). These channels share substantial similarities with IP_3 receptors because they are transmembrane proteins with large cytoplasmic domains and a quatrefoil arrangement of four identical subunits (Fleischer and Inui, 1989). A key property of ryanodine receptors, first recognized from studies on isolated sarcoplasmic reticulum, is that they can be activated by Ca^{2+}, thus providing a molecular explanation for the phenomenon of CICR. This property was later confirmed by channel isolation and reconstitution in artificial bilayers (Fleischer and Inui, 1989). Ryanodine receptors have now been shown to be widespread in many cell types, and have been incorporated into many models to explain the spatial and temporal complexity of cellular Ca^{2+} signals that are often oscillatory (Fleisher and Inui, 1989) and propagate as regenerative waves or sometimes as spiral waves (Lechleiter *et al.*, 1991), phenomena collectively referred to as Ca^{2+} excitability.

Four different kinds of ryanodine receptors have been documented to date. The first one to be purified and characterized in molecular terms was the Type 1 ryanodine receptor from skeletal muscle sarcoplasmic reticulum, where it plays a major role in excitation–contraction coupling (Fill and Coronado, 1988). These receptors are massive structures, consisting of four identical transmembrane subunits each of ~560 kDa. The purified receptors could be visualized as quatrefoil or clover-leaf structures by electron microscopy (Fleischer and Inui, 1989). The cytoplasmic domain is very large. This domain is thought to interact with the dihydropyridine receptor in the sarcolemma, which is believed to confer voltage sensitivity onto Ca^{2+} release from the sarcoplasmic reticulum. This protein–protein interaction allows for rapid activation of the Ca^{2+} release, leading to a fast contractile response as demanded of skeletal muscle, without the apparent involvement of diffusable cytoplasmic messengers. Thus, the type 1 receptor may be atypical of ryanodine receptors because of this important specialization in its regulation.

Three other forms of ryanodine receptor have been sequenced. The type 2 receptor was characterized from cardiac muscle and is also thought to be found in the brain (Lai *et al.*, 1992). The widely distributed type 3 form was partially sequenced from an epithelial cell line, where its expression is governed by growth factors (Gianini *et al.*, 1992). This receptor was later sequenced from the brain in its entirety (Hakamata *et al.*, 1992). A fourth type of ryanodine receptor has been reported that is much smaller than the other three isoforms and is a brain specific transcript of the type 1 gene. This transcript encodes a protein of approximately 75 kDa, probably representing the transmembrane domain of the type 1 receptor (Takeshima et al., 1993).

The most commonly used method for demonstrating the functional presence of CICR in cells is based on the pharmacology of ryanodine receptors/Ca^{2+} channels. Caffeine and ryanodine are two known agonists for the channel. Caffeine is believed to act by sensitizing the channel to Ca^{2+} (Rousseau and Meissner, 1989). The action of ryanodine is more complicated. Low concentrations of ryanodine lock the Ca^{2+}-release channel into an open state with lower conductance, whereas higher concentrations close the channel (Smith et al., 1988). Two commonly used antagonists of the channel are ruthenium red and procaine (Fleischer and Inui, 1989).

Cyclic ADP–ribose has been shown to mobilize Ca^{2+} by a mechanism independent of the IP_3 receptor (Clapper et al., 1987; Dargie et al., 1990). Similarities between cADPR-sensitive and ryanodine/caffeine-sensitive Ca^{2+} release in sea urchin egg homogenates was examined using a pharmacological analysis. Ryanodine and caffeine were shown to release Ca^{2+} in egg homogenates in a concentration-dependent manner (Galione et al., 1991). The sarcoplasmic reticulum Ca^{2+}-release inhibitors, procaine and ruthenium red, selectively blocked Ca^{2+} release induced by caffeine but not by IP_3. Further, submaximal concentrations of caffeine sensitized the Ca^{2+}-release mechanism so addition of Ca^{2+} could induce further Ca^{2+} release, directly demonstrating the functional presence of CICR in egg microsomes. This phenomenon was later confirmed in a more detailed study (Lee, 1993a). The pharmacology of Ca^{2+} release from sea urchin egg microsomes was mirrored in studies in intact eggs. Microinjection of ryanodine into the intact eggs also could induce Ca^{2+} changes (Galione et al., 1991; Sardet et al., 1992). Also, cortical exocytosis in intact eggs, an index for intracellular Ca^{2+} changes, could be induced by treatments with ryanodine and caffeine and blocked by procaine and ruthenium red (Fujiwara et al., 1990). These results demonstrate that the CICR mechanism is present and functional in sea urchin eggs.

That cADPR-sensitive Ca^{2+} release is mediated by a CICR mechanism was shown by the fact that ruthenium red and procaine selectively blocked the cADPR- but not the IP_3-induced release in egg microsomes (Galione et al., 1991). Pretreatments with high concentrations of cADPR densitized the microsomes to subsequent additions of cADPR or the CICR agonists ryanodine and caffeine. The microsomes were, however, fully responsive to IP_3. High concentrations of caffeine and ryanodine also desensitized the microsomes to cADPR but not to IP_3. Additional support comes from the finding that low concentrations of cADPR that are not sufficient to release Ca^{2+} by themselves can, nevertheless, sensitize the Ca^{2+}-release mechanism so divalent cations such as Ca^{2+} and Sr^{2+}, but not Ba^{2+}, at 10- to 20-fold lower concentrations can induce further Ca^{2+} release (Lee, 1993a). In this respect, cADPR appears to be functioning very similarly to caffeine, except that it is 5 orders of magnitude more effective. Indeed, cADPR and caffeine can cross-sensitize each other. Thus, subthreshold concentrations of caffeine can reduce the half-maximal effective concentration of cADPR ~ 10-fold. The potency of caffeine can be similarly enhanced by low concentrations of cADPR (Lee, 1993a). These results provide convincing evidence that cADPR-sensitive Ca^{2+} release is mediated by a CICR mechanism.

Results from several other cellular systems have since confirmed the conclusion obtained with the egg microsomes. For examples, researchers showed that cADPR at nanomolar concentrations, but not IP_3, mimics caffeine in inducing oscillations in a Ca^{2+}-dependent ion current that is known to reflect intracellular Ca^{2+} oscillations in rat dorsal root ganglion cells (Currie et al., 1992). In three other microsomal systems derived from brain (Meszaros et al., 1993; Takasawa et al., 1993; White et al., 1993), pancreatic β cells (Taksawa et al., 1993), and cardiac cells (Mezsaros et al., 1993), the cADPR-induced Ca^{2+} release has been shown to be from a ryanodine-sensitive but IP_3-insensitive Ca^{2+} pool. Further, cADPR has been demonstrated to activate single cardiac ryanodine (type 2) receptors (Meszaros et al., 1993).

The finding that cADPR can sensitize the Ca^{2+}-release system to Ca^{2+} suggests that it may have two modes of action in cells (Lee, 1993a). The first would involve an agonist-induced production of intracellular cADPR, which binds to its receptor and activates Ca^{2+} release. This mode of action would be like the second messenger function of IP_3. The second mode would not require changes in cADPR levels but only its presence, when it can function in a permissive role by sensitizing the release system to Ca^{2+} so CICR can occur. In this case, the activator would be Ca^{2+} itself.

Investigators long have thought that CICR is a unique property of the ryanodine receptor. However, studies have shown that, under appropriate conditions, the IP_3 receptor also can function like a CICR channel. Pertinent to this idea is the observation that the IP_3-sensitive Ca^{2+} release is strongly Ca^{2+} dependent and, in fact, IP_3 and Ca^{2+} are coagonists of the IP_3 receptor (Bezprozvanny et al., 1991; Finch et al., 1991; Iino and Endo, 1992). Under conditions when IP_3 levels do not change, rapid elevations in the concentration of intracellular Ca^{2+} can trigger release of Ca^{2+} from intracellular stores (DeLisle and Walsh, 1992). Since in many cases this release is blocked by the IP_3 receptor antagonist heparin, researchers have suggested that the IP_3 receptor can be activated by a rise in Ca^{2+}, that is, that it has the property of CICR (Bezprozvanny et al., 1991; Finch et al., 1991; Iino and Endo, 1992).

Because of the findings just described, the understanding of Ca^{2+} mobilization mediated by IP_3 and ryanodine receptors has come full circle. Previously, investigators thought that the IP_3 receptor was a ligand-gated Ca^{2+} channel with no role in mediating CICR, whereas while the ryanodine receptor was solely responsible for CICR and was not gated by endogenous ligand. This dichotomy has now given way to a striking symmetry, in which both channels are involved in CICR and both have their own endogenous ligands. This functional symmetry is certainly harmonious with the similarity in molecule sizes and the sequence homology of the two receptors (Furuichi et al., 1989). However, the underlying mechanism of how these two channels are activated by their respective ligands may be quite different. In contrast to the IP_3 action, increasing evidence suggests that cADPR may not act directly on the ryanodine channel, but that its activating effect may be mediated by accessory proteins (Walseth and Lee, 1993; Walseth et al., 1993). This aspect of the cADPR action is discussed in more detail in Section V.

D. MAMMALIAN SYSTEMS

Since the discovery of cADPR and the detailed analysis of cADPR-induced Ca^{2+} release in sea urchin eggs (Clapper et al., 1987; Dargie et al., 1990), several studies have demonstrated cADPR-induced Ca^{2+} mobilization in various mammalian cell types. The first indication that cADPR may be a general Ca^{2+}-mobilizing agent came from the demonstration that cADPR was present in a wide range of mammalian tissue extracts (Walseth et al., 1991), and from the findings that ADP–ribosyl cyclase and cADPR hydrolyase are common enzymes in mammalian systems (Rusinko and Lee, 1989; Lee and Aarhus, 1993). Further, the demonstration that cADPR exerts its effects through a ryanodine-sensitive Ca^{2+}-release channel not only provided a *raison d'etre* for an additional Ca^{2+}-mobilizing messenger to IP_3, but also indicated that the most likely cells in which cADPR would be effective were those displaying ryanodine-sensitive Ca^{2+} release (Galione, 1992). The realization that ryanodine receptors are widespread offered a number of cellular systems for testing cADPR action. To date, five different mammalian cell preparations have been found to contain cADPR-sensitive Ca^{2+}-release activity in their microsomal fractions. Since cADPR is membrane impermeable, the majority of these studies have been performed on microsomes derived from these cells or in permeabilized cell suspensions, but one has been performed on intact cells with cADPR introduced by the whole-cell patch technique. In all these studies, the pharmacology of cADPR-induced Ca^{2+} release has been shown to be quite different from that of the IP_3 pathway. Where tested, cADPR-induced Ca^{2+} release appears to be via a ryanodine-sensitive mechanism, as demonstrated first in the sea urchin egg.

1. *Pituitary Cells*

The pituitary cell line GH_4C_1 contains at least two separate intracellular Ca^{2+} stores. One is sensitive to IP_3 and can be mobilized by the inositol lipid coupled cell-surface receptor for thyrotropin-releasing hormone (Wagner *et al.*, 1993), whereas the other is sensitive to caffeine and displays CICR, which is important for spontaneous and thyrotropin-releasing hormone-induced Ca^{2+} oscillations. cADPR has been shown to release Ca^{2+} in a dose-dependent manner from permeabilized GH_4C_1 cells at nanomolar concentrations (Koshiyama *et al.*, 1991). cADPR-induced Ca^{2+} release in these cells is via a mechanism independent of the IP_3 receptor since it persists after IP_3-sensitive stores are depleted and is unaffected by heparin. Further, GH_4C_1 cells appear to contain ADP–ribosyl cyclase since incubation of β-NAD^+ with an extract from pituitary cells for 15 min results in the formation of a Ca^{2+}-mobilizing activity that cross-desensitizes with authentic cADPR. Whether cADPR mobilizes Ca^{2+} from the caffeine-sensitive pool in these cells has yet to be demonstrated.

2. *Dorsal Root Ganglion Cells*

Rat dorsal ganglion neurons contain a caffeine-sensitive Ca^{2+} store that is relatively insensitive to IP_3. The presence of a robust caffeine-sensitive Ca^{2+}-release mechanism made this cell a perfect mammalian cell in which to test the generality of cADPR as a Ca^{2+}-mobilizing agent from caffeine- and ryanodine-sensitive stores.

Caffeine treatment of rat dorsal ganglion cells under whole-cell voltage clamp yields a stereotypic response of activating a series of oscillations of cationic currents across the plasma membrane (Currie *et al.*, 1992). These currents are exquisitely sensitive to increases in intracellular free Ca^{2+} concentrations and can be considered a physiological index of the intracellular free Ca^{2+} levels. Since these currents persist in the absence of extracellular Ca^{2+} and are abolished by intracellular application of the Ca^{2+} chelator EGTA, researchers concluded that oscillatory currents induced by caffeine reflect pulsatile Ca^{2+} release from intracellular stores. Intracellular application of IP_3 can occasionally give rise to a single short-lived activation of the cation conductance but rarely to oscillatory currents (Currie et al., 1992). Intracellular application of cADPR at pipette concentrations as low as 10 nM generated the Ca^{2+}-dependent oscillatory currents. This activity was enhanced by Ca^{2+} loading of the cells, consistent with the proposed enhancing effects of cADPR on CICR (Lee, 1993a). Ca^{2+} released by cADPR was also shown to be from the same intracellular pool as that released by caffeine, since the effects of cADPR were abolished in cells that had had their caffeine-sensitive pools emptied by prior incubation with caffeine in Ca^{2+}-free media. The Ca^{2+}-dependent currents induced by cADPR could be restored on reintroduction of Ca^{2+} into the extracellular solution. Interestingly, in these cells, the responses to caffeine and cADPR were not abolished by ruthenium red at 50 µM, which inhibits caffeine- and cADPR-induced Ca^{2+} release in sea urchin eggs.

3. Brain Microsomes

Three separate reports of cADPR-induced Ca^{2+} release from brain microsomes have been made, all indicating that cADPR is acting via a ryanodine-sensitive release mechanism. The first indication that cADPR might have a role in Ca^{2+} signaling in neural tissue came from the finding that nanomolar levels of cADPR were present in rat brain tissue (Walseth et al., 1991), and that brain tissue has the highest activities of ADP–ribosyl cyclase and cADPR hydrolase of any mammalian tissue examined to date (Lee and Aarhus, 1993).

Cerebellar microsomes have IP_3- and cADPR-sensitive Ca^{2+}-release mechanisms. Ca^{2+} release included by cADPR was not affected by heparin but was abolished by prior desensitization of the microsomes with high concentrations of ryanodine or cADPR, but not with IP_3. Maximal Ca^{2+} release occurred at ~400 nM cADPR, whereas maximal release by IP_3 was at 1 µM, although IP_3 released nearly twice as much Ca^{2+} (Takasawa et al., 1993). In another study (White et al., 1993), cADPR was demonstrated to release Ca^{2+} from whole brain microsomes in a dose-dependent manner. Release was abolished by heat inactivation of cADPR (Dargie et al., 1990). As for the cerebellar microsomes, release of Ca^{2+} by cADPR was abolished by prior treatment of the vesicles with ryanodine. Similar results were obtained when measuring Ca^{2+} release using the fluorescent indicator, fluo-3, or using $^{45}Ca^{2+}$ efflux.

In the third study, brain microsomes were passively loaded with

$^{45}Ca^{2+}$, which allows tight control over extra- and intravesicular Ca^{2+} concentrations (Meszaros et al., 1993). Here, cADPR increased the rate of $^{45}Ca^{2+}$ efflux, an effect that was abolished by ryanodine. The half-maximal effective concentration for cADPR-induced Ca^{2+} release was approximately 110 nM. In contrast to all other positive studies with cADPR using microsomal preparations, this study used an isolation medium based on sucrose rather than the N-methyl glucamine-containing medium originally designed to mimic intracellular ionic conditions (Clapper et al., 1987).

4. Pancreatic β Cells

Another study suggests that cADPR can mobilize Ca^{2+} from intracellular stores in pancreatic β cells and may have an obligatory role in stimulus-secretion coupling in this cell type (Takasawa et al., 1993). cADPR releases Ca^{2+} from rat β cell microsomes in a dose-dependent manner with a threshold dose of ~90 nM and a maximal release at 500 nM. The release mechanism desensitized in response to repeated applications of cADPR and showed cross-desensitization with ryanodine at 100 µM. Insulin secretion from permeabilized islets of Langerhans was stimulated by cADPR. The effect was abolished by EGTA, indicating that the effect was due to cADPR-dependent Ca^{2+} release. Further, incubation of intact islets with stimulatory concentrations of glucose resulted in the production of a metabolite that had Ca^{2+}-releasing properties identical to those of cADPR when tested with β cell microsomes. This result represented the first indication that an extracellular stimulus could stimulate the production of cADPR, consistent with its second messenger role.

How glucose stimulates cADPR production is not known. Interestingly, IP$_3$ neither mobilizes Ca^{2+} from microsomes, in contrast to other reports (Nilsson et al., 1987), nor stimulates insulin secretion from the permeabilized islets, in agreement with others (Wolf et al., 1985). The reason for the discrepancy in the Ca^{2+} release data is unclear. However, since glucose is effective in inducing insulin secretion in the absence of any change in inositol lipid turnover, considerable debate exists over whether release of Ca^{2+} from IP$_3$-sensitive stores plays a role in glucose-induced insulin secretion (Hallberg, 1986).

Studies using the diabetogenic β cytotoxins to uncouple stimulus–secretion coupling suggest an obligatory role for cADPR in stimulus–secretion coupling in β cells (Takasawa et al., 1993). Diabetogenic agents such as streptozotocin inhibit glucose-induced Ca^{2+} spiking (Grapengiesser et al., 1990) as well as insulin secretion in β cells and are used to produce experimental models of diabetes. These agents seem to be specific for β cells, although the reason for this underlying specificity is unknown (Okamoto, 1985). Researchers have proposed that the mode of action of these drugs is to induce DNA strand breakage, which leads to activation of nuclear poly(ADP–ribose) synthetases. Since the substrate for the synthetases is NAD^+, their activation could result in depletion of intracellular NAD^+ levels (Yamamoto et al., 1981). This possibility is supported by evidence that poly(ADP–ribose) polymerase inhibitors block the inhibitory effects of streptozotocin on insulin secretion from islets. Since NAD^+ is the precursor of cADPR synthesis by ADP–ribosyl cyclases, the effects of these agents on the production of a cADPR-like factor by glucose was examined (Takasawa et al., 1993). As discussed earlier, extracts of glucose-treated islets produced a metabolite that can release Ca^{2+} from cerebellar microsomes in a manner identical to the action of cADPR. This metabolite was not produced when the islets were also incubated with streptozotocin. The inhibitory action of streptozotocin could be reversed by the poly(ADP–ribose) polymerase inhibitors nicotinamide and 3-aminobenzamide.

In summary, cADPR appears to mobilize Ca^{2+} from a ryanodine-sensitive mechanism in β cell microsomes. Investigators have proposed cADPR as a second messenger for glucose-induced Ca^{2+} mobilization and as essential for stimulus–secretion coupling in β cells.

5. Cardiac SR Vesicles

The discovery that cADPR mobilizes intracellular Ca^{2+} in sea urchin eggs by activating a ryanodine-sensitive Ca^{2+}-release mechanism (Galione et al., 1991) immediately suggested experiments to examine effects of cADPR on Ca^{2+} release from sarcoplasmic reticulum, organelles with the highest densities of ryanodine receptors (Fleischer and Inui, 1989). The first such experiment on skeletal muscle sarcoplasmic reticulum involved the application of cADPR to rabbit skinned skeletal muscle fibers and the use of contraction of the fibers as an assay for Ca^{2+} release (Galione et al., 1991). The results of these experiments showed that cADPR did not induce contraction, although caffeine (50 mM) was effective. Two explanations for this effect are possible. Either cADPR did not activate the type 1 ryanodine receptor isoform found in skeletal muscle or, perhaps, the cADPR binding site was occluded. The latter possibility could occur, since the dihydropyridine receptor in the sarcolema is known to have high affinity for the ryanodine receptor; their interactions may be enhanced after removal of the sarcolemma by the chemical skinning process. However, experiments on sarcoplasmic reticulum vesicles derived from mammalian skeletal muscle, which should be free of sarcolemmal components, showed that cADPR did not induce Ca^{2+} release, suggesting that the type 1 ryanodine receptors are insensitive to cADPR.

In contrast, cADPR was able to induce Ca^{2+} release from passively loaded cardiac sarcoplasmic reticulum vesicles that contain the type 2 ryanodine receptor isoform. The rate of Ca^{2+} release from the vesicles was enhanced by 1 μM cADPR, an effect that was abolished by treating the vesicles with 50 μM ryanodine (Meszaros et al., 1993). Compelling data that cADPR is an endogenous activator of the type 2 ryanodine receptor come from reconstitution studies in which the receptors were incorporated into planar lipid bilayers. Here cADPR application to the cis or cytoplasmic side of the membrane greatly increased channel activity, with fast gating and numerous short-lived excursions into the open state (Meszaros et al., 1993).

A direct interaction between cADPR and type 2 ryanodine receptors is also supported by the finding that cADPR, like caffeine and Ca^{2+}, enhances [^{3}H]ryanodine binding to cardiac, but not skeletal, sarcoplasmic reticulum membranes. The effects of cADPR (1 μM) on the rate of Ca^{2+} release from, or in stimulating [^{3}H]ryanodine binding to, the sarcoplasmic reticulum membranes were Ca^{2+} sensitive, being most pronounced at submicromolar Ca^{2+} concentrations. At micromolar Ca^{2+} concentrations, which maximally activate CICR, cADPR had no additional effect, indicating that Ca^{2+} and cADPR were activating the same release mechanism (Meszaros et al., 1993).

These data, in conjunction with data from sea urchin microsomes demonstrating a clear potentiation of Ca^{2+} release by divalent cations or caffeine by submaximal concentrations of cADPR (Lee, 1993a), support the hypothesis that cADPR may sensitize CICR through ryanodine receptors. Since cADPR itself and ADP-ribosyl cyclase activity have been measured in mammalian heart (Rusinko and Lee, 1989; Walseth et al., 1991), the intriguing possibility is raised that cADPR is an important modulator of Ca^{2+} release by the cardiac sarcoplasmic reticulum and, subsequently, is a key regulator of cardiac excitability. The widespread distribution of type 2 ryanodine receptors between tissues, relative to the type 1 receptors (Lai et al., 1992), is consistent with the view that cADPR is a widespread activator of Ca^{2+} mobilization.

V. Cyclic ADP–Ribose Receptor

Knowledge of the proteins that interact with cADPR will be fundamental to the complete understanding of the mechanism by which cADPR regulates Ca^{2+} release. A variety of techniques, including binding studies, antagonist effects, and photoaffinity labeling, has been used to study cADPR binding protein(s).

A. SPECIFIC BINDING TO SEA URCHIN EGG MICROSOMES

Lee (1991) characterized the sea urchin egg cADPR receptor using conventional binding techniques. High specific activity [^{32}P]cADPR was synthesized from [^{32}P]NAD using brain microsomal ADP-ribosyl cyclase and was employed as ligand to probe cADPR binding in sea urchin egg microsomes. Microsomes from *S. purpuratus* showed specific binding of the ligand to a saturable site. The binding of [^{32}P]cADPR had a pH optimum of 6.7 and was localized to the microsomal fraction, as determined by Percoll density gradient fractionation. Binding to the mitochondrial fraction was very low. Specific binding of [^{32}P]cADPR was unaffected by micromolar concentrations of NAD^+ and ADP-ribose, but was completely eliminated by 0.3 μM cADPR. IP_3, at concentrations up to 10 μM, did not affect cADPR binding. Heparin, an antagonist of IP_3-induced Ca^{2+} release, also did not affect cADPR binding at concentrations that blocked IP_3 action. These data are additional indications that cADPR regulates Ca^{2+} release through a mechanism completely independent of the IP_3 receptor system. Scatchard analysis indicated a single class of binding sites with a binding affinity (K_d) of 17 nM and a capacity (B_{max}) of 25 fmol/mg protein. The affinity for binding is very similar to the half-maximal concentration of cADPR for Ca^{2+} release, indicating that the two activities are related.

Subsequent work on the microsomal binding revealed the presence of an inhibitor in the egg homogenates (H. C. Lee, unpublished data). The inhibitor is heat labile and is dialyzable, indicating that it is of low molecular weight. Although the inhibitor has not been characterized further, its properties suggest that it may be endogenous cADPR that is bound to microsomes. Removal of the inhibitor by dialysis allows specific microsomal binding to be detected much more readily. Figure 5A shows a Scatchard plot of the binding data obtained from dialyzed *S. purpuratus* microsomes (H. C. Lee, unpublished data). In contrast to the undialyzed microsomes (Lee, 1991), two classes of binding sites can be discerned from the Scatchard analysis. The high affinity sites have an affinity of ~0.7 nM and a density of ~11 fmol/mg. The density of the low affinity sites is much higher, at 600 fmol/mg with a K_d value of ~170 nM. Thus, the removal of the endogenous inhibitor appears to unmask some high affinity as well as low affinity sites. The functional significance of these two classes of binding site remains to be deter-

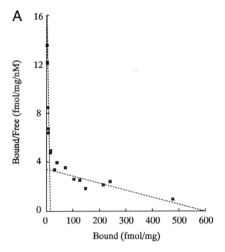
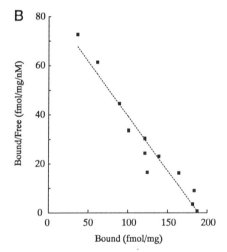

FIG. 5. Cyclic ADP–ribose binding sites in sea urchin egg microsomes. Scatchard plots of specific binding of [^{32}P]cyclic ADP–ribose to egg microsomes. (A) Two classes of binding sites are present in microsomes from *Strongylocentrotus purpuratus* eggs. The high affinity sites have an affinity of about 0.7 nM and a density of about 11 fmol/mg. The low affinity sites have a density of 600 fmol/mg with an affinity of about 170 nM. (B) Only one class of binding sites is present in microsomes from *Lytechinus pictus* eggs. The affinity of these sites is 1.6 ± 0.5 nM (n = 3, ± SE) and the density is 125 ± 41 fmol/mg protein.

mined. In this regard, two classes of IP_3 binding sites have been observed in hepatocytes; researchers thought that the low affinity sites correlated better with Ca^{2+} release activity (Mauger et al., 1989). Ca^{2+} in the physiological range can induce interconversion between the two classes of IP_3 binding site (Pietri et al., 1990).

Unlike *S. purpuratus* microsomes, dialyzed *L. pictus* microsomes display only a single class of binding, as shown in Fig. 5B. The affinity of these sites is 1.6 ± 0.5 nM ($n = 3$, $\pm$SE) and the density is 125 ± 41 fmol/mg protein (Walseth et al., 1993). To date, in all responsive systems, cADPR is active in the nanomolar concentration range, indicating that it is operating through a specific receptor. Demonstration of high affinity and specific binding in egg homogenates is consistent with such a notion. Direct binding studies with [^{32}P]cADPR will serve an important role in the purification and characterization of cADPR binding proteins in sea urchin eggs as well as in other systems.

B. Cyclic ADP–Ribose Antagonists

Walseth and Lee (1993) have described several 8-substituted analogs of cADPR that antagonize the actions of cADPR. 8-Amino-, 8-bromo-, and 8-azido-derivatives of cADPR were synthesized by incubating the appropriately substituted NAD^+ analogs with the purified ADP–ribosyl cyclase from *Aphysia* ovotestis. The general synthetic scheme is shown in Fig. 6. The 8-substituted NAD^+ analogs can be synthesized by chemically coupling their respective 8-substituted AMP analogs to nicotinamide mononucleotide (NMN) using a carbodiimide reagent (EDC) (Prescott and McLennan, 1990). The only precursor compound that is not commercially available is 8-amino-AMP, which can be easily synthesized by reducing 8-azido-AMP with dithiothreitol. The *Aplysia* cyclase is quite tolerant to the substitutions at the 8-position of NAD^+ and readily cyclizes these analogs.

Studies using these analogs indicate that the 8-position of cADPR is critically important for its Ca^{2+}-releasing activity, since none of the analogs had the ability to release Ca^{2+} from sea urchin egg microsomes (Walseth and Lee, 1993). However, all three block cADPR from releasing Ca^{2+}; 8-amino-cADPR is the most effective. These molecules are, therefore, specific antagonists of the cADPR-dependent Ca^{2+}-release mechanism. The antagonistic effects of these cADPR analogs is competitive since high concentrations of cADPR can overcome the inhibition, suggesting that the analogs may bind to the same site as cADPR itself. This possibility has proved to be true since 8-amino-cADPR also inhibits [^{32}P]cADPR binding to microsomes, as does cAD-

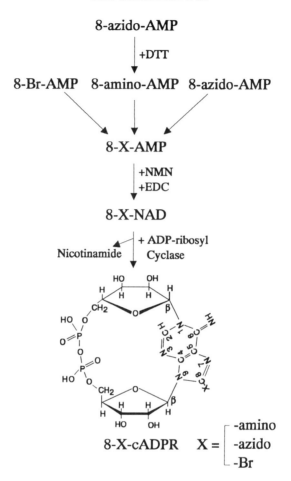

FIG. 6. Schematic for synthesis of 8-substituted analogs of cyclic ADP–ribose. The groups substituted at the 8 position of the adenine ring are denoted by X, which can be amino-, azido- or Br. Abbreviations: NMN, nicotinamide mononucleotide; EDC, 1-ethyl-3(3-dimethyl-amino-propyl)-carbodiimide-HCl; DTT, dithiothreitol.

PR itself, an indication that the cADPR receptor has similar affinities for cADPR and 8-amino-cADPR. 8-Bromo- and 8-azido-cADPR also inhibit cADPR binding, but not as effectively as 8-amino-cADPR (Walseth and Lee, 1993). These findings provide a good explanation for the antagonistic effect of the 8-substituted analogs of cADPR. These analogs can bind to the cADPR receptor site with good affinity, but cannot activate the Ca^{2+}-release mechanism because of the substitutions at the 8-position. By competitively displacing cADPR from its receptor, these analogs can block the Ca^{2+}-releasing activity of cADPR. An implication of these results is that occupation of the receptor site does not necessarily lead to Ca^{2+} release. Appropriate interaction between the receptor and the 8-position of the ligand is also required.

An interesting question concerning the mechanism of Ca^{2+} release induced by cADPR is whether the continuous occupation of the receptor site by cADPR is required to sustain the Ca^{2+} release. The availability of these competitive antagonists allows this question to be addressed. Researchers found that 8-amino-cADPR added immediately after cADPR can displace cADPR from its receptor and block further Ca^{2+} release from sea urchin egg microsomes. The extent of inhibition is dependent on the concentration of 8-amino-cADPR used; at sufficiently high concentration, immediate cessation of the Ca^{2+} release results (Walseth and Lee, 1993). Collectively, these results indicate that occupation of the cADPR receptor is certainly a necessary condition for Ca^{2+} release, but is not a sufficient one. The presence of an appropriate group at the 8-position of the ligand is also needed.

These novel antagonists of cADPR not only provide useful tools for studying the mechanism of the cADPR-dependent Ca^{2+} release, but also are useful in determining the physiological role of the cADPR system. The approach is to introduce these antagonists into cells to inhibit the cADPR-dependent Ca^{2+}-release pathway and then to determine which cellular functions are blocked by the treatment. This approach is analogous to using heparin to block the IP_3 pathway. Compared with heparin, the antagonists of cADPR should be much more selective. Although heparin has been shown to be a competitive inhibitor of IP_3 binding to its receptor (Ghosh et al., 1988), it is a polyanion known to have affinity for a variety of proteins. Considerable effort has been expended to synthesize specific antagonists for the IP_3 receptor; however, to date all the analogs reported are weak agonists (Hirata et al., 1989,1990). An alternative approach is to use, instead, inhibitory antibodies against the IP_3 receptor, as has been done in hamster eggs (Miyazaki et al., 1992).

The usefulness of the cADPR antagonists has been demonstrated in a study of cADPR function during sea urchin egg fertilization. Microinjection of heparin to inhibit the IP_3 pathway is known to only delay the onset of the fertilization-associated Ca^{2+} increase but not to block it (Rakow and Shen, 1990; Crossley et al., 1991; Lee et al., 1993a), indicating that Ca^{2+} mobilization in sea urchin egg can proceed via multiple mechanisms. This result is in contrast with results in vertebrate eggs, in which fertilization-induced Ca^{2+} changes can be totally inhibited if the IP_3 pathway is blocked (Miyazaki et al., 1992; Galione *et al.*, 1993a). Preloading sea urchin eggs with 8-amino-cADPR inhibits subsequent microinjection of cADPR from elevating internal Ca^{2+}, indicating that the antagonist is effective in the intact cell as well as in homogenates (Lee *et al.*, 1993a). The antagonist, however, does not block the fertilization-associated Ca^{2+} transient. The only condition that results in inhibition of the fertilization response is the co-injection of 8-amino-cADPR and heparin into sea urchin eggs (Lee *et al.*, 1993a). Similar results were obtained by co-injection of heparin and ruthenium red, an inhibitor of the ryanodine receptor/Ca^{2+} channel that also blocks the cADPR-sensitive Ca^{2+} release (Galione *et al.*, 1991, 1993a). Therefore, both the IP_3 and the cADPR release system are involved in mediating the Ca^{2+}-mobilization response during fertilization. These results provide the first direct evidence of a physiological role for cADPR in cell functions. The antagonists of cADPR promise to be important in unraveling the Ca^{2+}-release mechanism of cADPR.

Similar experimental paradigms should be useful in examining the role of cADPR in other cellular systems. For example, antagonists such as 8-amino-cADPR could be introduced into cells to determine which cellular function(s) are affected by the presence of the antagonists. Indeed, several methods are available for introducing small molecules into cells by permeabilizing them transiently. After resealing, cells can recover and function normally. For example, electroporation permeabilizes sea urchin eggs to molecules at least as large as ATP. Afterward, the eggs can be fertilized and develop normally (Sweze and Epel, 1989). Various cultured cells can be scrape-loaded with macromolecules, yet maintain viability and growth (McNeil *et al.*, 1984). A commercially available lipid-based system has been shown to be able to permeabilize transiently a variety of mammalian cells (Karger *et al.*, 1990); ATP^{4-} has been shown to induce reversible permeabilization in mast cells and other transformed cell lines (Gomperts and Fernandez, 1985). These procedures can be used to introduce the antagonists into cells and to examine changes in cellular functions after resealing and recovery. The advantage of introducing the antagonists instead of cADPR is that biological effects of the antagonists can be assessed after recovery whereas the agonist effects must be determined at the time of permeabilization. This approach could provide clues to the type of cellular functions that may be regulated by cADPR.

The success of synthesizing a series of 8-substituted antagonists illustrates a useful approach, that is, taking advantage of the substrate tolerance of the ADP–ribosyl cyclase. In principle, cADPR analogs with modifications on other position of the molecules can be synthesized with this approach. Indeed, various positions of the NAD^+ have been extensively modified (Woenckhaus and Jeck, 1987) and can potentially be used as substrates for the cyclase to produce other cADPR analogs with interesting biological properties. An analog having a reactive group that can be linked to a macromolecule would be useful for developing specific antibodies against cADPR. As discussed in Section II,C, an immunoassay for cADPR would be of great value in determining the endogenous levels of cADPR in cells and tissues.

C. Photoaffinity Labeling of cADPR Binding Sites

Among the cADPR antagonists synthesized, the 8-azido-cADPR is photoactive because of the azido group and thus can be used as a photoaffinity probe for identifying the cADPR receptor(s). Two different methods have been used to synthesize ^{32}P-labeled 8-azido-AMP with high specific activity and then chemically to convert the molecule to 8-azido-[^{32}P]NAD^+ using the scheme outlined in Fig. 6 (Walseth et al., 1993). The first method uses (γ-^{32}P)ATP and the enzyme adenosine kinase (Miller et al., 1979) to phosphorylate 8-azido-adenosine to 8-azido-5'[^{-32}P]AMP. In the second method, 3'-8-azido-AMP can be phosphorylated by polynucleotide kinase (Walseth and Johnson, 1979) and [γ-^{32}P]ATP to produce 8-azido-5'-[^{32}P],3'-ADP, and then dephosphorylated with nuclease P1 (Walseth and Johnson, 1979) to 5'-^{32}P-labeled 8-azido-AMP. The specific activity of the product synthesized with either method should be the same as that of [γ-^{32}P]ATP.

Employing this radiolabeled photoactive probe in sea urchin egg preparations has led to the identification of two specifically labeled proteins having molecular masses of 140 kDa and 100 kDa. The labeling of these two proteins is specific for cADPR, since it is not affected by micromolar concentrations of ADP–ribose, AMP, ADP, ATP, cAMP, or IP_3, but is inhibited half-maximally by about 10 nM cADPR and completely by 500 nM cADPR.

As discussed in Section IV,C, cADPR releases Ca^{2+} through the ryanodine receptor/CICR system. A 380-kDa protein in sea urchin egg that cross-reacts with antibodies against skeletal ryanodine receptor has been identified by Western analyses (McPherson et al., 1992). The cADPR binding proteins identified by photolabeling are certainly much smaller than 380 kDa, suggesting that cADPR may not interact directly with the ryanodine receptor, but may instead exert its effect on the ryanodine receptor through intermediate proteins (the 140-kDa and 100-kDa proteins). This possibility is supported by the finding that

8-amino-cADPR, although effectively inhibiting Ca^{2+} release induced by cADPR, does not block the Ca^{2+} release induced by either caffeine or ryanodine. The cADPR antagonist, however, can completely eliminate the potentiating effects of cADPR on the Ca^{2+}-releasing activity of caffeine (Walseth and Lee, 1993). Thus, the cADPR system may be more complicated than the IP_3 system, which is known to function by IP_3 binding directly to the IP_3 receptor/Ca^{2+}-release channel to induce Ca^{2+} release. Interestingly caffeine, an agonist of the ryanodine receptor, preferentially inhibits the photolabeling of the 100-kDa protein relative to the 140-kDa protein (Walseth et al., 1993). This effect could account for the desensitizing effects of caffeine on the cADPR-dependent Ca^{2+}-release system described in Section IV,C.

Alternatively, the 140-kDa and/or 100-kDa proteins identified by photoaffinity labeling could be unknown variants of the ryanodine receptor. A 75-kDa variant of the ryanodine receptor has been described in brain (Takeshima et al., 1993). Determining whether or not the 140-kDa and 100-kDa proteins represent smaller variants of the ryanodine receptor will require further characterization. The success of specifically labeling cADPR binding sites with ^{32}P-labeled 8-azido-cADPR represents a major step toward the eventual purification and identification of these proteins.

VI. PHYSIOLOGICAL ROLES OF CYCLIC ADP–RIBOSE

A. FERTILIZATION

cADPR was first discovered as a Ca^{2+}-mobilizing agent in sea urchin eggs (Clapper et al., 1987). To date, the sea urchin egg remains the most extensively studied system with respect to cADPR physiology. Much of current knowledge about its mechanism of action, metabolism, and role is derived from studies on this cell type.

As described in Section II,A, microinjection of cADPR into intact eggs can induce a large Ca^{2+} transient, trigger the cortical reaction, and parthenogenically activate the eggs to undergo multiple cycles of nuclear envelope breakdown and DNA synthesis (Dargie et al., 1990). This effect is independent of external Ca^{2+}, consistent with its major action in mobilizing Ca^{2+} from internal stores. A key question is whether the cADPR signaling pathway is activated at fertilization in the sea urchin egg.

Fertilization reactions in most eggs are characterized by a very large Ca^{2+} transient that is independent of the influx of extracellular Ca^{2+}

(Whitaker and Swann, 1993) and is initiated at the site of sperm–egg interaction, which then sweeps across the egg as a propagating wave (Jaffe, 1991). Two studies demonstrate a redundancy in the mechanism of Ca^{2+} mobilization at fertilization (Galione et al., 1993a; Lee et al., 1993a). In the first study, researchers showed that ryanodine receptors and IP_3 receptors both needed to be blocked by the pharmacological inhibitors ruthenium red and heparin before the wave was abolished (Galione et al., 1993a). This result indicates that ryanodine and IP_3 receptors are both recruited by the fertilizing sperm to propagate the Ca^{2+} signals across the egg. Direct demonstration that the cADPR pathway is involved in fertilization comes from the use of a novel specific antagonist of the cADPR-binding site, 8-amino-cADPR (Walseth and Lee, 1993). These authors showed that activation of the sea urchin egg on fertilization could be blocked only if the IP_3 receptor antagonist heparin and 8-amino-cADPR were co-injected into the egg (Lee et al., 1993a). Inhibition of either system was not sufficient to block the fertilization response. Collectively, these studies provide direct evidence that cADPR has a role in fertilization and that this role is mediated by the ryanodine receptor. Whether cADPR plays a permissive role, where it acts as a coagonist for Ca^{2+} mobilization, or a true second messenger role, where its levels rise on fertilization, remains to be determined.

The striking redundancy serves to highlight further the similarity in function of the two classes of intracellular release channels, IP_3 and ryanodine receptors, which mirrors their structural homologies (Berridge, 1993). These channels can be considered to be primarily CICR channels that can be modulated by specific intracellular messengers, namely cADPR and IP_3. Whereas these two types of Ca^{2+}-release mechanism can give very similar responses in the sea urchin egg, and may both be employed at fertilization to insure that the important transition from dormancy to activation is made, in other cells their singular activation may give rise to different or spatially restricted Ca^{2+} signals, resulting in a different cellular message.

B. INSULIN SECRETION

Electrophysiological studies of β cells have emphasized the importance of plasma membrane channels in the secretory response of cells to nutrients. Glucose induces oscillations in the membrane potential. Much emphasis has been placed on the interactions between ATP-dependent potassium conductances and L-type Ca^{2+} channels in stimulus–response coupling (Ashcroft, 1988). In these models, metabolism of glucose leads to enhanced intracellular ATP levels that close the ATP-controlled potassium channels. The resulting membrane depolarization then activates the L-type Ca^{2+} channels. Influx of Ca^{2+} triggers two responses: first, exocytosis of insulin-containing granules and, second, opening of the Ca^{2+}-activated potassium channels. The latter event is responsible for repolarization of the membrane. The positive and negative interactions between the channels in the plasma membrane generate slow waves of membrane potential.

Imaging of single pancreatic β cells has revealed pronounced spatiotemporal complexity of intracellular Ca^{2+} signals (Gylfe et al., 1991; Theler et al., 1992). The physiological importance of these signaling patterns is only just being appreciated. For example, the frequency of Ca^{2+} oscillations and that of the pulsatile insulin secretion are very similar. Both are highly sensitive to nutrient concentration and are abolished by drugs that impair insulin secretion. Intracellular Ca^{2+} oscillations and propagating Ca^{2+} waves are now a well-established feature of Ca^{2+} signaling in many cells, and are thought to depend critically on the properties of Ca^{2+}-release channels located on the endoplasmic reticulum. This relationship has renewed interest in the properties and control of these channels in the endoplasmic reticulum of the β cell, and in the mechanism by which a rise in blood glucose leads to their activation. One obvious mechanism of activation is that the glucose-induced influx of Ca^{2+} during plasma membrane depolarization leads to mobilization of Ca^{2+} from internal stores, possibly by CICR (Roe et al., 1993). A CICR mechanism has been demonstrated in β cells and cell lines that are sensitive to caffeine, but relatively insensitive to ryanodine (Islam et al., 1992; Hellman et al., 1992; Roe et al., 1993). In addition, in permeabilized β cells and cell lines, or in microsomes derived from these cells, an IP_3-sensitive release mechanism that could also be activated by CICR has been reported (Iino and Endo, 1992).

As discussed in Section IV,D, Okamoto and his colleagues have reported a possible role of cADPR in insulin secretion (Takasawa et al., 1993). This report is important for several reasons. First, this report demonstrates that cADPR can mobilize Ca^{2+} from β cell microsomes via a ryanodine-sensitive mechanism. This situation is the first clear demonstration of a ryanodine-sensitive channel in this cell, although it appears to contradict the previous finding of insensitivity of the glucose-induced oscillations in β cells to ryanodine (Hellman et al., 1993). Second, this report suggests that glucose enhances the synthesis of cADPR and thus implicates cADPR as a second messenger of glucose. Third, this report suggests an obligatory role for cADPR and, hence, Ca^{2+} mobilization in insulin secretion. Finally, this report provides an elegant mechanism by which impairing the cADPR signaling pathway with various diabetogenic drugs can induce experimental diabetes.

The finding that drugs that inhibit glucose-induced insulin secretion also impair the ability of glucose to stimulate the production of a cADPR-like activity in islets suggests a key role for the cADPR signaling pathway in stimulus–secretion coupling in the pancreatic β cells (Takasawa et al., 1993). The requirement for a glucose-induced Ca^{2+} influx across the plasma membrane as well as glucose-enhanced cADPR synthesis for insulin secretion suggests that a CICR mechanism may operate to stimulate the secretory response. A rise in cADPR levels would sensitize the CICR mechanism so influx of Ca^{2+} across the plasma membrane would trigger additional Ca^{2+} release from intracellular stores (Galione, 1993). This CICR model for stimulus–secretion coupling in the β cell is illustrated in Fig. 7. Regenerative Ca^{2+} release from stores would then result in large Type A cytosolic Ca^{2+} oscillations (Hellman et al., 1992), which may control the pulsatile insulin release from the pancreas. Further, regenerative Ca^{2+} release also has a spatial aspect; the Ca^{2+} signals can propagate as waves that could help synchronize clusters of cells in the pancreas and coordinate their activities.

Impairment of Ca^{2+}-release mechanisms would result in the abolition of cytosolic Ca^{2+} oscillations as well as propagation of Ca^{2+} signals between cells. Researchers have proposed that abolition of pulsatile insulin secretion, which reflects Type A Ca^{2+} oscillations, is an early indicator of Type II (insulin-resistant) diabetes (O'Rahilly et al., 1988). Since streptozotocin can induce defects similar to those seen in Type II diabetes and abolishes Ca^{2+} oscillations (Grapengiesser et al., 1990), insulin secretion, and glucose-dependent increases in cADPR, the primary lesion in Type II diabetes may be a defect in the cADPR signaling pathway.

How glucose stimulates cADPR synthesis remains to be resolved. However, one intriguing possibility is that cGMP might be a coupling factor. As described in Section III,D, ADP–ribosyl cyclase in sea urchin eggs may be regulated by cGMP-dependent phosphorylation (Galione et al., 1993b). cGMP has been proposed as a mediator of insulin secretion from rat islets (Laychock, 1981), since glucose causes an increase in cGMP levels in islets and since agents that elevate cGMP in islets, such as ascorbic acid and sodium nitroprusside, enhance insulin secretion. Further, LY83583, an inhibitor of guanylyl cyclase, abolishes glucose-induced insulin secretion (Laychock et al., 1991). Nitric oxide,

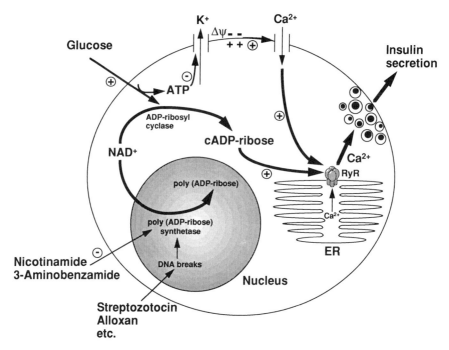

FIG. 7. Possible scheme for the involvement of cyclic ADP–ribose in stimulus–secretion coupling in the pancreatic β cell. Glucose stimulates both Ca^{2+} influx and cADPR synthesis, leading to calcium-induced calcium release (CICR) from intracellular stores. This event may give rise to Ca^{2+} spiking in the cell, leading to exocytosis of insulin-containing granules. Streptozotocin induces DNA strand breakage, which activates poly(ADP–ribose) synthetase in the nucleus, resulting in the depletion of cellular NAD^+. This event compromises the ADP–ribosyl cyclase pathway, since substrate for cADPR synthesis is reduced, leading to inhibition of Ca^{2+} mobilization and insulin secretion. Inhibitors of poly(ADP–ribose) synthetase, such as 3-aminobenzamide, protect the cell against NAD^+ depletion, so the pathway for stimulus–secretion coupling functions normally even in the presence of streptozotocin.

an activator of soluble guanylyl cyclase, can also enhance insulin secretion from pancreatic β cells as well as increase islet cGMP levels (Schmidt et al., 1992). In the presence of glucose or the oral hypoglycemic drug tolbutamide, treatments with L-arginine, the cellular precursor for nitric oxide production by nitric oxide synthases, enhance insulin secretion and can stimulate nitric oxide production in β cells (Schmidt et al., 1992). Whether cGMP/nitric oxide-enhanced insulin secretion is mediated by increasing cADPR synthesis warrants investigation.

VII. Conclusion

Figure 8 summarizes the current information on the various components known to be involved in the cADPR-dependent Ca^{2+}-mobilization system. The metabolic pathway of cADPR consists of synthesis from NAD^+ by ADP-ribosyl cyclase and degradation by the cADPR hydrolase to ADP-ribose. Bifunctional enzymes, such as CD38 on the surface of lymphocytes and a 39-kDa splenic protein, can catalyze both reactions. CD38-like bifunctional enzymes are likely to be

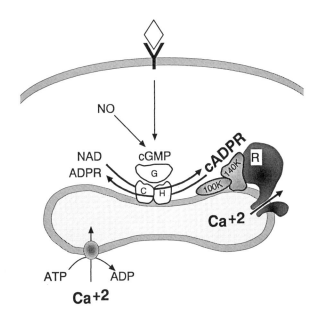

FIG. 8. Model of the cyclic ADP-ribose-dependent Ca^{2+}-mobilization system, depicting various components known to be involved in the system. The metabolic pathway of cADPR consists of synthesis from NAD^+ by ADP-ribosyl cyclase (C) and degradation by the cADPR hydrolase (H) to ADP-ribose. Bifunctional enzymes, such as CD38, can catalyze both reactions. Elevation of cytoplasmic cADPR level elicits CA^{2+} release from internal stores via a ryanodine receptor-like (R) Ca^{2+} channel. The receptor(s) for cADPR may not be the channel itself but, instead, may be the 140-kDa and/or the 100-kDa proteins identified by photoaffinity labeling. The released Ca^{2+}, after serving its function, is eventually sequestered by an ATP-dependent Ca^{2+} pump. The whole system appears to be turned on by a cGMP-dependent mechanism, which acts by stimulating the ADP-ribosyl cyclase to produce cADPR. Researchers postulated that any signaling mechanisms that elevate cGMP, either through surface receptor activation or diffusable messengers such as nitric oxide (NO), would trigger the cADPR-dependent Ca^{2+}-release pathway.

responsible for regulating the cellular concentration of cADPR. Elevation of cytoplasmic cADPR levels elicits Ca^{2+} release from internal stores via a mechanism very similar to the ryanodine receptor/Ca^{2+} channel of the sarcoplasmic reticulum. However, the receptor(s) for cADPR may not be the channel itself, but may be the 140-kDa or the 100-kDa protein identified by photoaffinity labeling. The released Ca^{2+}, after serving its function, is eventually sequestered by an ATP-dependent Ca^{2+} pump. The whole system appears to be turned on by a cGMP-dependent mechanism, which acts by stimulating the ADP-ribosyl cyclase to produce cADPR. Any signaling mechanisms that elevate cGMP, through surface receptor activation or through diffusable messengers such as nitric oxide, would trigger the cADPR-dependent Ca^{2+} release pathway. The cADPR system thus fulfills most of the criteria of a second messenger system.

Since the discovery of cADPR in 1987, important advances have helped clarify its physiological functions and mechanism of action. Its cyclic structure has been elucidated and its natural occurrence demonstrated. The enzymatic pathways of the metabolism of cADPR have been identified. Increasing evidence shows that this molecule is an agonist for the Ca^{2+}-induced Ca^{2+} release mechanism and that it is active in mobilizing Ca^{2+} in invertebrate and mammalian cellular systems. Participation of cADPR in mediating the Ca^{2+} changes associated with fertilization of sea urchin eggs has been directly demonstrated. Technological advances in synthesizing antagonists, and use of a photoaffinity label of cADPR, have set the stage for identifying its receptor and for understanding the mechanism of its action in molecular terms. That CD38 is an ecto-enzyme catalyzing the synthesis and the degradation of cADPR raises the possibility that cADPR may have extracellular functions and points to a whole new frontier that must be explored. The finding that the synthesis of cADPR is regulated by a cGMP-dependent mechanism promises to usher in a novel concept of integrating three messenger pathways—nitric oxide, cGMP, and cADPR—into a Ca^{2+}-mobilization mechanism. The implication of these findings is likely to be far reaching. Cyclic ADP–ribose, an unexpected discovery, promises to bring forth many more unexpected findings.

ACKNOWLEDGMENTS

We thank David Levitt of the Department of Physiology, University of Minnesota for sharing the unpublished results on crystalization of the *Aplysia* ADP–ribosyl cyclase and Richard M. Graeff for critical reading of the manuscript. We are also grateful for support from the National Institutes of Health (H.C.L. and T.F.W.), the Medical Research Council (A.G.), and the Beit Memorial Fellowship for Medical Research (A.G.).

REFERENCES

Ashcroft, F. M. (1988). Adenosine triphosphate sensitive K$^+$ channels. *Annu. Rev. Neurosci.* **11**, 97–118.
Berridge, M. J. (1993). Inositol trisphosphate and calcium signalling. *Nature (London)* **361**, 315–325.
Bezprozvanny, I., Watras, J., and Ehrlich, B. E. (1991). Bell-shaped calcium-response curves of Ins(1,4,5)P$_3$- and calcium-gated channels from endoplasmic reticulum of cerebellum. *Nature (London)* **351**, 751–754.
Bredt, D. S., and Snyder, S. H. (1990). Isolation of nitric oxide synthase, a calmodulin-requiring enzyme. *Proc. Natl. Acad. Sci. USA* **89**, 5159–5162.
Bredt, D. S., Mourey, R. J., and Snyder, S. H. (1989). A simple, sensitive, and specific radioreceptor assay for inositol 1,4,5-trisphosphate in biological tissues. *Biochem. Biophys. Res. Commun.* **159**, 976–982.
Breitman, T. R., Selonick, S. E., and Collins, S. J. (1980). Induction of differentiation of the human promyelocytic leukemia cell line (HL-60) by retinoic acid. *Proc. Natl. Acad. Sci. USA* **77**, 2936–2940.
Burgoyne, R. D., Cheek, T. R., Morgan, A., O'Sullivan, A. J., Moreton, R. B., Berridge, M. J., Mata, A. M., Colyer, J., Lee, A. G., and East, J. M. (1989). Distribution of two distinct Ca^{2+}–ATPase-like proteins and their relationships to the agonist-sensitive calcium store in adrenal chromaffin cells. *Nature (London)* **342**, 72–74.
Campbell, A. K. (1983). "Intracellular Calcium: Its Universal Role as Regulator." Academic Press, New York.
Carafoli, E. (1987). Intracellular calcium homeostasis. *Annu. Rev. Biochem.* **56**, 395–433.
Caterina, M. J., and Devreotes, P. N. (1991). Molecular insights into eukaryotic chemotaxis. *FASEB J.* **5**, 3078–3085.
Chen, Q., and van Breeman, C. (1992). Function of smooth muscle sarcoplasmic reticulum. *Adv. Second Mess. Phosphorylation Res.* **26**, 335–350.
Clapper, D. L., and Lee, H. C. (1985). Inositol trisphosphate induces Ca^{2+}-release from nonmitochondrial stores in sea urchin egg homogenates. *J. Biol. Chem.* **260**, 13947–13954.
Clapper, D. L., Walseth, T. F., Dargie, P. J., and Lee, H. C. (1987). Pyridine nucleotide metabolites stimulate calcium release form sea urchin egg microsomes desensitized to inositol trisphosphate. *J. Biol. Chem.* **262**, 9561–9568.
Collins, S. J. (1987). The HL-60 promyelocytic leukemia cell line: proliferation, differentiation, and cellular oncogene expression. *Blood* **70**, 1233–1244.
Crossley, I., Whally, T., and Whitaker, M. (1991). Guanosine 5'-thiotriphophate may stimulate phosphoinositide messenger production in sea urchin eggs by a different route than the fertilizing sperm. *Cell Reg.* **2**, 121–133.
Currie, K., Swann, K., Galione, A., and Scott, R. H. (1992). Activation of Ca^{+2}-dependent currents in cultured neurons by a sperm factor and cyclic ADP-ribose. *Mol. Biol. Cell* **3**, 1415–1422.
Cusack, N. J., and Hourani, S. M. (1990). Subtypes of P2-purinoceptors. Studies using analogues of ATP. *Ann. N.Y. Acad. Sci.* **603**, 172–181.
Dargie, P. J. Agre, M. C., and Lee, H. C. (1990). Comparison of Ca^{+2} mobilizing activities of cyclic ADP-ribose and inositol trisphosphate. *Cell Reg.* **1**, 279–290.
DeLisle, S., and Walsh, M. J. (1992). Inositol trisphosphate is required for the propagation of calcium waves in *Xenopus* oocytes. *J. Biol. Chem.* **267**, 7963–7966.
Dicker, P., Heppel, L. A., and Rozengurt, E. (1980). Control of membrane permeability by external and internal ATP in 3T6 cells grown in serum-free medium. *Proc. Natl. Acad. Sci. USA* **77**, 2103–2107.

Edwards, F. A., Gibb, A. J., and Colquhoun, D. (1992). ATP receptor-mediated synaptic currents in the central nervous system. *Nature (London)* **359,** 144–147.

Epel, D. (1980). Experimental Analysis of the role of intracellular calcium in the activation of the sea urchin egg at fertilization. *In* "The Cell Surface: Mediator of Developmental Processes" (S. Subtelny and N. Wessels, eds.), pp. 169–185. Academic Press, Orlando, Florida.

Ferris, C. D., and Snyder, S. H. (1992). Inositol 1,4,5-trisphosphate-activated calcium channels. *Annu. Rev. Biochem.* **54,** 469–488.

Fill, M., and Coronado, R. (1988). Ryanodine receptor channel of sarcoplasmic reticulum. *Trends Neurosci.* **11,** 453–457.

Finch, E. A., Turner, T. J., and Goldin, S. M. (1991). Calcium as a coagonist of inositol 1,4,5-trisphosphate-induced calcium release. *Science* **252,** 443–446.

Gleischer, S., and Inui, M. (1989). Biochemistry and biophysics of excitation-contraction coupling. *Annu. Rev. Biophys. Chem.* **18,** 333.

Fujiwara, A., Taguchi, K., and Yasumasu, I. (1990). Fertilization membrane formation in sea urchin eggs induced by drugs known to cause Ca^{+2} release from isolated sarcoplasmic reticulum. *Dev. Growth Diff.* **32,** 303–314.

Furuichi, T., Yoshikawa, S., Miyawaki, A, Wada, K., Maeda, N., and Mikoshiba, K. (1989). Primary structure and functional expression of the inositol 1,4,5-trisphosphate-binding protein P400. *Nature (London)* **342,** 32–38.

Galione, A. (1992). Ca^{2+}-induced Ca^{2+} release and its modulation by cyclic ADP-ribose. *Trends Pharmacol. Sci.* **13,** 304–306.

Galione, A. (1993). Cyclic ADP-ribose, a new way to control calcium. *Science* **259,** 325–326.

Galione, A., Lee, H. C., and Busa, W. B. (1991). Ca^{+2}-induced Ca^{+2} release in sea urchin egg homogenates and its modulation by cyclic ADP-ribose. *Science* **253,** 1143–1146.

Galione, A., McDougall, A., Busa, W. B., Willmott, N., Gillot, I., and Whitaker, M. (1993a). Redundant mechanism of calcium-induced calcium release underlying calcium waves during fertilization of sea urchin eggs. *Science* **261,** 348–352.

Galione, A., White, A., Willmott, N., Turner, M., Potter, B.V.L., and Watson, S. P. (1993b). cGMP mobilizes intracellular Ca^{+2} in sea urchin eggs by stimulating cyclic ADP-ribose synthesis. *Nature (London)* **365,** 456–459.

Ghosh, T. K., Eis, P. S., Mullaney, J. M., Ebert, C. L., and Gill, D. L. (1988). Competitive, reversible, and potent antagonism of inositol 1,4,5-trisphosphate-activated calcium release by heparin. *J. Biol. Chem.* **263,** 11075–11079.

Gianini, G., Clementi, E., Ceci, R., Marziali, G., and Sorrentino, V. (1992). Identification of a novel and broadly expressed, ryanodine-receptor/Ca^{+2} channel and its regulation by TGFβ. *Science* **257,** 91–94.

Gill, D. L. (1989). Receptor kinships revealed. *Nature (London)* **342,** 16–18.

Glick, D. L., Hellmich, M. R., Beushausen, S., Tempst, P., Bayley, H., and Strumwasser, F. (1991). Primary structure of a molluscan egg-specific NADase, a second-messenger enzyme. *Cell Reg.* **2,** 211–218.

Gomperts, B. D., and Fernandez, J. M. (1985). Techniques for membrane permeabilization. *Trends Biol. Sci.* **10,** 414–417.

Grapengiesser, E., Gylfe, E., and Hellman, B. (1990). Disappearance of glucose-induced oscillations of cytoplasmic Ca^{+2} in pancreatic cells exposed to streptozotocin or alloxan. *Toxicology* **63,** 263–271.

Gylfe, E., Grapengiesser, E., and Hellman, B. (1991). Propagation of cytoplasmic Ca^{+2} oscillations in clusters of pancreatic β-cells. *Cell Calcium* **12,** 229–240.

Hakamata, Y., Nakai, J., Takeshima, H., and Imoto, K. (1992). Primary structure and distribution of a novel ryanodine receptor/calcium release channel from rabbit brain. *FEBS Lett.* **312**, 229–235.

Hallberg, A. (1986). Dissociation of phosphatidylinositol hydrolysis and insulin secretion of cultured mouse pancreatic islets. *Acta Physiol. Scand.* **128**, 267–276.

Harada, N., Santos-Argumedo, L., Chang, R., Grimaldi, J. C., Lund, F. E., Brannan, C. I., Copeland, N. G., Jenkins, N. A., Heath, A. W., Parkshouse, R.M.E., and Howard, M. (1993). Expression cloning of a cDNA encoding a novel murine B cell activation marker. Homology to human CD38. *J. Immunol.* **151**, 3111–3118.

Hellman, B., Gylfe, E., Grapengiesser, E., Lund, P.-E., and Bers, A. (1992). Cytoplasmic Ca^{2+} oscillations in pancreatic β cells. *Biochim. Biophys. Acta* **1113**, 295–305.

Hellmich, M. R., and Strumwasser, F. (1991). Purification and characterization of a molluscan egg-specific NADase, a second-messenger enzyme. *Cell Reg.* **2**, 193–202.

Hemmi, H., and Breitman, T. R. (1982). Induction by retinoic acid of NAD^+-glycohydrolase activity of myelomonocytic cell lines HL-60, THP-1 and U-193, and fresh human acute promyelocytic leukemia cells in primary culture. *Biochem. Biophys. Res. Commun.* **109**, 669–774.

Hirata, M., Watanabe, Y., Ishimatsu, T., Ikebe, T., Kimura, Y., Yamaguchi, K., Ozaki, S., and Koga, T. (1989). Synthetic inositol trisphosphate analogs and their effects on phosphatase, kinase, and the release of Ca^{+2}. *J. Biol. Chem.* **264**, 20303–20308.

Hirata, M., Yanaga, F., Koga, T., Ogasawara, T., Watanabe, Y., and Osaki, S. (1990). Stereospecific recognition of inositol 1,4,5-trisphosphate analogs by the phosphatase, kinase, and binding proteins. *J. Biol. Chem.* **265**, 8404–8407.

Hofmann, F., Dostmann, W., Keilbach, A., Landgraf, W., and Ruth, P. (1992). Structure and physiological role of cGMP-dependent protein kinase. *Biochim. Biophys. Acta* **1135**, 51–60.

Howard, M., Grimaldi, J. C., Bazan, J. F., Lund, F. E., Santos-Argumedo, L., Parkhouse, R.M.E., Walseth, T. F., and Lee, H. C. (1993). Formation and hydrolysis of cyclic ADP-ribose catalyzed by lymphocyte antigen CD38. *Science* **262**, 1056–1059.

Iino, M., and Endo, M. (1992). Calcium-dependent immediate feedback control of inositol 1,4,5 trisphosphate-induced calcium release. *Nature (London)* **360**, 76–78.

Islam, Md.S., Rorsman, P., and Berggren, P.-O. (1991). Ca^{+2}-induced Ca^{+2} release in insulin-secreting cells. *FEBS Lett.* **296**, 287.

Jackson, D. G., and Bell, J. I. (1990). Isolation of a cDNA encoding the human CD38 (T10) molecule, a cell surface glycoprotein with an unusual discontinuous pattern of expression during lymphocyte differentiation. *J. Immunol.* **144**, 2811–2815.

Jaffe, J. A., and Terasaki, M. (1993). Structural changes of the endoplasmic reticulum of sea urchin eggs during fertilization. *Dev. Biol.* **156**, 566–573.

Jaffe, L. F. (1991). The path of calcium in cytosolic calcium oscillations, a unifying hypothesis. *Proc. Natl. Acad. Sci. USA* **88**, 9883–9887.

Karger, B. D., Sykes-Saloranta, A. G., Shonk, L. A., and Henrich, C. J. (1990). The TransPort™ transient cell permeabilization kit: An update. *Focus* **12**, 110–111.

Kim, H., Jacobson, E. L., and Jacobson, M. K. (1993a). Synthesis and degradation of cyclic ADP-ribose by NAD glycohydrolases. *Science* **261**, 1330–1333.

Kim, H., Jacobson, E. L., and Jacobson, M. K. (1993b). Position of cyclization in cyclic ADP-ribose. *Biochem. Biophys. Res. Commun.* **194**, 1143–1147.

Koshiyama, H., Lee, H. C., and Tashijian, A. H., Jr. (1991). Novel mechanism of intracellular calcium release in pituitary cells. *J. Biol. Chem.* **266**, 16985–16988.

Kontani, K., Nishina, H., Ohoka, Y., Takahashi, K., and Katda, T. (1993). NAD glycohydrolase specifically induced by retinoic acid in human leukemic HL-60 cells: Identification of the NAD glycohydrolase as leukocyte cell surface antigen CD38. *J. Biol. Chem.* **268**, 16895–16898.

Kume, S., Muto, A., Aruga, J., Nakagawa, T., Michikawa, T., Furuichi, T., Nakade, S., Okano, H., and Mikoshiba, K. (1993). The *Xenopus* IP$_3$ receptor: Structure, function and localization in oocytes and eggs. *Cell* **73**, 555–570.

Lai, F. A., Dent, M., Wickenden, C., Xu, L., Kumari, G., Misra, M., Lee, H. B., Sar, M., and Meissner, G. (1992). Expression of a cardiac Ca^{2+} release channel isoform in mammalian brain. *Biochem. J.* **288**, 553–564.

Laychock, S. G. (1981). Evidence for guanosine 3′,5′monophosphate as a putative mediator of insulin secretion from isolated rat islets. *Endocrinology* **108**, 1197–1205.

Laychock, S. G., Modica, M. E., and Cavanaugh, C. T. (1991). L-Arginine stimulates cyclic guanosine 3′,5′monophosphate formation in rat islets of Langerhans and RINm5F insulinoma cells. Evidence for L-arginine, nitric oxide synthase. *Endocrinology* **129**, 3043–3052.

Lechleiter, J., Girard, S., Peralta, E., and Clapham, D. (1991). Spiral calcium wave propagation and annihilation in *Xenopus laevis* oocytes. *Science* **251**, 123–126.

Lee, H. C. (1991). Specific binding of cyclic ADP-ribose to calcium-storing microsomes from sea urchin eggs. *J. Biol. Chem.* **266**, 2276–2281.

Lee, H. C. (1993a). Potentiation of calcium- and caffeine-induced calcium release by cyclic ADP–ribose. *J. Biol. Chem.* **268**, 293–299.

Lee, H. C. (1994). Cyclic ADP-ribose: A calcium mobilizing metabolite of NAD$^+$. *Mol. Cell. Biochem.* (in press).

Lee, H. C., Aarhus, R., and Levitt, D. (1994). The crystal structure of cyclic ADP-ribose. *Nature Structural Biology* **1**, 143–144.

Lee, H. C., and Aarhus, R. (1991). ADP-ribosyl cyclase: An enzyme that cyclizes NAD$^+$ into a calcium-mobilizing metabolite. *Cell Reg.* **2**, 203–209.

Lee, H. C., and Aarhus, R. (1993). Wide distribution of an enzyme that catalyzes the hydrolysis of cyclic ADP-ribose. *Biochem. Biophys. Acta* **1164**, 68–74.

Lee, H. C., Walseth, T. F., Bratt, G. T., Hayes, R. N., and Clapper, D. L. (1989). Structural determination of a cyclic metabolite of NAD$^+$ with intracellular Ca^{+2} mobilizing activity. *J. Biol. Chem.* **264**, 1608–1615.

Lee, H. C., Aarhus, R., and Walseth, T. F. (1993a). Calcium mobilization by dual receptors during fertilization of sea urchin eggs. *Science* **261**, 352–355.

Lee, H. C., Zocchi, E., Guida, L., Franco, L., Benatti, U., and De Flora, A. (1993b). Production and hydrolysis of cyclic ADP-ribose at the outer surface of human erythrocytes. *Biochem. Biophys. Res. Commun.* **191**, 639–645.

Lin, S. H. (1989). Localization of the ecto-ATPase (ecto-nucleotidase) in the rat hepatocyte plasma membrane. *J. Biol. Chem.* **264**, 14403–14407.

Lowenstein, C. J., and Snyder, S. H. (1992). Nitric oxide, a novel biologic messenger. *Cell* **70**, 705–707.

Malavasi, F., Funaro, A., Alessio, M., DeMonte, L. B., Ausiello, C. M., Dianzani, U., Lanza, F., Magrini, E., Momo, M., and Roggero, S. (1992). CD38: A multi-lineage cell activation molecule with a split personality. *Int. J. Clin. Lab. Res.* **22**, 73–80.

Mauger, J.-P., Claret, M., Pietri, F., and Hilly, M. (1989). Hormonal regulation of inositol 1,4,5-trisphosphate receptor in rat liver. *J. Biol. Chem.* **264**, 8821–8826.

Matsuuchi, L., Gold, M. R., Travis, A., Grosschedl, R., DeFranco, A. L., and Kelley, R. B. (1992). The membrane IgM-associated proteins MB-1 and Ig-β are sufficient to promote surface expression of a partially functional B-cell antigen receptor in a nonlymphoid cell line. *Proc. Natl. Acad. Sci. USA* **89**, 3404–3408.

McNeil, P. L., Murphy, R. F., Lanni, F., and Taylor, D. L. (1984). A method of incorporating macromolecules into adherent cells. *J. Cell Biol.* **98,** 1556–1564.

McPherson, P. S., and Campbell, K. P. (1993). The ryanodine receptor/Ca^{2+} release channel. *J. Biol. Chem.* **268,** 13765–13768.

McPherson, S. M., McPherson, P. S., Mathews, L., Campbell, K. P., and Longo, F. J. (1992). Cortical localization of a calcium release channel in sea urchin eggs. *J. Cell Biol.* **116,** 1111–1121.

Meldolesi, J., Madeddu, L., and Pozzan, T. (1990). Intracellular storage organelles in nonmuscle cells: heterogeneity and functional assignment. *Biochim. Biophys. Acta* **1055,** 130–140.

Meszaros, L. G., Bak, J., and Chu, A. (1993). Cyclic ADP-ribose as an endogenous regulator of the non-skeletal type ryanodine receptor Ca^{+2} channel. *Nature (London)* **364,** 76–79.

Miller, R. L., Adamczyk, D. L., and Miller, W. H. (1979). Adenosine kinase from rabbit liver. I. Purification by affinity chromatograhy and properties. *J. Biol. Chem.* **254,** 2339–2345.

Miyazaki, S., Yuzaki, M., Nakada, K., Shirakawa, H., Nakanishi, S., Nakade, S., and Mikoshiba, K. (1992). Block of Ca^{+2} wave and Ca^{+2} oscillation by antibody to the inositol 1,4,5-trisphosphate receptor in fertilized hamster eggs. *Science* **257,** 251–255.

Nilsson, T., Arkhammar, P., Hallberg, A., Hellman, B., and Berggren, P.-O. (1987). Characterization of inositol 1,4,5-trisphosphate-induced Ca^{+2} release in pancreatic β cells. *Biochem. J.* **248,** 329–336.

Okamoto, H. (1985). Molecular basis of experimental diabetes, Degeneration, oncogenesis and regeneration of pancreatic β-cells of islets of Langerhans. *Bioessays* **2,** 16–20.

O'Rahilly, S., Turner, R. C., and Matthews, D. R. (1988). Impaired pulsatile secretion of insulin in relatives of patients with non-insulin-dependent diabetes. *New Engl. J. Med.* **318,** 1225–1230.

Osipchuk, Y., and Cahalan, M. (1992). Cell-to-cell spread of calcium signals mediated by ATP receptors in mast cells. *Nature (London)* **359,** 241–244.

Pearson, R. B., and Kemp, B. E. (1991). Protein kinases and phosphorylation site sequences. *Meth. Enzymol.* **200,** 62–81.

Pekala, P. H., and Anderson, B. M. (1978). Studies of bovine erythrocyte NAD glycohydrolase. *J. Biol. Chem.* **253,** 7453–7459.

Pietri, F., HIlly, M., and Mauger, J-P (1990). Calcium mediates the interconversion between two states of the liver inositol 1,4,5-trisphosphate receptor. *J. Biol. Chem.* **265,** 17478–17485.

Pilkis, S. J., ed. (1990). "Fructose-2,6-Bisphosphate." CRC Press, Boca Raton, Florida.

Pilkis, S. J., and Granner, D. K. (1992). Molecular physiology of the regulation of hepatic gluconeogenesis and glycolysis. *Annu. Rev. Physiol.* **54,** 885–909.

Poenie, M., Alderston, J., Tsien, R. Y., and Steinhardt, R. A. (1985). Changes in free calcium levels with stages of the cell cycle. *Nature (London)* **315,** 147–149.

Prescott, M., and McLennan, A. G. (1990). Synthesis and applications of 8-azido photoafinity analogs of P1,P3-bis(5'-adenosyl)triphosphate and P1,P4-bis(5'-adenosyl)-tetraphosphate. *Anal. Biochem.* **184,** 330–337.

Price, S. R., and Pekala, P. H. (1987). Pyridine nucleotide-linked glycohydrolase. *In* "Coenzymes and Cofactors" (D. Dolphin, R. Poulson, and O. Avramovic, eds.), Vol. 2, pp. 513–548.

Publicover, N. G., Hammond, E. M., and Sanders, K. M. (1993). Amplification of nitric oxide signaling by interstitial cells isolated from canine colon. *Proc. Natl. Acad. Sci. USA* **90**, 2087–2091.

Rakow, T. L., and Shen, S. S. (1990). Multiple stores of calcium are release in the sea urchin egg during fertilization. *Proc. Natl. Acad. Sci. USA* **87**, 9285–9289.

Roe, M. W., Lancaster, M. E., Mertz, R. J., Worley, J. F., III, and Dukes, I. D. (1993). Voltage-dependent intracellular calcium release from mouse islets stimulated by glucose. *J. Biol. Chem.* **268**, 9953–9956.

Rousseau, E., and Meissner, G. (1989). Single cardiac sarcoplasmic reticulum Ca^{+2}-release channel: activation by caffeine. *Am. J. Physiol.* **256**, H328–333.

Rusinko, N., and Lee, H. C. (1989). Widespread occurrence in animal tissues of an enzyme catalyzing the conversion of NAD^+ into a cyclic metabolite with intracellular Ca^{+2} mobilizing activity. *J. Biol. Chem.* **264**, 11725-11731.

Sanderson, M. J., Charles, A. C., and Dirksen, E. R. (1990). Mechanical stimulation and intercellular communication increases intracellular Ca^{+2} in epithelial cells. *Cell Reg.* **1**, 585–596.

Sardet, C., Gillot, I., Ruscher, A., Payan, P., Girard, J-P., and de Renzis, G. (1992). Ryanodine activates sea urchin eggs. *Dev. Growth Diff.* **34**, 37–42.

Schmidt, H.H.H.W., Lohmann, S. M., and Walter, U. (1993). The nitric oxide and cGMP signal transduction, regulation and mechanism of action. *Biochim. Biophys. Acta* **1178**, 153–175.

Schmidt, H.H.H.W., Warner, T. D., Ishii, K., Sheung, H., and Murad, F. (1992). Insulin secretion from pancreatic B cells caused by D-arginine-derived nitrogen oxides. *Science* **255**, 721–723.

Schmit, T., Patton, C., and Epel, D. (1982). Is there a role for the Ca^{+2} influx during fertilization of the sea urchin egg? *Dev. Biol.* **58**, 185–196.

Schulz, S., Chinkers, M., and Garbers, D. L. (1989). The guanylate cyclase/receptor family of proteins. *FASEB J.* **3**, 2026–2035.

Sharp, A. H., McPherson, P. S., Dawson, T. M., Aoki, C., Campbell, K. P., and Snyder, S. H. (1993). Differential immunohistochemical localization of inositol 1,4,5-trisphosphate- and ryanodine-sensitive Ca^{+2} release channels in rat brain. *J. Neurosci.* **13**, 3051–3063.

Shen, S. S., and Buck, W. R. (1993). Sources of calcium in sea urchin eggs during the fertilization response. *Dev. Biol.* **157**, 157–169.

Sorrentino, V., and Volpe, P. (1993). Ryanodine receptors, how many, where and why? *Trends Pharmacol. Sci.* **14**, 98–103.

Smith, J. S., Imagawa, T., Ma, J., Fill, M., Campbell, K. P., and Coronado, R. (1988). Purified ryanodine receptor from rabbit skeletal muscle is the calcium-release channel of sarcoplasmic reticulum. *J. Gen. Physiol.* **92**, 1–26.

States, D. J., Walseth, T. F., and Lee, H. C. (1992). Similarities in amino acid sequences of *Aplysia* ADP-ribosyl cyclase and human lymphocyte antigen CD38. *Trends Biochem. Sci.* **17**, 495.

Steinhardt, R. A., and Alderton, J. M. (1982). Calmodulin confers calcium sensitivity on secretory exocytosis. *Nature (London)* **295**, 154–155.

Swann, K., and Whitaker, M. (1986). The part played by inositol trisphosphate and calcium in the propagation of the fertilization wave in sea urchin eggs. *J. Cell Biol.* **103**, 2333–2342.

Swezey, R. R., and Epel, D. (1989). Stable, resealable pores formed in sea urchin eggs by electric discharge (electroporation) permit substrate loading for assay of enzymes in vivo. *Cell Reg.* **1**, 65–74.

Takasawa, S., Nata, K., Yonekura, H., and Okamoto, H. (1993). Cyclic ADP-ribose in insulin secretion from pancreatic β cells. *Science* **259**, 370–373.

Takeshima, H., Nishimura, S., Nishi, M., Ikea, M., and Sugimoto, T. (1993). A brain-specific transcript from the 3'-terminal region of the skeletal muscle ryanodine receptor gene. *FEBS Lett.* **322**, 105–110.
Terasaki, M., and Jaffe, L. A. (1991). Organization of the sea urchin egg endoplasmic reticulum and its reorganization at fertilization. *J. Cell Biol.* **114**, 929–940.
Terasaki, M., and Sardet, C. (1991). Demonstration of calcium uptake and release by sea urchin cortical endoplasmic reticulum. *J. Cell Biol.* **115**, 1031–1037.
Theler, J-M., Mollard, P., Guerineau, N., Vacher, P., Pralong, W. F., Schlegel, W., and Wollheim, C. B. (1992). Video-imaging of cytosolic Ca^{2+} in pancreatic β cells stimulated by glucose, carbachol, and ATP. *J. Biol. Chem.* **267**, 18110–18117.
Tiff, C. J., Proia, R. L., and Camerini-Otero, R. D. (1992). The folding and cell surface expression of CD4 requires glycosylation. *J. Biol. Chem.* **267**, 3268–3273.
Vacquier, V. D. (1975). The isolation of intact cortical granules from sea urchin eggs: Calcium ions trigger granule discharge. *Dev. Biol.* **43**, 62–74.
Volpe, P., Krause, K., Hashimoto, S., Zorzato, F., Pozzan, T., Meldolesi, J., and Lew, D. P. (1988). "Calciosome", a cytoplasmic organelle: The inositol 1,4,5 trisphosphate-sensitive Ca^{+2} store of nonmuscle cells? *Proc. Natl. Acad. Sci. USA* **85**, 1091–1095.
Wagner, K. A., Yacono, P. W., Golan, D. E., and Tashjian, A. H., Jr. (1993). Mechanism of spontaneous intracellular calcium fluctuations in single GH4C1 rat pituitary cells. *Biochem. J.* **292**, 175–182.
Walseth, T. F., and Johnson, R. A. (1979). The enzymatic preparation of [α-^{32}P]nucleoside triphosphate, cyclic [^{32}P]AMP, and cyclic [^{32}P]GMP. *Biochim. Biophys. Acta* **562**, 11–31.
Walseth, T. F., and Lee, H. C. (1993). Synthesis and characterization of antagonists of cyclic ADP–ribose. *Biochim. Biophys. Acta* **1178**, 235–242.
Walseth, T. F., Aarhus, R., Zeleznikar, R. J., Jr., and Lee, H. C. (1991). Determination of endogenous levels of cyclic ADP-ribose in rat tissues. *Biochim. Biophys. Acta* **1094**, 113–120.
Walseth, T. F., Aarhus, R., Kerr, J. A., and Lee, H. C. (1993). Identification of cyclic ADP-ribose binding proteins by photoaffinity labeling. *J. Biol. Chem.* **268**, 26686–26691.
Whalley, T., McDougall, A., Crossley, I., Swann, K., and Whitaker, M. (1992). Internal calcium release and activation of sea urchin eggs by cGMP are independent of the phosphoinositide signaling pathway. *Mol. Biol. Cell* **3**, 373–383.
Whitaker, M., and Irvine, R. F. (1984). Inositol 1,4,5-trisphosphate microinjection activates sea urchin eggs. *Nature (London)* **312**, 636–639.
Whitaker, M. J., and Swann, K. (1993). Lighting the fuse at fertilization. *Development* **117**, 1–12.
White, A., Watson, S. P., and Galione, A. (1993). Cyclic ADP-ribose-induced calcium release from brain microsomes. *FEBS Lett.* **318**, 21–24.
Woenckhaus, C., and Jeck, R. (1987). Preparation and properties of NAD^+ and $NADP^+$ analogs. *In* "Coenzymes and Cofactors" (D. Dolphin, R. Poulson, and O. Avramovic, eds.), Vol. 2, pp. 449–568. Wiley, New York.
Wolf, B. A., Comens, P. G., Ackerman, K. E., Sherman, W. R., and McDaniel, M. L. (1985). The digitonin-permeabilized pancreatic islet model. Effect of myo-inositol 1,4,5-trisphosphate on Ca^{2+} mobilization. *Biochem. J.* **227**, 965–969.
Yamamoto, H., Uchigata, Y., and Okamoto, H. (1981). Streptozotocin and alloxan induce DNA strand breaks and poly(ADP-ribose) synthetase in pancreatic islets. *Nature (London)* **294**, 284–286.

A Critical Review of Minimal Vitamin B_6 Requirements for Growth in Various Species with a Proposed Method of Calculation

STEPHEN P. COBURN

Biochemistry Department
Fort Wayne State Developmental Center
Fort Wayne, Indiana 46835

I. Characteristics of Vitamin B_6 Metabolism
II. Vitamin B_6 Requirements for Growth
   A. Rodents
   B. Rabbits
   C. Carnivores
   D. Swine
   E. Horses
   F. Ruminants
   G. Nonhuman Primates
   H. Birds
   I. Aquatic Species
   J. Exotic Animals
III. Discussion
IV. Summary
   References

I. Characteristics of Vitamin B_6 Metabolism

Vitamin B_6 is associated with a greater variety of reactions than most other vitamins (Zubay, 1988). This molecule is also interconverted to seven common forms (pyridoxine, pyridoxal, pyridoxamine, their 5'-phosphates, and 4-pyridoxic acid) in most species, as well as to additional forms in some species (e.g., N-methyl pyridoxine in cats; Coburn and Mahuren, 1987). Pyridoxal 5'-phosphate is the most thoroughly studied cofactor form of this vitamin and is listed in *Enzyme Nomenclature* (Nomenclature Committee of the International Union of Biochemistry and Molecular Biology, 1992) as a component of 103 enzymes spanning all six major categories except ligases (Table I). Only 42 of the 71 aminotransferases are specifically listed as pyridoxal phosphate enzymes. Some of the others (e.g., 2.6.1.40 and 2.6.1.62) are known to contain pyridoxal phosphate whereas some definitely do not (e.g., 2.6.1.30 and 2.6.1.31). Therefore, assuming that all aminotransferases contain pyridoxal phosphate is not appropriate. The

TABLE I
NUMBER OF ENZYMES LISTED AS CONTAINING PYRIDOXAL PHOSPHATE
IN ENZYME NOMENCLATURE[a]

Group	Type	Number
1.4.4	Amine dehydrogenases	1
2.1.2	Formyl and hydroxymethyl transferases	2
2.3.1	Acyl transferases	3
2.4.1.1	Phosphorylases	1
2.6	Aminotransferases	42
3.7	Carbon–carbon hydrolases	1
4.1.1	Decarboxylases	16
4.1.2	Aldehyde lyases	3
4.1.99	Other carbon–carbon lyases	2
4.2.1	Hydrolyases	8
4.2.99	Other carbon–oxygen lyases	4
4.3.1	Carbon–nitrogen lyases	2
4.4.1	Carbon–sulfur lyases	10
4.5.1	Carbon–halide lyases	2
5.1.1	Racemases acting on amino acids	5
5.4.3	Intramolecular transfer of NH_2 groups	1
	Total	103

[a] Nomenclature Committee of the International Union of Biochemistry and Molecular Biology (1992).

major vitamin B_6 pool in animals is in muscle, where pyridoxal phosphate is bound primarily to glycogen phosphorylase. Interestingly, in this case the phosphate rather than the aldehyde may participate in the catalytic process (Helmreich, 1992). In addition to functioning as an enzyme cofactor, pyridoxal 5'-phosphate may have a role in regulating steroid hormone receptors through a strictly chemical Schiff base mechanism (Allgood and Cidlowski, 1991). The possibility of an interaction between vitamin B_6 and steroid hormone action is strengthened by the observation that pyridoxine kinase activity is higher in the adrenal gland than in any other tissue examined (Coburn et al., 1981).

Although pyridoxal 5'-phosphate has been the most extensively studied form of vitamin B_6, pyridoxamine 5'-phosphate is present in liver, heart, brain, and kidney in amounts comparable to pyridoxal 5'-phosphate (Coburn et al., 1988b). The role of pyridoxamine 5'-phosphate and the kinetics of the interchange between the aldehyde and amine forms in vivo remain to be studied. In view of the importance of vitamin B_6 in amino acid metabolism, the fact that the nervous system, which utilizes a variety of amino acid derivatives as neurotransmitters, and erythropoietic systems, which must synthesize large quantities of protein, are readily affected by vitamin B_6 deficiency is not surprising. The nervous system is also subject to vitamin B_6 toxicity (Bendich and Cohen, 1990).

The literature contains many reports in which 4′-deoxypyridoxine is administered in conjunction with a vitamin B_6-deficient diet. Often, the use of 4′-deoxypyridoxine is mentioned only in the methods section. Results obtained with 4′-deoxypyridoxine do not necessarily reflect those obtained with dietary deficiency alone (Coburn, 1981). Therefore, caution should be used in extrapolating such results to a simple dietary deficiency.

Assessment of vitamin B_6 requirements remains a challenging task. The task is complicated by lack of consensus on the best criteria for assessing vitamin B_6 status. In young animals, growth rate is a useful parameter. However, in adult animals some other indicator is required. Concentration of vitamin B_6 compounds, particularly pyridoxal phosphate, in plasma has been shown to be a valid measure under controlled conditions in rats (Lumeng *et al.*, 1978) but may be misleading under other circumstances (Barnard *et al.*, 1987; Raiten *et al.*, 1991). Urinary excretion of 4-pyridoxic acid is also widely used in human nutrition (Leklem, 1990a). Other common assessment techniques are urinary excretion of various tryptophan metabolites after a tryptophan load (Brown, 1981) and aminotransferase activity in erythrocytes before and after addition of pyridoxal 5′-phosphate *in vitro* (Skala *et al.*, 1981). Use of urinary excretion of metabolites of sulfur amino acids has also been suggested (Linkswiler, 1981). Any of these techniques is likely to be valid under laboratory conditions when the groups being compared differ only in vitamin B_6 intake. The interpretation of results obtained from free-living organisms with uncontrolled diets is more difficult. Evidence suggests that protein intake (Shultz and Leklem, 1981), as well as exercise and carbohydrate intake (Manore *et al.*, 1987), can influence vitamin B_6 metabolism. The protein interaction presumably reflects the importance of vitamin B_6 as a cofactor in amino acid metabolism. The effect of carbohydrate may reflect interactions with metabolism of body tissues and intestinal microflora (Mickelsen, 1956). The interactions between vitamin B_6 and riboflavin, which in the form of flavin mononucleotide (FMN) is a cofactor for pyridoxine phosphate oxidase, have been reviewed (McCormick, 1989). Vitamin B_6 requirements are also influenced by natural (e.g., linatine; Klosterman, 1979) or synthetic (e.g., isoniazid, D-penicillamine; Bhagavan, 1985) agents that alter vitamin B_6 metabolism.

Considering that the vitamin B_6 requirements for rats and humans remain a subject of debate despite hundreds of studies, obviously the data for most other species must be viewed as preliminary. Short-term requirements for meat production—for which rate of gain, feed efficiency, and market quality are of primary concern—can be assessed quite readily and are probably accurate. However, minimal requirements over an entire life-time remain uncertain for any species. Caution must be used in transferring biochemical techniques between species. For example, although 4-pyridoxic acid is a major urinary metabolite in many species, it is a minor urinary metabolite in cats (Coburn and Mahuren, 1987). At least a 50-fold variation in plasma pyridoxal phosphate concentrations occurs among species (Coburn et al., 1984). Also, marked interspecies variation is seen in the metabolism of the vitamin B_6 antagonist 4'-deoxypyridoxine (Coburn and Mahuren, 1979).

Several factors make quantitative measurement of vitamin B_6 in biological samples difficult. Vitamin B_6 exists in multiple forms and no method responds to all forms equally. The phosphorylated forms are usually protein bound and must be released for assay. In addition, in many methods the phosphate group must be removed prior to assay. Finally, some forms, particularly pyridoxal phosphate, are subject to photodecomposition (Ang, 1979). Microbiological assay with the yeast *Saccharomyces uvarum* has been among the most common methods of measuring total vitamin B_6 (Polansky, 1981). The radiometric microbiological assay using *Kloeckera apiculata* (Guilarte, 1986) avoids interference from turbidity in the sample, is easily automated, and can be used for biotin, folate, niacin, pantothenic acid, and vitamin B_{12} in addition to vitamin B_6. Microbiological assay currently remains the most reliable assay for vitamin B_6 in most samples of plant origin. The chromatograms given by extracts of such samples are so complex that the identification of vitamin B_6 compounds is questionable (Toukairin-Oda et al., 1989). In fact, interfering peaks have been encountered even in the plasma of animals consuming forage products (Coburn et al., 1984). Plant products may contain significant amounts of pyridoxine glucoside and other derivatives not normally found in animal tissues. One review tabulates the glycosylated vitamin B_6 content of a variety of foods, including previously unpublished data (Leklem, 1990b). The analyst must decide whether such derivatives should be included in the data for a given study. In addition to the microbiological assay for total vitamin B_6, enzymatic assays for pyridoxal phosphate have been widely employed and have shown good interlaboratory reproducibility (Reynolds, 1983). However, in liver, kidney, heart, and brain, pyridoxamine derivatives may be present in concentrations equal to or exceeding the concentration of pyridoxal phosphate (Coburn et al., 1988b). Also, under some circumstances such as pregnancy, a shift in the balance between pyridoxal and pyridoxal phosphate may occur (Barnard et al., 1987). Therefore, the use of chromatographic techniques to quantitate all the individual vitamers in animal tissues during studies of vitamin B_6 requirements is highly recommended.

When human volunteers are placed on a low vitamin B_6 intake diet, their urinary excretion of 4-pyridoxic acid rapidly declines until it approximates the intake (Kelsay *et al.*, 1968; Coburn *et al.*, 1991). The net loss of vitamin B_6 before reaching the new steady state is only about 4% of the body pool, and no significant decrease in the vitamin B_6 content of the muscle is detected (Coburn *et al.*, 1991). Rats (Black *et al.*, 1978) and swine (Russell *et al.*, 1985a) also maintain relatively normal pyridoxal phosphate concentrations in muscle during low vitamin B_6 intake. Although researchers frequently assume that water-soluble vitamins must be replaced daily, these observations suggest that existing tissues might be able to conserve vitamin B_6 very efficiently. If so, the vitamin B_6 requirement might reflect primarily special physiological demands such as the needs of new tissue during growth and pregnancy or losses during lactation. Data from rats (Coburn *et al.*, 1988b), swine (Coburn *et al.*, 1985), and humans (Coburn *et al.*, 1988a) suggest that 15 nmol/g is a reasonable average of vitamin B_6 content for the total body. Therefore, a reasonable estimate of the vitamin B_6 requirement of a growing animal might be obtained by multiplying the daily weight gain in grams by 15 nmol vitamin B_6/g (Coburn, 1990). This concept is referred to throughout this chapter as the calculated requirement.

Ongoing efforts to understand better the regulation of vitamin B_6 metabolism have led to several noteworthy conclusions. One is that the ability of cells to conserve vitamin B_6 cannot be explained solely by enzyme kinetics because the excretion of pyridoxic acid drops much more than the tissue concentrations of B_6 vitamers. This result reemphasizes the role of protein binding in controlling vitamin B_6 metabolism. Second, the ability to conserve vitamin B_6 suggests that the turnover may be at least partially dependent on the intake. In studies in which the turnover was estimated at about 1%/day, the intake was about 1% of the body pool (Johansson *et al.*, 1966). In subjects with a vitamin B_6 intake of only 1.8 μmol (0.4 mg)/day, the turnover was greatly reduced (Pauly *et al.*, 1991). Data from swine (Coburn *et al.*, 1985) and humans (Johansson *et al.*, 1966), as well as the differing pool sizes calculated from oral and intravenous doses (Johansson *et al.*, 1966; Tillotson *et al.*, 1966), demonstrate that the two-pool model suggested by Johansson *et al.* (1966) is not appropriate. We are attempting to develop a more versatile model (Coburn and Townsend, 1992).

Researchers often assume that the liver is the center of vitamin B_6 interconversions. However, data from mice (Sakurai *et al.*, 1988, 1992) and swine (S. P. Coburn *et al.*, unpublished data) indicate that small doses of pyridoxine and pyridoxamine can be completely converted to pyridoxal by the intestinal wall before being released into the blood, but larger doses result in the appearance of pyridoxine or pyridoxamine in the portal plasma. Thus, the role of the liver might not be as important as previously thought under conditions of normal vitamin B_6 intake. Pyridoxal phosphate appears to enter the plasma primarily in association with proteins secreted by the liver (Lumeng *et al.*, 1974). However, the relative importance of pyridoxal phosphate and pyridoxal as transport forms of vitamin B_6 in plasma remains to be determined, as does the possible role of the erythrocytes. These observations indicate that the importance of the various pathways of vitamin B_6 metabolism probably shifts depending on the vitamin B_6 intake.

In general, the minimum vitamin B_6 intake that produces maximum growth is lower than the intake that produces maximum activity of aminotransferases or minimal excretion of tryptophan metabolites after a load. Therefore, recommended intakes may vary considerably based on the criteria used. Some authors have suggested distinguishing between minimal and optimal requirements (Roth-Maier and Kirchgessner, 1977; Hemilä, 1991). A key question that remains unanswered in many cases is whether the additional intake needed to meet the optimal requirements produces any significant health benefits to the organism. Minimum requirements based solely on growth may be less than optimal for reproduction. The requirements for reproduction are usually higher than those needed to supply the estimated 15 nmol vitamin B_6/g fetal tissue. Plasma vitamer concentrations are likely to become the key factor in insuring adequate placental transport of vitamin B_6; the intake required to produce an adequate plasma concentration is likely to be greater than the absolute amount of vitamin B_6 needed by the fetal tissue.

Processing of vitamin B_6 by mammary tissue poses another challenging problem. The inverse relationship between pyridoxal phosphate concentrations and alkaline phosphatase activity in the milk of a variety of monogastric and ruminant mammals indicates that, in most species, pyridoxal phosphate is a significant fraction of the secreted vitamin B_6 (Coburn *et al.*, 1992b). However, the very low levels of pyridoxal phosphate and alkaline phosphatase in human milk suggest that human mammary tissue uses a markedly different mechanism for secreting vitamin B_6.

A continuing point of confusion in the literature is the indiscriminate and interchangeable use of the terms vitamin B_6 and pyridoxine, without specifying whether the data are calculated in terms of the free base, the hydrochloride form, or a mixture of the vitamers. In some articles, the terms pyridoxine and pyridoxine hydrochloride are used interchangeably to refer to the same data. A diet containing 2 mg pyridoxine/kg diet is not identical to a diet containing 2 mg pyridoxine hydrochloride/kg diet. This potential source of confusion and miscalculation could be eliminated if data were stated in molar units rather than in mass units. Use of molar units also simplifies comparisons of metabolites and calculations of overall balance. Throughout this chapter, we assume that, unless the authors specified otherwise, the terms vitamin B_6 and pyridoxine refer to pyridoxine hydrochloride.

Another point worth noting is that fresh animal tissues normally contain mainly pyridoxal phosphate and pyridoxamine phosphate, with very little pyridoxal or pyridoxine (Coburn et al., 1988b). Plants may contain significant amounts of glycosylated vitamers (Leklem, 1990b). These various forms of vitamin B_6 are not metabolized identically (Wozenski et al., 1980; Trumbo and Gregory, 1989; Gregory et al., 1991; Szadkowska et al., 1993). Since the vast majority of studies of vitamin B_6 requirements have utilized pyridoxine as the source of vitamin B_6, the possibility exists that optimal vitamin B_6 intakes obtained from natural diets may need to be slightly different than the intakes obtained using pyridoxine hydrochloride in conjunction with purified diets.

Finally, a common experimental design involves placing two or more groups of growing animals on varying vitamin B_6 intakes for a given time, followed by sacrifice and analysis of tissues. The lower intake groups usually have lower vitamin B_6 concentrations at the end of the experiment. Investigators frequently assume that the tissues have lost vitamin B_6. However, the total vitamin B_6 content of the tissues is rarely compared with the initial values. If vitamin B_6 is conserved by existing tissues, the lower final concentrations in low intake groups might reflect dilution because of additional growth in the absence of adequate vitamin B_6 intake rather than actual loss of vitamin B_6. In one study in which such a comparison was made (Sampson and O'Connor, 1989), researchers found that, although the pyridoxal phosphate concentration in muscle declined almost 50% over 2 wk, the total amount remained essentially unchanged. Variations in the abilities of tissues to conserve vitamin B_6 may be important in the development of deficiency symptoms. For example, the fact that skin and bone marrow cells are constantly reproducing may limit their ability to conserve vitamin B_6 and, thus, may contribute to their susceptibility to vitamin B_6 deficiency.

II. Vitamin B_6 Requirements for Growth

A. Rodents

1. Laboratory Rat (Rattus)

Not surprisingly, the vitamin B_6 requirement has probably been studied more thoroughly for the rat than for any other species. The data provide good examples of the difficulty in determining optimal vitamin B_6 intake. Physical symptoms of deficiency include symmetrical scaling dermatitis on the tail, feet, face, and ears; microcytic anemia; and hyperexcitability (Sherman, 1954). Convulsions may occur in severely deficient young animals, but generally do not occur readily in older rats placed on vitamin B_6-deficient diets. Chen and Marlatt (1975) concluded that male weanling rats needed intakes of about 50 nmol/day to maintain normal average growth of 3 g/day for 16 wk. Ingesting about 10 g of a diet containing 4.9 μmol (1 mg) vitamin B_6/kg diet would provide about 50 nmol, which would meet the requirement of 45 nmol/day predicted by our calculations to be needed daily to supply 3 g of new tissues. Using a liquid diet with a constant vitamin B_6 intake, Lumeng et al. (1978) found that 140 nmol/day produced maximal growth. An intake of 71 nmol/day sustained maximal growth for about 6 wk. However, by 9 wk, this group weighed slightly but significantly ($p < 0.05$) less than the group receiving 140 nmol/day. Since the growth of animals receiving 71 nmol/day was only slightly reduced, the minimal intake needed to achieve maximal growth probably lies between 71 and 140 nmol/day. This use of a constant vitamin B_6 intake is not directly comparable to studies with ad libitum feeding, since food consumption and, thus, vitamin B_6 intake may change with growth. With constant intake, the intake expressed as g/body wt will steadily decline. Thus an intake of 71 nmol/day might be adequate when the animals weigh 50 g but apparently is not quite sufficient when the animals weigh 250 g. Therefore, these data cannot be converted directly to units of μmol/kg diet. However, 71 nmol/day would be equivalent to 1–2 mg/kg diet. Roth-Maier and Kirchgessner (1981) achieved maximal growth for 3 wk with 7.3 μmol (1.5 mg)/kg diet. Van den Berg et al. (1982) reported maximal growth for 4 wk at 7.8 μmol (1.6 mg)/kg, but not at 3.4 μmol (0.7 mg)/kg.

Although a vitamin B_6 level of 4.9 μmol (1 mg)/kg diet maintained growth, higher intakes were needed to achieve saturation of enzymes and tissues. Aminotransferase activity is a commonly used indicator of vitamin B_6 status. Chen and Marlatt (1975) concluded that 19 μmol (4 mg)/kg diet was needed to maintain normal activity of liver, serum, and erythrocyte alanine aminotransferase. Beaton and Cheney (1965) concluded that 29–34 μmol (6–7 mg)/kg diet was needed to maintain erythrocyte alanine aminotransferase activity. Lumeng et al. (1978) found that erythrocyte alanine aminotransferase activity increased steadily with increasing vitamin B_6 intake, whereas aspartate aminotransferase activity started to plateau at an intake of 142 nmol/day. The liver activity of these two enzymes and of tryosine aminotransferase approached maximum activity with an intake of just 24 nmol/day. However, the activity of serine dehydratase in the liver increased with each increase in vitamin B_6 intake. Compared with rats receiving 28.2 μmol (5.8 mg) vitamin B_6/kg diet, animals receiving 5.8 μmol (1.2 mg) weighed slightly less after 9 wk, had significantly smaller livers, and had significantly less absolute protein synthesis (Sampson et al., 1988). However, the fractional synthesis in the liver was similar in both groups. Therefore, although it supported essentially normal overall growth, the lower vitamin B_6 intake was associated with statistically significant alterations in at least one organ. The key question, which cannot be answered at this time, is how such changes affect the long-term health of the organism. Maximal growth is not necessarily associated with maximal longevity. These situations re-emphasize the need for improved methods of assessing health. At this point, we may need to extend our experiments over the entire lifespan rather than look at very isolated segments. Such experiments would be more expensive and time consuming. However, the lengthy controversy over the 10th edition of the *Recommended Daily Allowances* clearly demonstrated the need for more definitive data on nutrient requirements.

Use of tissue composition as a measure of vitamin B_6 status is complicated by the fact that tissues vary in their response to alterations in vitamin B_6 intake. The pyridoxal phosphate content of brain and liver of growing rats seems to reach a plateau with an intake of 71 nmol/day (Lumeng et al., 1978) to 177 nmol/day (Driskell et al., 1973). However, the pyridoxal phosphate content of muscle did not reach a plateau (Black et al., 1977; Lumeng et al., 1978). This observation caused speculation that vitamin B_6 supplementation might stimulate an undesirable increase in muscle glycogen phosphorylase (Anonymous, 1978). However, these data were obtained in growing animals and might not be replicated in mature animals. In fact, we found no statistically significant increase in vitamin B_6 after adult male human volunteers consumed 0.98 mmol vitamin B_6/day for 6 wk (Coburn et al., 1991). Driskell et al. (1973) reported that, although an intake of 180 nmol/day produced normal growth, about 250 nmol/day was needed to maintain normal behavior in sexually mature rats. Alton-Mackey and Walker (1978) examined the effect of graded pyridoxine during lactation. Their data must be evaluated with care. These investigators designated their intakes as percentages of the National Research Council (NRC) recommendations and cited the 1972 edition of *Nutrient Requirements of Laboratory Animals* (Subcommittee on Laboratory Animal Nutrition, 1978). However, that edition recommends 34 μmol (7 mg)/kg whereas Alton-Mackey and Walker (1978) state that 0.4 mg pyridoxine hydrochloride/kg diet is 100% of the requirement. Therefore, the amount they cite as 400% of the recommendation is only 7.7 μmol, which is much less than the 34 μmol (7 mg)/kg actually recommended. As a result, whether vitamin B_6 levels greater than 7.7 μmol (1.6 mg)/kg would have yielded different results is not clear.

Although a vitamin B_6 intake of about 75 nmol/day (1 mg/kg diet) or more may maintain maximal growth in weanling male rats for at least a few weeks, the requirement for normal reproduction in females is higher. Intakes of 200–300 nmol/day (2–3 mg/kg diet) produced offspring of near normal weight but slightly altered brain composition (Kirksey et al., 1975). Because the vitamin B_6 content of milk and the activity of erythrocyte aminotransferase were decreased with diets below 23.3 μmol (4.8 mg)/kg diet, the NRC Subcommittee on Laboratory Animal Nutrition concluded that diets should contain at least 24 μmol (5 mg)/kg for growth and reproduction and 34 μmol (7 mg)/kg for maintenance of normal aminotransferase activity (Subcommittee on Laboratory Animal Nutrition, 1978).

A report by Kirksey et al. (1975) is often cited as indicating that brain tissue did not reach saturation levels of vitamin B_6 even when the diet contained 93 μmol (19.2 mg)/kg. The analyses supporting this statement were obtained from fetal brains obtained by Cesarean section at 21 days of gestation. The placental transport of vitamin B_6 and the relationship of dietary vitamin B_6 intake to fetal metabolism have not been thoroughly studied. Even at the highest vitamin B_6 intake (93 μmol/kg), the concentration of vitamin B_6 in fetal brain was only 6 nmol/g (Kirksey et al., 1975) compared with 10 nmol/g in deficient 30-day-old pups receiving a diet containing only 3 μmol/kg (Groziak and Kirksey, 1987). Since the highest vitamin concentration in fetal brain was only 60% of the concentration in vitamin B_6-deficient pups, vitamin B_6 concentrations in at least some fetal tissues may normally be significantly lower than the concentrations found after birth. Until fetal requirements are better understood, data from fetal tissues cannot be used reliably in determining vitamin B_6 requirements.

This entire discussion has dealt with animals subject to the stress of growth and/or pregnancy. Few data are available on long-term requirements of adult animals not subject to physiological stress. Black et al. (1978) found that the vitamin B_6 content of muscle was maintained during vitamin B_6 deficiency until the deficiency became so severe that weight loss was encountered. Using 12-wk-old female Long–Evans rats, Schaeffer and Kretsch (1987) observed a depression of growth and detected gait abnormalities within 9 day on an intake of 4 nmol/day or about 0.016 nmol/g body wt. Canham et al. (1966) observed that men receiving about 0.02 nmol/g body wt showed transient electroencephalographic changes, suggesting that they were adjusting to the low intake. Some women receiving < 0.05 mg vitamin B_6/day developed altered electroencephalograms that were returned to normal by increasing the vitamin B_6 intake to 0.5 mg/day (Kretsch et al., 1991). Therefore, we conclude that the minimal maintenance requirement in unstressed adults is slightly more than 0.02 nmol/g body wt or about 1 µmol (0.2 mg)/kg diet. However, in the Schaeffer and Kretsch study, the *ad libitum* control rats receiving 408 nmol/day were still growing about 1 g/day over 10 wk. Therefore, the daily intake of these animals probably should have been an additional 15 nmol/day for a total of 2 µmol/kg or about 0.4 mg/kg diet. Although such low intakes would be accompanied by declines in some biochemical indicators of vitamin B_6 status, we are unaware of any data on the long-term health of adult animals receiving such intakes. Under some circumstances, dietary restriction may actually increase longevity (Snyder and Wostmann, 1989).

Although intestinal microorganisms no doubt synthesize vitamin B_6, the fact that rats readily develop a vitamin B_6 deficiency suggests that the availability of microbially produced vitamin B_6 to the body may be limited. Comparisons between conventional and germ-free animals receiving adequate nutrition confirmed this idea (Coburn et al., 1989). However, some indication exists that conventional animals are slightly more resistant than germ-free rats to dietary deficiency of vitamin B_6 (Ikeda et al., 1979a,b). This consequence may be the result of increased coprophagy, as has been reported in thiamine deficiency (Wostmann et al., 1962). The complexity of interactions with the intestinal flora is illustrated by the fact that significant amounts of microbially produced folate seem to be available without coprophagy, whereas coprophagy is required for utilization of pantothenate (Daft et al., 1963). Daft et al. also commented on the need to consider the role of the oral flora and microbial growth in the food and water (Daft et al., 1963). We also have encountered problems with microbial growth in water, in animal studies (Coburn et al., 1989) and in analytical situations (Coburn, 1983).

2. Mouse (Mus)

Miller and Baumann (1945) and Morris (1947) concluded that 1 mg pyridoxine/kg diet would support normal growth in mice. If mice grow as much as 1.3 g/day (Canolty and Koong, 1976) and consume 3.5 g feed/day (Troelsen and Bell, 1963), a diet containing 4.9 μmol (1 mg) vitamin B_6/kg would provide about 17 nmol/day compared to our calculated requirement (1.3 g gain × 15 nmol vitamin B_6/g) of 19.5 nmol. Since the 1.3 g/day growth rate is unusually high (Canolty and Koong, 1976), a diet containing 4.9 μmol (1 mg)/g should be adequate for most mice.

Large variations in protein intake can influence the vitamin B_6 requirement. Miller and Baumann (1945) found that a vitamin B_6 concentration of 2.4 μmol (0.5 mg)/kg diet supported normal growth at 20% casein, whereas 9.7 μmol (2 mg)/kg was required to achieve maximal growth with 60% casein. However, the maximal growth rate achieved over 18 day with 60% casein was only about half the growth rate reported with 20% casein. No improvement was obtained, even when the vitamin B_6 content was increased to 49 μmol (10 mg)/kg. Therefore, although large increases in protein intake do increase the vitamin B_6 requirement, such diets apparently induce an abnormal metabolic state that is less than optimal, even when supplemented with vitamin B_6.

Note that the requirement of 4.9 μmol (1 mg)/kg suggested earlier is a general one. Many inbred strains of mice have altered biochemistry and may, therefore, have altered nutritional requirements. For example, the I/St² strain mouse has an increased sensitivity to vitamin B_6 deficiency (Lyon et al., 1958). Absorption and metabolic conversion of vitamin B_6 to cofactor were normal (Bell and Haskell, 1971). Urinary losses during deficiency were increased although aldehyde oxidase activity was below that of controls (Bell et al., 1971). Therefore, the cause of the sensitivity to vitamin B_6 depletion remains obscure. Whereas Lyon et al. (1958) concluded that the vitamin B_6 requirement of this strain was about 10 μmol (2 mg)/kg diet, the methods section of their paper indicates that the diet contained 100 μmol (20 mg)/kg, thus raising questions about which number is correct.

Beck et al. (1950) reported that vitamin B_6 deficiency in mice was accompanied by poor growth, posterior paralysis, alopecia, and necrotic degeneration of the tail. Hematological changes included a progressive microcytic anemia with hypersideremia and increased reticulocyte count (Keyhani et al., 1974).

3. Gerbil (Gerbillus)

We have not located any specific studies of the vitamin B_6 requirement of gerbils. Levels of 19–107 μmol (4–22 mg)/kg diet have been used (Subcommittee on Laboratory Animal Nutrition, 1978). Assuming gains of 1 g/day and food intake of about 5 g/day (Subcommittee on Laboratory Animal Nutrition, 1978), a vitamin B_6 concentration of 3 μmol (0.6 mg)/kg diet should be adequate.

4. Guinea Pig (Caviidae)

The only detailed studies of vitamin B_6 requirements in guinea pigs appear to be those of Reid (1954,1964). This investigator concluded that 9.7–14.6 μmol (2–3 mg)/kg diet was needed to achieve maximal growth and optimal fur condition with a 30% protein diet. With growth of about 7 g/day (Reid and Briggs, 1953), consumption of 8–12 g/day of such diets should meet the requirements for growth. We commented earlier on the ability of the body to conserve vitamin B_6. Therefore, the relatively mature state of guinea pigs at birth may make them fairly resistant to the effects of vitamin B_6 deficiency. For example, although growth was reduced about 50%, 7 of 12 animals were still alive after 6 wk on a vitamin B_6-deficient diet (Reid, 1954). Deficiency symptoms included poor growth, anorexia, weakness, enlargement of adrenals, atrophy of sex organs, and disturbances of the gastrointestinal tract such as hemorrhagic condition of the cecum (Reid, 1954). The animals did not develop scaly dermatitis. No biochemical and histological findings were reported.

5. Hamster (Cricetus)

Diets containing 29–97 μmol (6–20 mg) vitamin B_6/kg have supported normal growth in hamsters. Maximum growth seems to be slightly more than 2 g/day (subcommittee on Laboratory Animal Nutrition, 1978). With a food intake ≥ 5 g/day (Subcommittee on Laboratory Animal Nutrition, 1978), a diet containing 9.7 μmol (2 mg)/kg should be adequate. Deficiency was associated with marked atrophy of the thymus, which may have been partly the result of reduced food intake (Schwartzman and Strauss, 1949).

6. Ground Squirrel (Spermophilus richardsonii)

Munger and Holmes (1988) reported a gain of about 43% of the initial wt (306 ± 96 g) in 23 day with a diet containing 107 μmol (22 mg) pyridoxine hydrochloride/kg compared with a loss of about 6% on a pyridoxine-deficient diet. With a mean gain of about 6 g/day and food intake of about 25 g/day, the minimum requirement predicted by our guideline of 15 nmol/g gain would have been about 3.6 μmol (0.7 mg)/kg diet. Animals that were infected with *Trypanosoma ostospermophili* after receiving a vitamin B_6-deficient diet for 10 day showed reasonably normal growth for 10 day, then lost weight in parallel with the noninfected vitamin B_6-deficient animals. The increased initial weight gain was associated with increased food intake in the *ad libitum* group. The increase in food intake and weight gain did not occur in animals receiving a restricted intake of the vitamin B_6-deficient diet nor in animals receiving the complete diet. Also, animals receiving restricted intake of the vitamin B_6-deficient diet maintained their body weight somewhat better than those receiving the same diet *ad libitum*. Since the two *ad libitum* groups contained a total of only 9 animals and the weight gains were significantly different at only 2 of 13 dates, we conclude that further studies are needed before we can be confident that these data are reproducible.

7. *Meadow Vole (Microtus pennsylvanicus)*

Meadow voles have been used to compare forages (Shenk, 1976). No studies of vitamin B_6 requirements have been reported. Diets containing 97 µmol (20 mg)/kg provide good growth, but the true requirement is probably lower. Changes in the carbohydrate:protein ratio influenced growth although the vitamin B_6 intake was high (Shenk *et al.*, 1970).

B. RABBITS

Although coprophagy supplies some vitamins, vitamin B_6 deficiency in rabbits (*Lagomorpha*) will cause decreased growth, dermatological symptoms, and seizures. Intakes of about 190 nmol (39 µg)/day or 4.9 µmol (1 mg)/kg diet appeared adequate (Hove and Herndon, 1957). Assuming a maximum relative gain of 2%/day and food intake of 114 g/day for a 2-kg animal (Subcommittee on Rabbit Nutrition, 1966), our calculated requirement would be 40 g × 15 nmol/g or 600 nmol/day or 5 µmol (1 mg)/kg diet. The 1977 edition of *Nutrient Requirements of Rabbits* (Subcommittee on Rabbit Nutrition, 1977; p. 8) erroneously states that Hove and Herndon found the requirement to be 39 µg/g diet rather than 39 µg/rabbit/day. Therefore, the requirement of 39 mg/kg diet in Table 1 (p. 14) of that document is also erroneously high. A diet containing 15 µmol (3 mg)/kg was adequate with up to 50% protein. Lower vitamin B_6 content was not tested (Hove and Herndon, 1957).

C. CARNIVORES

1. *Cat (Felis catus)*

Cats have rather unusual vitamin B_6 metabolism. Plasma concentrations of pyridoxal 5'-phosphate are higher than in any other species examined, averaging from 1000 to over 2000 nmol/liter (Coburn et al., 1984; Bai et al., 1989). Whereas 4-pyridoxic acid is the major urinary vitamin B_6 metabolite in many species, in cats it is a minor metabolite; pyridoxal 3-sulfate, pyridoxine 3-sulfate, and N-methylpyridoxine are the major urinary forms of vitamin B_6 (Coburn and Mahuren, 1987). In addition to the common abnormalities of growth, nerve function (Buckmaster et al., 1993), and hematology, vitamin B_6 deficiency in cats is associated with oxalate nephrocalcinosis (Carvalho da Silva et al., 1959; Gershoff et al., 1959; Blanchard et al., 1991). Gershoff et al. (1959) and Bai et al. (1989) estimated the vitamin B_6 requirement of kittens to be greater than 4.9 μmol (1 mg)/kg diet but less than 9.7 μmol (2 mg)/kg. In additional studies (Bai et al., 1991), growth of vitamin B_6-depleted kittens with a diet containing 30% casein and 9.7 μmol (2 mg) vitamin B_6/kg was significantly ($p < 0.05$) greater than that of kittens with a diet containing 60% casein and the same vitamin B_6 content. The authors again concluded that the requirement was greater than 4.9 μmol/kg diet, but could not determine whether it was higher than 9.7 μmol/kg. Using our assumed maintenance requirement of 0.02 nmol/g body wt and growth requirement of 15 nmol/g, 1500-g kittens growing 20 g/day would need 30 nmol vitamin B_6 for maintenance and 300 nmol for growth, yielding a total requirement of 330 nmol (70 μg)/day. With food intake of about 60 g/day, the vitamin B_6 content would need to be about 5.5 μmol (1.1 mg)/kg diet. In the protein study, the two groups with the best growth each consumed 27.2 nmol vitamin B_6/g gain whereas the two groups with lesser growth consumed 12.9 and 17.5 nmol/g gain. Food consumption in the four groups ranged from 45 to 81 g/day with food consumption lowest with the 60% casein diets. Whether the lower consumption of the 60% casein diet reflects a metabolic and/or a palatability effect is not clear. In addition, the kittens had been placed on a vitamin B_6-deficient diet for 6 wk prior to the assessment of vitamin B_6 requirements. All these factors may have contributed to the apparent increase in the vitamin B_6 requirement above the proposed 15 nmol/g gain.

The literature just cited deals with domestic cats. A single urine specimen from a bobcat (*Felis rufus*) suggested that the unusual urinary vitamin B_6 metabolites found in domestic cats also occur in at least some other feline species (Coburn and Mahuren, 1987).

Gershoff et al. (1959) reported that the histological appearance of the bone marrow in vitamin B_6-deficient cats showed a deficiency of normoblasts compared with the hyperplasia reported in monkeys (Rinehart and Greenberg, 1949; Poppen et al., 1952). In cats, hemosiderin deposits were preponderant in hepatic cells whereas in swine (Wintrobe et al., 1943) the spleen was most involved. Dermatological changes, fatty liver, changes in lymphoid and adrenal tissue, and histological abnormalities in the nervous system were not found in the vitamin B_6-deficient cats. Vitamin B_6 supplements of 4.9 μmol (1 mg)/kg diet prevented convulsions in 3 of 4 cats and also corrected the anemia. With 9.7 μmol (2 mg)/kg diet, growth was comparable to a stock diet containing 19 μmol (4 mg)/kg. However, urinary oxalate excretion was reduced almost 50% by increasing the vitamin B_6 content from 9.7 μmol to 19 μmol/kg. Additional study of the interactions between vitamin B_6, protein, growth, oxalate, and other aspects of feline metabolism is needed to determine conclusively the optimal vitamin B_6 intake.

2. Dog (Canis familiaris)

Michaud and Elvehjem (1944) found that 24 nmol (5 μg)/kg body wt/day did not permit survival in growing puppies. According to our calculations, an intake of 292 nmol (60 μg)/kg body wt (Subcommittee on Dog Nutrition, 1985) would support growth at a rate of 2% of the body wt/day. This intake seems to be adequate. Because dogs adjust their food intake to meet energy requirements and because commercial feeds range from low moisture to high moisture, a general recommendation for the vitamin B_6 content/kg feed cannot be made. The suggested intake of 292 nmol/kg body wt would be met by a diet containing 4.9 μmol (1 mg)/kg diet if the dog consumed 60 g diet/kg body wt.

3. Fox (Vulpes)

A concentration of 10 μmol (2 mg)/kg diet was reported to prevent symptoms of vitamin B_6 deficiency in foxes (Schaefer et al., 1947). This experiment provided evidence against the concept that water-soluble vitamins are rapidly excreted and must be supplied on a daily basis. The foxes on the vitamin B_6-deficient diet stopped growing after 5 wk. At that time, they were given a single dose of 12 μmol (2.5 mg) vitamin B_6 which sustained growth for another 3 wk. Although our calculations would suggest that 12 μmol should have supported about 800 g growth, the foxes only grew about 400 g. Some of the dose would obviously be required by existing depleted tissues and some would be lost by excretion. Animals receiving the complete diet grew about 25 g/day, which would suggest a requirement of 375 nmol (77 μg)/day.

4. Mink (Mustela vision)

Sterility was observed in mink fed low vitamin B_6 plus deoxypridoxine (Helgebostad et al., 1968). The symptoms were prevented by an intake of 3 μmol (0.6 mg)/day. Leoschke (1960) recommended 9.7 μmol (2 mg)/kg feed. Bowman et al. (1968) found that 3.6 μmol (0.75 mg)/kg diet supported near normal growth but did not completely eliminate deficiency symptoms. The next level tested [7.3 μmol (1.5 mg)/kg diet] corrected both biochemical and physical deficiency symptoms. Assuming relative growth of 2.5% of the body wt/day and food intake of about 80 g/kg body wt, our calculation would suggest a requirement of about 4.7 μmol (1 mg)/kg diet. Bowman et al. (1968) also concluded that tryptophan metabolism in mink may be limited by picolinic carboxylase. Unfortunately, these researchers made no direct measurements of vitamin B_6 metabolites.

D. SWINE

Although the effects of vitamin B_6 deficiency in swine (Sus) have been studied since 1940, the requirement remains uncertain, at least in part because of the use of different assessment criteria. Husbandry practices may also influence the results. Whereas some workers apparently achieved vitamin B_6 deficiency easily, Hughes and Squibb (1942) were unsuccessful until they routinely cleaned and disinfected the floor of the pens. Apparently, their animals were able to ingest adequate microbially generated vitamin B_6 by licking the floor. Another indication of the complexities of assessing requirements is provided by the data of Russell et al. (1985b). Rats receiving 0.018 nmol pyridoxine hydrochloride/g body wt showed alterations in tryptophan metabolism and gait after 9 day (Schaeffer and Kretsch, 1987), yet swine reportedly receiving a similar intake grew almost 300 g/day for 6 mo. The only physical evidence of vitamin B_6 deficiency was slight hair loss (Russell et al., 1985b). According to our proposed method of calculating requirements, such growth would require an intake of 4.5 μmol/day compared with the estimated dietary intake of 2.2 μmol/day. Since the 4 animals tested for 6 mo weighed an average of 122 kg at the start of the experiment, even if they received only 2 μmol/day, at the end of the experiment their body content of vitamin B_6 would average over 12 nmol/g compared with our estimate of 15 nmol/g for normal animals. This difference might have been too small to detect in this study. In addition, data supplied by the author (Russell, personal communication) revealed that, although total body weight increased about 60% over 6 mo, the weight of the semimembranous muscle increased only about 30%, even in the group receiving 400 μmol vitamin B_6/day. If the amount of muscle deposited was a decreasing fraction of the weight gain, our formula might overestimate the requirements. This result is further evidence of the complexities encountered in estimating vitamin B_6 requirements after the weaning period when the relative changes in body weight and total vitamin content of the body are reduced.

The percentage stimulation of erythrocyte aspartate aminotransferase activity increased from about 20% at day 0 to 80% after 6 wk and to about 130% after 12 wk in swine receiving 2.2 μmol/day (Russell et al., 1985b). This effect was not influenced by giving 7.3 μmol/day for the last 3 wk of the study. Muscle aspartate aminotransferase also dropped to 25% of the initial activity in the group receiving 2.2 μmol/day. Since growth was unaffected despite these changes, aminotransferase may reflect vitamin B_6 intake but not necessarily absolute requirement.

The 1988 NRC recommendation for 1- to 5-kg swine (Subcommittee on Swine Nutrition, 1988) recommendation was 2.4 μmol (0.5 mg)/day. Using our estimate of 15 nmol/g gain and their estimated gain of 200 g/day, the requirement would be about 3 μmol (0.6 mg)/day. Assuming their estimated feed efficiency of 1.25 g feed/g gain, the vitamin B_6 content would need to be 12 μmol (2.5 mg)/kg. Based on growth alone, Kösters and Kirchgessner (1977) estimated the daily requirement of early weaned swine to be 2.4 μmol (0.5 mg)/day, corresponding to 9.7 μmol (2.0 mg)/kg feed. Kösters and Kirchgessner (1976) found that minimal xanthurenic acid excretion after a tryptophan load was achieved with 13.6 μmol (2.8 mg)/kg, but maximal aspartate aminotransferase activity required 17 μmol (3.5 mg)/kg. Although the minimal requirements based on our calculations are usually lower than recommendations in the literature, in this case the calculated requirement is slightly higher than the amount recommended in *Nutrient Requirements of Swine* (Subcommittee on Swine Nutrition, 1988) although it is comparable to the data of Kösters and Kirchgessner (1976). Perhaps increases in feed intake and decreases in relative growth occur rapidly enough in weanling swine to prevent a deficiency from developing at slightly lower vitamin B_6 intakes. Alternatively, the swine may benefit from microbial vitamin B_6 in their environment, as described earlier.

Although little need may exist to supplement normal corn and soybean rations with pyridoxine for growth (Easter et al., 1983), supplements have produced slight but possibly significant improvements in litter size (Richie et al., 1960; Easter et al., 1983). As noted earlier in the section on rats, few data are available on placental transport of vitamin B_6. Swine have the lowest plasma pyridoxal phosphate concentrations of any species examined (Coburn et al., 1984). A slight increase during pregnancy may be beneficial. If the swine gain about

260 g/day during gestation and weigh about 150 kg, our calculations suggest that an intake of about 7 µmol/day or 3.7 µmol (0.8 mg)/kg diet should be adequate if food intake is 1.9 kg/day. A corn–soybean diet was estimated to contain about 14 µmol (3 mg)/kg (Easter et al., 1983), yet supplementing with 4.9 µmol (1 mg)/kg diet increased litter size.

The problem of anemia in swine has been studied extensively. Cartwright and Wintrobe (1948) concluded that vitamin B_6 deficiency impaired the synthesis of protoporphyrin. Petri and Petri (1970) published an interesting series of papers on a phenomenon they called gastroprival pellagra. Basically, these investigators reported that partial or total gastrectomy in swine was associated with degeneration of the central nervous system that could be corrected by administration of vitamin B_6. The condition was associated with severe anemia, probably because of the effect of achlorhydria on iron absorption (Murray and Stein, 1970). The importance of iron in the brain has been reviewed (Beard et al., 1993). However, why gastrectomy should influence vitamin B_6 metabolism is not clear. Observations on the effect of gastrectomy or atrophic gastritis on vitamin B_6 metabolism in humans have yielded conflicting results (Brummer and Markkanen, 1963; Ribaya-Mercado et al., 1987; Turkki et al., 1992). Initial attempts in our laboratory to reproduce the results of Petri et al. using partial gastrectomy were unsuccessful (Coburn and Van Vleet, unpublished results). In fact, in their last report (Petri et al., 1980), these investigators also failed to see any degeneration of the nervous system. However, such observations may underscore the need for better understanding of the interactions between metabolism of vitamin B_6, iron, hemoglobin, and erythrocytes. Since the half-life of plasma pyridoxal phosphate is about 17 min (Coburn et al., 1992a) and pyridoxal is metabolized even more quickly, the erythrocytes may play a significant role in the transport and distribution of vitamin B_6. Erythrocytes in swine metabolized pyridoxal about twice as rapidly as erythrocytes in goats (S. P. Coburn et al., 1992a).

E. HORSES

Pyridoxine is produced in the lower digestive tract of horses (*Equidae*) (Carroll et al., 1949), but no definitive studies of requirements are available. Carroll (1950) reported that a diet deficient in B vitamins reduced muscle stores of riboflavin, pantothenic acid, nicotinic acid, biotin, and folate but not pyridoxine. This result agrees with observations in rats (Black et al., 1978), swine (Russell et al., 1985a), and humans (Coburn et al., 1991), that muscle content of vitamin B_6 is not readily altered by vitamin B_6 deficiency. Assuming that the maximum relative rate of gain is about 0.5%, our calculations would suggest that the vitamin B_6 requirement for growth would be about 75 nmol (0.015 mg)/kg body weight. Thus, a 275-kg colt growing 1.3 kg/day would have a requirement of about 20 μmol (4 mg) pyridoxine hydrochloride/day. Since oats, alfalfa meal, and various other common feeds contain over 10 μmol (2 mg)/kg (Subcommittee on Horse Nutrition, 1989), an intake of 20 μmol should be readily achieved. Using data from *Nutrient Requirements of Horses* (Subcommittee on Horse Nutrition, 1989), if a 400-kg mare produces 12 kg milk containing 1400 nmol vitamin B_6/kg, the foal would receive about 17 μmol (3.5 mg)/day. Therefore, these estimates based on either calculation or observation suggest a vitamin B_6 requirement for the horse of about 20 μmol (4 mg)/day. The horse would be a good species in which to examine the interaction between vitamin B_6 metabolism and exercise. However, we are unaware of any such studies.

F. RUMINANTS

Researchers generally assume that the microflora in the rumen can provide an adequate supply of B vitamins including vitamin B_6. Although the production of vitamins in the rumen can be altered experimentally (Pavel et al., 1968), that this would be a significant consideration for vitamin B_6 nutrition under normal circumstances seems unlikely. Therefore, concern about deficiencies is limited primarily to the initial postnatal period, when the animals are receiving only milk or milk replacements and the rumen is not fully functional, or when the animal is subjected to unusual stresses such as shipping (Zinn et al., 1987; Dubeski et al., 1993).

1. *Dairy Cattle (Bos)*

The fact that dairy calves are usually removed from the dam shortly after birth and placed on milk substitutes requires particular attention to their nutritional requirements. Johnson et al. (1950) found that calves placed on a deficient ration generally developed anorexia in 2–4 wk and seizures after 12–14 wk. No hematological changes were seen in these animals. Vitamin B_6 excretion in the urine declined markedly from 9 to 15 days, suggesting that depletion of the rapid turnover pools was associated with the development of anorexia. The observation that another 10–12 wk were needed before seizures appeared indicates that the remaining vitamin B_6 was efficiently conserved.

The observation that pyridoxic acid excretion in the urine did not appear to decline was probably the result of interference in the Huff and Perlzweig method (1944). Controls fed 0.32 µmol (65 µg)/kg body wt developed normally. The most rapid relative growth rate shown in *Nutrient Requirements for Dairy Cattle* (Subcommittee on Dairy Cattle Nutrition, 1988) is 2%/day for a 40-kg veal calf. Our calculations (15 nmol/g × 800 g/day) suggest an intake of 12 µmol/day or 0.3 µmol/kg body wt, which is comparable to the value suggested by Johnson et al. (1950).

One of the nicest demonstrations of B vitamin production in the rumen is presented in the report by Virtanen (1963) on the performance of dairy cows receiving a purified diet devoid of water-soluble vitamins for 1.5 yr. The vitamin content of the milk was unaffected by the lack of vitamins in the diet. The vitamin B_6 concentrations found by microbiological assay were generally comparable to the values found using cation-exchange chromatography (Coburn et al., 1992b). At concentrations of 2-8 µmol/liter even with milk production as low as the 6.7 kg/day reported by Virtanen, vitamin B_6 secretion in the milk would be 13 µmol (2.7 mg)/day or more. A cow producing 30-40 kg milk/day could be secreting 60-100 µmol (12-20 mg) vitamin B_6/day. Examining the kinetics of the synthesis, transport, and metabolism of this large amount of vitamin B_6 would be interesting.

2. Beef Cattle (Bos)

The requirements for beef calves would be similar to those of dairy calves. Maximum relative gain after weaning is about 1%/day in 150-kg calves (Subcommittee on Beef Cattle Nutrition, 1984). Using our calculation, this rate of growth would require 1500 g × 15 nmol/g or 22.5 µmol/day for growth plus about 3 µmol for maintenance for a total of 25.5 µmol (5 mg)/day. This amount appears to be readily supplied by the rumen. Zinn et al. (1987) found that essentially all supplemented vitamin B_6 passed through the rumen and appeared in the duodenum. These researchers estimated the net microbial synthesis of vitamin B_6 appearing in the duodenum to be 27 µmol (5.6 mg)/kg digestible organic matter. This level of synthesis would support the output mentioned earlier for lactating dairy cattle.

4. Sheep (Ovis)

We have found no specific studies of vitamin B_6 deficiency in sheep. Maximum relative gain shown in *Nutrient Requirements of Sheep* (Subcommittee on Sheep Nutrition, 1985) is 2.5%/day for 10-kg early weaned lambs. Using our calculation (15 nmol/g × 250 g/day), the requirement would be 3.75 μmol/day. Assuming food intake of 0.6 kg/day, the vitamin B_6 content would need to be 6.25 μmol (1.3 mg)/kg diet if the requirement were not met by the rumen. Odynets et al. (1975) reported that vitamin B_6 supplementation did increase energy retention in sheep and suggested that this treatment would be beneficial in cold weather.

4. *Goats (Capra)*

As would other ruminants, goats would not be expected to have a dietary vitamin B_6 requirement after weaning. Since kids are usually fed goat milk rather than milk replacement, vitamin B_6 deficiency is unlikely to be a problem. We are unaware of any specific studies of vitamin B_6 requirements in goats.

G. NONHUMAN PRIMATES

The growth rate for rhesus monkeys (*Macaca mulatta*) is about 4 g/day for at least the first year (van Wagenen and Catchpole, 1956). Therefore, we predict that the intake for growth should be at least 60 nmol (12 μg)/day. In a 500-g monkey, this amount would be equivalent to 0.12 μmol (25 μg)/kg body wt. Using the yeast assay, Greenberg and Rinehart (1949) reported vitamin B_6 concentrations of 300–1500 nmol/liter in the blood of control monkeys supplemented with 34 μmol vitamin B_6/wk for an average of almost 5 μmol/day. The authors noted that, when the animals were received and before being fed the supplement, the vitamin B_6 concentration in blood was about 150 nmol/liter. This result tends to confirm the observation, based on our estimated requirement of 60 nmol/day, that an average intake of 5 μmol/day is more than necessary. On withdrawal of vitamin B_6, the plasma concentration leveled off at 50–150 nmol/liter although body weight declined markedly. Rinehart and Greenberg (1956) estimated the requirement for maximum growth to be 0.25 μmol (51 μg)/kg/day. Emerson et al. (1960) concluded that the requirement was 2.4 μmol (500 μg)/kg/day. The reasons that these estimates are so much higher than the 60 nmol/day that was calculated to be needed for growth of new tissue is not clear. However, a requirement of 0.25 μmol/kg body wt/day is comparable to values for several other species discussed earlier. Krishnaswamy and Rao (1977) used a control diet that was estimated to provide an intake of about 10 μmol (2 mg)/day for 6-kg adult males (*Macaca radiata*). This amount yielded vitamin B_6 concentrations of about 270 nmol/liter in whole blood. Reducing the intake to about 2.6 μmol (0.5 mg)/day combined with 5 mg 4'-deoxypyridoxine/kg body wt/day reduced the vitamin B_6 concentration in blood to 41 nmol/liter. Vitamin B_6 concentrations in liver were about 29 and 15 nmol/g in the control and deficient animals, respectively. Although the experiment lasted for 14 mo, no significant differences in weight were detected in the two groups. The deficient group did show some mild physical symptoms and a slight but statistically significant drop in hemoglobin. Dzhelieva *et al.* (1974) found 24 μmol (5 mg)/kg diet to be adequate for baboons, rhesus monkeys, and green marmosets receiving a diet based on sunflower seed, dry milk, and sugar.

Although the reported total vitamin B_6 values for whole blood are quite high, Boxer *et al.* (1957) reported that, in most rhesus monkeys, pyridoxal phosphate concentrations in whole blood were less than 40 nmol/liter. Using a 1-mg intramuscular dose of [^{3}H]-pyridoxine, Greenberg and Peng (1965) estimated the average half-life to be 16.5 day.

Marsh *et al.* (1955) found that, on removing vitamin B_6 from the diet, blood vitamin B_6 concentrations dropped to about 50 nmol/liter within 2 wk and remained at that concentration. Aminotransferase activity tended to decline more slowly but increased promptly in response to supplements. These investigators determined that aminotransferase activity was related to vitamin B_6 intake by the equation

$$\text{activity} = 364(\log \mu g\ B_6/\text{kg body wt}) - 446$$

Since activity was expressed in units related to the specific method, we converted the equation to percentage change using the activity at an intake of 4.86 μmol (1000 μg)/kg as 100% and changed the intake to μmol/kg body wt. The equation then becomes

$$\%\ \text{activity} = 56.4(\log \mu\text{mol}\ B_6/\text{kg}) + 61.3$$

Determining whether this relationship has broader applications to other situations will be interesting.

H. BIRDS

1. *Chicken (Gallus domesticus)*

Chicks with severe vitamin B_6 deficiency showed severe neurological changes (e.g., ataxia, intermittent hyperactivity, convulsions) and died within 6–10 days with no obvious pathological lesions (Gries and Scott, 1972). Milder deficiencies were associated with severe perosis. The changes in cartilage structure associated with vitamin B_6 deficiency have been further examined by Massé et al. (1990). Pyridoxal concentrations in plasma seem to be the most sensitive indicators of vitamin B_6 intake (Heard and Annison, 1986; Massè et al., 1989). Comparison of the various forms of vitamin B_6 indicated that pyridoxine, pyridoxal, and pyridoxamine were equally effective in diets based on autoclaved starch (Waibel et al., 1952). The decreased effectiveness of pyridoxal and pyridoxamine in diets containing glucose or sucrose was attributed to alterations in the intestinal microflora resulting in increased microbial metabolism of the aldehyde and amine vitamers. Rogerson and Singsen (1976a,b) developed an equation to describe the interactions of high intakes of magnesium and vitamin B_6. However, even their best feed efficiencies were less than those that can be achieved with more normal intakes of these nutrients.

Our proposed estimation of vitamin B_6 requirements appears applicable to most of the data in chicks, even at very high feed efficiencies. For example, Gries and Scott (1972) reported a feed efficiency of 1.0 with a diet containing 13.6 μmol (2.8 mg) vitamin B_6/kg. This diet would have provided about 14 nmol vitamin B_6/g gain. Even so, 20% of this group did have reduced growth and developed peropsis. One significant exception to our estimated requirement of 15 nmol vitamin B_6/g gain was a group in their studies that achieved normal growth and an initial feed efficiency of 1.3 g feed/g gain for 7 day with a diet containing 22% protein and only 6.8 μmol (1.4 mg) vitamin B_6/kg. This diet would have provided only 8.8 nmol vitamin B_6/g gain. Yen et al. (1976) reported similar results for 9 day. Determining whether the vitamin B_6 content of the muscle of these young chicks averaged less than our estimate of 15 nmol/g would be interesting.

The importance of the interaction between vitamin B_6 requirement and protein is indicated by the fact that chicks receiving 31% protein and 9.2 μmol (1.9 mg) vitamin B_6/kg diet died within 2 wk although their vitamin B_6 intake was 21 nmol/g gain (Gries and Scott, 1972). With 31% protein, normal growth was achieved with 16.5 μmol (3.4 mg) vitamin B_6/kg diet; the vitamin B_6 intake was still 21 nmol/g gain. With the increased protein intake, a higher level of aminotransferase activity may become the critical factor. Saroka and Combs (1986) also found a requirement of about 24 nmol/g gain over 3 wk on a low methionine diet. Boxer et al. (1957) noted that the pyridoxal phosphate concentrations in whole blood in pigeons, chicks, and turkeys were above 800 nmol/liter, most of which was in the erythrocytes. These investigators reasoned that the composition of the nucleated erythrocytes of birds would be expected to be fairly similar to that of other body tissues. Since a 60-g egg contains about 300 nmol (66 μg) vitamin B_6 Orr, 1969) and feed consumption of laying hens exceeds 100 g/day (Subcommittee on Poultry Nutrition, 1984), feed containing sufficient vitamin B_6 for maximum growth should also be adequate for laying hens.

2. Turkey (Meleagrididae)

Sullivan et al. (1967) found that vitamin B_6 concentrations ≤ 11 μmol (2.3 mg)/kg resulted in 100% mortality in 25 day. These researchers concluded that the requirement was between 19 μmol (3.9 mg) and 21 μmol (4.4 mg)/kg feed. Waldroup et al. (1976) reported feed efficiencies of 1.44 g feed/g gain. Our calculations would suggest a requirement of 10 μmol (2 mg)/kg diet to meet the growth needs. Waldroup et al. used a natural diet based on corn and soybean meal plus other ingredients. Neither the vitamin B_6 nor the protein content was specified. These investigators found that adding 24 μmol (5 mg) vitamin B_6/kg diet did improve growth.

3. Duck (Anatidae)

Based on the growth and feed consumption data in *Nutrient Requirements of Poultry* (Subcommittee on Poultry Nutrition, 1984), our calculations would yield a maximal requirement of 15 μmol (3 mg)/kg feed. This amount is comparable to the 12.6 μmol (2.6 mg) to 15 μmol (3 mg)/kg recommended in *Nutrient Requirements of Poultry* (Subcommittee on Poultry Nutrition, 1984).

4. Japanese Quail (Coturnix japonica)

Based on growth and survival of 3-day chicks for 35 day, Mak and Vohra (1982) concluded that the requirement was 7.3 μmol (1.5 mg)/kg. Growth averaged 2.3 g/day, which according to our assumptions would require an intake of 35 nmol/day. This amount, in turn, would require a feed intake of 5 g/day of a diet containing 7.3 nmol/kg.

5. Hoatzin (Opisthocomus hoazin)

Our discussion should not close without mentioning the report of foregut fermentation in the hoatzin (Grajal et al., 1989). Such activity could influence a variety of nutritional requirements in that species.

I. AQUATIC SPECIES

An important consideration in evaluating the vitamin requirements of aquatic species is loss due to leaching from the food prior to consumption. Therefore, the amount of vitamin received by the fish is highly dependent on the procedures used to prepare the diet and feed the fish. Another noteworthy point is that most fish diets are approximately 50% protein. The effect of high protein intake on vitamin B_6 requirements has been discussed several times in this review. As noted for other species, the vitamin B_6 intake needed to achieve maximal aminotransferase activity in fish is usually greater than the intake required for maximal growth (Kissil et al., 1981). However, in several studies normal growth has been achieved with diets containing 5–10 μmol (1–2 mg) vitamin B_6/kg. This amount would seem to be less than would be required by mammals with a 50% protein diet. Knowing more about the biochemical adjustments encountered with high protein diets in various species would be interesting. Unfortunately, much of the work on vitamin requirements of fish is reported in specialized publications that we have been unable to locate. Therefore, some of our comments are based on secondary sources.

Disturbingly, a number of studies report the vitamin B_6 content of the vitamin B_6-deficient diets to be 5–15 μmol (1–3 mg)/kg diet. In our experience, rodent diets based on vitamin-free casein contain less than 0.48 μmol (0.1 mg) vitamin B_6/kg. Therefore, whether the higher values found by others indicate less efficient extraction of vitamin B_6 during preparation of the vitamin-free casein or inaccurate analyses is unclear. In either case, these high reported values in the basal diets complicate the interpretation of the results.

Fish can often achieve feed efficiencies of about 1 g/g gain. Our calculated vitamin B_6 requirements under such conditions would be 15 μmol (3 mg)/kg diet. This amount is compatible with much of the data discussed in this section. As was found in other species, vitamin B_6 deficiency symptoms in fish include seizures and anemia. In addition, at least one species of fish (*Channa punctatus*) develops opacity of the lens of the eye (Agrawal and Mahajan, 1983). Degeneration of the proximal tubules in the kidney has been observed in rainbow trout (Smith et al., 1974) and Atlantic salmon (Herman, 1985).

1. Cold Water Fish

a. *Trout and Salmon*

1. Rainbow trout (*Oncorhynchus mykiss*, formerly *Salmo gairdneri*). Early studies by McLaren et al. (1947) reported normal growth with 4.9 μmol (1 mg) and 49 μmol (10 mg)/kg diet. A concentration of 100 μmol (20 mg)/kg diet reduced growth. Halver (1972) estimated the requirement at 50–75 μmol (10–15 mg)/kg diet. However, other results (Woodward, 1990) again yielded maximal growth with 9.7 μmol

(2 mg)/kg. Maximal alanine and aspartate aminotransferase activities were achieved with levels of 14.6 μmol (3 mg) and 29 μmol (6 mg)/kg diet, respectively. Studies of other aspects of vitamin B_6 deficiency in rainbow trout have been reported by Hardy et al. (1987) and Jurss et al. (1988).

2. Other trout. Halver (1972) suggested requirements of 50–75 μmol (10–15 mg)/kg diet for brook trout (*Salmo fontinalis*) and brown trout (*Salmo trutta*). However, the requirement for these species is likely to be comparable to that for the rainbow trout, as discussed earlier, and is probably around 15 μmol (3 mg)/kg diet.

3. Atlantic salmon (*Salmo salar*). The histopathology of vitamin B_6 deficiency was reported by Herman (1985). Lall and Weerakoon (1990) found that 12 μmol (2.5 mg)/kg diet prevented clinical and histological symptoms of deficiency but 24 μmol (5 mg)/kg (the next level tested) was needed to achieve maximal growth and activity of alanine aminotransferase in the liver.

4. Chinook salmon (*Oncorhynchus tshawytscha*). Halver (1957) demonstrated a qualitative requirement for vitamin B_6 and later estimated it at 75–100 μmol (15–20 mg)/kg diet (Halver, 1972). Our estimate would be about 15 μmol (3 mg)/kg.

5. Coho salmon (*Oncorhynchus kisutsch*). Coates and Halver (1958) demonstrated a qualitative requirement and Halver (1972) estimated the requirement at 75–100 μmol (15–20 mg)/kg diet. Again, our estimate would be much lower—about 15 μmol (3 mg)/kg.

b. Herring (Clupea harengus). Blaxter et al. (1974) concluded that 2.9 μmol (0.6 mg)/kg body wt/day would be adequate. Our calculations would suggest that this intake would support a relative daily gain of almost 20%. Therefore, such an intake should be very adequate.

2. Warm Water Species

a. Fresh Water Fish

1. Carp (*Cyprinus carpio*). Ogino (1965) found that vitamin B_6 intake had little effect on growth of fingerlings for 4 wk. Maximal growth was achieved with a vitamin B_6 content of 26 μmol (5.4 mg)/kg diet. This value is higher than the 15 μmol (3 mg)/kg we calculated for an efficiency of 1 g feed/g gain.

2. Channel catfish (*Ictalurus punctatus*). Andrews and Murai (1979) reported normal growth of channel catfish fingerlings for 4 wk on an unsupplemented purified diet, reportedly containing 5.8 μmol (1.2 mg) pyridoxine hydrochloride/kg. This amount seems unusually high for a purified diet. All members of this group died by 12 wk. Supplementa-

tion with 4.9 μmol (1 mg) pyridoxine hydrochloride prevented death but yielded reduced growth. Maximum growth was achieved with a supplement of 9.7 μmol (2 mg)/kg. Since we suspect that the vitamin B_6 content of the basal diet was overestimated, the requirement may be closer to the 10 μmol (2 mg)/kg supplement than to the authors' estimate of 15 μmol (3 mg)/kg that includes the basal diet. Unexpectedly, pyridoxine hydrochloride supplements of 100 μmol (20 mg)/kg or more significantly reduced the hematocrit and hemoglobin values.

3. Eels (*Anguilla japonica*). Arai et al. (1972) observed deficiency symptoms in eels deprived of vitamin B_6. No symptoms were seen in animals receiving 195 μmol (40 mg)/kg diet. With a growth rate of 36 mg/day, the calculated requirement would be 0.5 nmol (111 ng)/day.

b. *Marine Fish.*

1. Gilthead bream (*Sparus aurata*). Supplementation of a basal diet with 4 μmol (0.85 mg)/kg permitted normal growth and prevented deficiency symptoms (Kissil et al., 1981). The vitamin B_6 content of the vitamin-free casein was given as 11 μmol (2.3 mg)/kg. Again, this value is very high in our experience. Assuming that the feed contained at least 5 μmol (1 mg), the small fish (2.67 g initial wt) consumed 18–20 nmol/g gain whereas the large group (69 g initial wt) consumed only about 10 nmol/g gain. The authors did not comment on why the larger fish achieved much better feed efficiency.

2. Red sea bream (*Chrysopheys major*). Takeda and Yone (1971) recommended about 26 μmol (5–6 mg)/kg diet.

3. Turbot (*Scophthalamus maximum*). Maximum growth was achieved at concentrations above 4.9 μmol (1 mg)/kg feed whereas maximum aminotransferase activity required 12 μmol (2.5 mg)/kg feed (Adron et al., 1978). At an average growth of 0.4 g/day, we would predict a requirement of only 6 nmol (1.2 μg)/day. With such slow growth, even the most deficient fish grew normally for 6–8 wk. The authors noted that some pyridoxine probably leached out of the feed pellets during feeding.

4. Yellowtail (*Seriola quinqueradiata*). Sakaguchi et al. (1969) reported deficiency symptoms but did not determine the requirement.

c. *Crustaceans.* Prawn (*Penaeus japonicus*). The unusually high requirement of 583 μmol (120 mg)/kg diet reported by Deshimaru and Kuroki (1979) may reflect the leaching problem mentioned earlier since the diet was cut into 3-mm cubes for feeding and prawn were fed so no feed remained the following morning. Prawn eat much more slowly than fish. Although the best gain was achieved with 583 μmol (120 mg)/kg, the authors did not indicate whether the improvement over the 292 μmol (60 mg)/kg group was statistically significant. Whereas the mammalian species examined have averaged about 15 nmol (3 μg) vitamin B_6/g body wt, the prawn contained only about 2.4 nmol (0.5 μg)/g body wt. Since the authors did not describe their procedure for preparing the tissue, determining whether the cause of this discrepancy is an analytical problem or a real difference in body composition, possibly due to the presumably low vitamin B_6 content of the shell, is difficult. Using their body content, the prawn should require an intake of only 0.05 nmol (10 ng) vitamin B_6/day/animal. Increasing the pyridoxine content of the diet to 1167 μmol (240 mg)/kg reduced growth.

J. Exotic Animals

In general, the nutrient requirements of exotic species have not been examined rigorously. The usual approach is simply to insure an excess of nutrients. The data in Table II show that 10 μmol (2 mg) vitamin B_6/kg feed would be adequate for growth and maintenance of most nonruminant species. Increasing the level to 25 μmol (5 mg)/kg during pregnancy and lactation would be prudent.

III. Discussion

Because of limitations of space, this review has been selective rather than exhaustive. Other useful reviews include the *Nutrient Requirements of Domestic Animals* series published by the National Academy Press (Washington, D.C.). Some of the earlier work in a variety of species is summarized by Fuller (1964). Although the data discussed thus far clearly demonstrate that much remains to be learned about vitamin B_6 metabolism and requirements, our current impressions can be summarized as follows. Although considerable interspecies variation occurs in the concentration and distribution of B_6 vitamers in plasma (Coburn et al., 1984) and in the end products of vitamin B_6 metabolism (Coburn and Mahuren, 1979, 1987; Mahuren et al., 1991), relatively little interspecies variation exists in vitamin B_6 requirements. This result suggests that the intermediate intracellular processing of vitamin B_6 as an enzyme cofactor is probably quite similar among species. Our proposal that vitamin B_6 is efficiently conserved and that requirements can therefore be estimated by assuming that 15

TABLE II
SUMMARY OF VITAMINE B_6 REQUIREMENTS FOR NORMAL GROWTH BASED ON *IN VIVO*
STUDIES WITH COMPARISON TO ESTIMATES CALCULATED ASSUMING 15 NMOL VITAMIN B_6
FOR EACH GRAM OF GAIN

Species	Vitamin B_6 requirement (μmol/kg feed)	
	Literature	Calculated[a]
Birds		
Chicken (*Gallus domesticus*)	14.5	12
Duck (*Anatidae*)	12.6	—[b]
Turkey (*Meleagridiae*)	19.4	10
Fish and crustaceans		
Cold water		
Brook trout (*Salmo fontinalis*)	50–75	—
Brown trout (*Salmo trutta*)	50–75	—
Rainbow trout (*Oncorhynchus mykiss*)	14.6	15
Atlantic salmon (*Salmo salar*)	12–24	15
Chinook salmon (*Oncorhynchus tshawytscha*)	75–100	15
Coho salmon (*Oncorhynchus kisutsch*)	75–100	15
Herring (*Clupea harengus*)	(2.9)[c]	—
Warm water		
Fresh water		
Carp (*Cyprinus carpio*)	26	—
Channel catfish (*Ictalurus punctatus*)	15	—
Eels (*Anguilla japonica*)	195[d]	—
Marine		
Gilthead bream (*Sparus aurata*)	9.6	7.5
Red sea bream (*Chrysopheys major*)	26	—
Turbot (*Scophthalmus maximum*)	4.9	—
Prawn (*Penaeus japonicus*)	583	—
Mammals		
Calf (*Bos*)	(0.3)	(0.3)
Cat (*Felis catus*)	4.9–9.7	5.5
Dog (*Canis familiaris*)	(0.3)	(0.3)
Fox (*Vulpes*)	10	8.2
Gerbil (*Gerbillus*)	19[d]	3.0
Ground squirrel (*Spermophilus richardsonii*)	3.6	—
Guinea pig (*Caviidae*)	15	10.5
Hamster (*Cricetus*)	30[d]	7
Horse (*Equidae*)	—	3.6
Mink (*Mustela vison*)	3.6–7.3	4.7
Mouse (*Mus*)	4.9	5.6
Nonhuman primate	(0.25–2.1)	(0.12)
Rabbit (*Lagomorpha*)	4.9	4.5
Rat (*Rattus*)	5	5
Swine (*Sus*)	7.2	12

[a] Calculated minimal requirement for normal growth (μmol/kg feed) = [(15 nmol vitamin B_6/g tissue) × (g tissue gained/day)]/(g feed eaten/day). Higher intakes may be needed to maximize biochemical parameters and certain measures of reproductive performance.
[b] Insufficient information for calculation.
[c] μmol/kg body wt/day.
[d] Lower concentrations were not tested.

nmol is needed for each g gain seems to be applicable to most young animals (Table II). Prawn were the major exception, possibly because of loss of vitamin B_6 from the diet by leaching. Unusually high vitamin B_6 requirements have also been reported for nonhuman primates, but further studies are needed to evaluate minimal requirements more thoroughly. A diet that contains sufficient vitamin B_6 to support maximal growth will usually also support the adult animal.

Investigators have shown repeatedly that the minimal intake that produced maximal growth was not necessarily associated with maximal activity of aminotransferases or minimal excretion of metabolites in response to a tryptophan load. Roth-Maier and Kirchgessner (1977) did find about a 2-fold variation in the vitamin B_6 content of various lots of feedstuffs. This variance would be another reason to plan for the optimal rather than the minimal requirement.

The vitamin B_6 requirement during pregnancy was generally greater than the requirement we would have predicted solely on the basis of fetal growth. We suspect that the plasma concentration of vitamin B_6 is the crucial factor in providing adequate placental transport of vitamin B_6. In the human, pyridoxal is the major form extracted from the maternal circulation by the placenta (Schenker et al., 1992). Both pyridoxal and pyridoxal phosphate are released into the fetal circulation.

Although increasing protein intake can increase vitamin B_6 requirements above our estimated 15 nmol/g gain, species such as carnivores and fish, which usually consume high protein diets, do not seem to have higher vitamin B_6 requirements than other species. This result suggests that the increased vitamin B_6 requirement sometimes observed with marked increases in protein intake may reflect the metabolic demand of disposing of excess amino acids and/or of utilizing amino acids as an energy source. A discussion of the high protein content of commercial infant formulas compared with breast milk suggested that the higher plasma urea concentrations in formula-fed babies indicated that the protein intake was in excess of metabolic needs (Hanning et al., 1992). Birds seem to be most sensitive to the interactions between vitamin B_6 and protein intake. Several aquatic species show reduced growth at vitamin B_6 intakes above the optimal level. These observations warrant further study.

Vitamin B_6 exists in multiple forms. The major form of vitamin B_6 in skeletal muscle is pyridoxal phosphate. In heart, liver, kidney, and brain, the concentration of pyridoxamine phosphate may equal or exceed the concentration of pyridoxal phosphate (Coburn et al., 1988b). Except under conditions of very high intakes, pyridoxine is rarely found in more than trace amounts in animal tissues. Studies in humans showed that pyridoxal is metabolized more rapidly than pyridoxine or pyridoxamine (Wozenski et al., 1980; Szadkowska et al., 1993). At this time, whether these differences would influence the minimum requirements for growth is unclear since most studies of vitamin B_6 requirements use pyridoxine as the only source of vitamin B_6 although natural diets will contain significant amounts of the aldehyde forms.

Plants contain additional vitamin B_6 derivatives. Pyridoxine B-glucoside is the most common and may account for over half of the vitamin B_6 (Gregory and Ink, 1987). Whether the glucoside form is detected by microbiological assay depends on the method of sample preparation. Determining whether current tables of composition of feed components include the glucoside may be difficult. The bioavailability of the glucoside can vary significantly among species. This form is utilized more completely by the human (Gregory et al., 1991) than by the rat (Trumbo and Gregory, 1989). Data are not available in other species. Therefore, at this time, whether incomplete bioavailability of pyridoxine glucoside is of nutritional importance under field conditions is not clear. Failure to identify significant bioavailability problems over the past 50 years suggests that any species that must normally rely on the glucoside as a major source of vitamin B_6 have developed the capability of utilizing it. However, the occurrence and bioavailability of the glucoside should be considered when formulating diets in which the amount and/or type of plant products is significantly different from the diet that the species would consume under natural conditions. The fact that new vitamin B_6 derivatives have been identified in both plants and animals in recent years further emphasizes the value of expressing all vitamin B_6 data in terms of moles rather than mass. Mass values have little meaning unless the chemical form, including any accompanying ions, is explicitly stated. All too often, such information is not provided.

IV. SUMMARY

We conclude that vitamin B_6 is efficiently conserved by the tissues and that the major dietary requirement is to meet physiological stresses such as growth, pregnancy, lactation, vigorous exercise, and/or disease. A reasonable approximation of the minimal requirement for growth in most species can be obtained by assuming 15 nmol vitamin B_6 for each g gain in body weight. For birds and fish, which can achieve 1 g gain/g feed, the vitamin B_6 content would need to be about 15 μmol (3 mg)/kg feed. Such an intake is achievable with a variety of natural diets and would also be adequate for most mammalian species, which have lower feed efficiencies. Although well-balanced natural diets are usually adequate, supplementation with 5–10 μmol (1–2 mg)/kg diet is inexpensive and would provide a safety factor to allow for variations in feed composition and physiological stresses such as pregnancy, disease, parasites, and so on. At this time, whether the metabolic changes associated with increases in vitamin B_6 intake above the minimum confer long-term health benefits on the organisms is not clear. Excessive vitamin B_6 intake can be toxic. Fish and prawn seem to be particularly sensitive to this toxicity. Additional work is needed to clarify the interactions between vitamin B_6 requirements and protein intake.

ACKNOWLEDGMENTS

The kinetic studies and modeling efforts that led to our proposal of 15 nmol vitamin B_6/g gain were supported in part by Grants 85-CRCR-1-1554 and 88-37200-3666 from the USDA Competitive Grant Program.

REFERENCES

Adron, J. W., Knox, D., Cowey, C. B., and Ball, G. T. (1978). Studies on the nutrition of marine flatfish. The pyridoxine requirement of turbot (*Scophthalmus maximus*). *Br. J. Nutr.* **40,** 261–268.

Agrawal, N. K., and Mahajan, C. L. (1983). Pathology of vitamin B-6 deficiency in *Channa punctatus* Bloch. *J. Fish Dis.* **6,** 439–450.

Allgood, V. E., and Cidlowski, J. A. (1991). Novel role for vitamin B-6 in steroid hormone action: A link between nutrition and the endocrine system. *J. Nutr. Biochem.* **2,** 523–534.

Alton-Mackey, M. G., and Walker, B. L. (1978). The physical and neuromotor development of progeny of female rats fed graded levels of pyridoxine during lactation. *Am. J. Clin. Nutr.* **31,** 76–81.

Andrews, J. W., and Murai, T. (1979). Pyridoxine requirements of channel catfish. *J. Nutr.* **109,** 533–537.

Ang, C. Y. W. (1979). Stability of three forms of vitamin B-6 to laboratory light conditions. *J. Assoc. Off. Anal. Chem.* **62,** 1170–1173.

Anonymous (1978). Phosphorylase response to vitamin B-6 feeding. *Nutr. Rev.* **36,** 55–57.

Arai, S., Nose, T., and Hashimoto, Y. (1972). Qualitative requirements of young eels *Anguilla japonica* for water soluble vitamins and their deficiency symptoms. *Bull. Freshwater Fish. Res. Lab.* **22,** 69–83.

Bai, S. C., Sampson, D. A., Morris, J. G., and Rogers, Q. R. (1989). Vitamin B-6 requirement of growing kittens. *J. Nutr.* **119,** 1020–1027.

Bai, S. C., Sampson, D. A., Morris, J. G., and Rogers, Q. R. (1991). The level of dietary protein affects the vitamin B-6 requirement of cats. *J. Nutr.* **121,** 1054–1061.

Barnard, H. C., de Kock, J. J., Vermaak, W.J.H., and Potgieter, G. M. (1987). A new perspective in the assessment of vitamin B-6 nutritional status during pregnancy in humans. *J. Nutr.* **117,** 1303–1306.

Beard, J. L., Connor, J. R., and Jones, B. C. (1993). Iron in the brain. *Nutr. Rev.* **51,** 157–170.

Beaton, G. H., and Cheney, M. C. (1965). Vitamin B-6 requirement of the male albino rat. *J. Nutr.* **87,** 125–132.

Beck, E. M., Fenton, P. F., and Cowgill, G. R. (1950). The nutrition of the mouse. IX. Studies on pyridoxine and thiouracil. *Yale J. Biol. Med.* **23**, 190–194.

Bell, R. R., and Haskell, B. E. (1971). Metabolism of vitamin B-6 in the I-strain mouse. Absorption, excretion and conversion of vitamin to enzyme cofactor. *Arch. Biochem. Biophys.* **147**, 588–601.

Bell, R. R., Blanchard, C. A., and Haskell, B. E. (1971). Metabolism of vitamin B-6 in the I-strain mouse II. Oxidation of pyridoxal. *Arch. Biochem. Biophys.* **147**, 602–611.

Bendich, A., and Cohen, M. (1990). Vitamin B-6 safety issues. *Ann. N.Y. Acad. Sci.* **585**, 321–330.

Bhagavan, H. N. (1985). Interaction between vitamin B-6 and drugs. In "Vitamin B-6: Its Role in Health and Disease" (R. D. Reynolds and J. E. Leklem, eds.), pp. 401–415. Liss, New York.

Black, A. L., Guirard, B. M., and Snell, E. E. (1977). Increased muscle phosphorylase in rats fed high levels of vitamin B-6. *J. Nutr.* **107**, 1962–1968.

Black, A. L., Guirard, B. M., and Snell, E. E. (1978). The behavior of muscle phosphorylase as a reservoir for vitamin B-6 in the rat. *J. Nutr.* **108**, 670–677.

Blanchard, P. C., Bai, S. C., Rogers, Q. R., and Morris, J. G. (1991). Pathology associated with vitamin B-6 deficiency in growing kittens. *J. Nutr.* **121**, S77–S78.

Blaxter, J.H.S., Roberts, R. J., Balbontin, F., and McQueen, A. (1974). B-group vitamin deficiency in cultured herring. *Aquaculture* **3**, 387–394.

Bowman, A. L., Travis, H. F., Warner, R. G., and Hogue, D. E. (1968). Vitamin B-6 requirements of the mink. *J. Nutr.* **95**, 554–562.

Boxer, G. E., Pruss, M. P., and Goodhart, R. S. (1957). Pyridoxal-5' phosphoric acid in whole blood and isolated leukocytes of man and animals. *J. Nutr.* **63**, 623–636.

Brown, R. R. (1981). The tryptophan load test as an index of vitamin B-6 nutrition. In "Methods in Vitamin B-6 Nutrition" (J. E. Leklem and R. D. Reynolds, eds.), pp. 321–340. Plenum Press, New York.

Brummer, P., and Markkanen, T. K. (1963). Urinary excretion of vitamin B-6 and folic acid in achlorhydria and after partial gastrectomy. *Acta Med. Scand.* **173**, 495–497.

Buckmaster, P. S., Holliday, T. A., Bai, S. C., and Rogers, Q. R. (1993). Brainstem auditory evoked potential interwave intervals are prolonged in vitamin B-6-deficient cats. *J. Nutr.* **123**, 20–26.

Canham, J. E., Baker, E. M., Raica, N., and Sauberlich, H. E. (1966). Vitamin B-6 requirement of adult men. *Proc. 7th Internat. Congr. Nutr.* **5**, 558–562.

Canolty, N. L., and Koong, L. J. (1976). Utilization of energy for maintenance and for fat and lean gains by mice selected for rapid post-weaning growth rate. *J. Nutr.* **106**, 1202–1208.

Carroll, F. D. (1950). B vitamin content in the skeletal muscle of the horse fed a B vitamin low diet. *J. Anim. Sci.* **9**, 139–142.

Carroll, F. D., Goss, H., and Howell, C. E. (1949). The synthesis of B vitamins in the horse. *J. Anim. Sci.* **8**, 290–299.

Cartwright, G. E., and Wintrobe, M. M. (1948). Studies on free erythrocyte protoporphyrin plasma copper and plasma iron in normal and in pyridoxine deficient swine. *J. Biol. Chem.* **172**, 557–565.

Carvalho da Silva, A., Fajer, A. B., de Angelis, R., Pontes, M. A., Giesbrecht, A. M., and Fried, R. (1959). The domestic cat as a laboratory animal for experimental nutrition studies VII. Pyridoxine deficiency. *J. Nutr.* **68**, 213–229.

Chen, L. H., and Marlatt, A. L. (1975). Effects of dietary vitamin B-6 levels and exercise on glutamic-pyruvic transaminase activity in rat tissues. *J. Nutr.* **105**, 401–407.

Coates, J. A., and Halver, J. E. (1958). Water soluble vitamin requirements of silver salmon. *U.S. Fish Wildlife Serv. Bureau Sport Fish. Wild. Spec. Sci. Rep. Fisheries Ser.* **281**, 1–9.

Coburn, S. P. (1981). "The Chemistry and Metabolism of 4'-Deoxypyridoxine." CRC Press, Boca Raton, Florida.

Coburn, S. P. (1983). More on contaminated water purification cartridges. *Clin. Chem.* **29**, 1872.

Coburn, S. P. (1990). Location and turnover of vitamin B-6 pools and vitamin B-6 requirements of humans. *Ann. N.Y. Acad. Sci.* **585**, 76–85.

Coburn, S. P., and Mahuren, J. D. (1979). Major urinary metabolites of 4'- and 5'-deoxypridoxines in various species. *IRCS Med. Sci.* **7**, 556.

Coburn, S. P., and Mahuren, J. D. (1987). Identification of pyridoxine 3-sulfate, pyridoxal 3-sulfate and N-methylpyridoxine as major urinary metabolites of vitamin B-6 in domestic cats. *J. Biol. Chem.* **262**, 2642–2644.

Coburn, S. P., and Townsend, D. W. (1992). A self-regulating model of vitamin B-6 metabolism based on enzyme and binding kinetics. *FASEB J.* **6**, A1373.

Coburn, S. P., Mahuren, J. D., Schaltenbrand, W. E., Wostmann, B. S., and Madsen, D. (1981). Effects of vitamin B-6 deficiency and 4'-deoxypyridoxine on pyridoxal phosphate concentrations, pyridoxine kinase and other aspects of metabolism in the rat. *J. Nutr.* **111**, 391–398.

Coburn, S. P., Mahuren, J. D., and Guilarte, T. R. (1984). Vitamin B-6 content of plasma of domestic animals determined by HPLC, enzymatic and radiometric microbiologic methods. *J. Nutr.* **114**, 2269–2273.

Coburn, S. P., Mahuren, J. D., Szadkowska, Z., Schaltenbrand, W. E., and Townsend, D. W. (1985). Kinetics of vitamin B-6 metabolism examined in miniature swine by continuous administration of labelled pyridoxine. *In* "Mathematical Models in Experimental Nutrition" (N. L. Canolty and T. P. Cain, eds.), pp. 99–111. University of Georgia, Athens.

Coburn, S. P., Lewis, D. N., Fink, W. J., Mahuren, J. D., Schaltenbrand, W. E., and Costill, D. L. (1988a). Human vitamin B-6 pools estimated through muscle biopsies. *Am. J. Clin. Nutr.* **48**, 291–294.

Coburn, S. P., Mahuren, J. D., Kennedy, M. S., Schaltenbrand, W. E., Sampson, D. A., O'Connor, D. K., Snyder, D. L., and Wostmann, B. S. (1988b). B-6 vitamer content of rat tissues measured by isotope tracer and chromatographic methods. *BioFactors* **1**, 307–312.

Coburn, S. P., Mahuren, J. D., Wostmann, B. S., Snyder, D. L., and Townsend, D. W. (1989). Role of intestinal microflora in the metabolism of vitamin B-6 and 4'-deoxypyridoxine examined using germfree guinea pigs and rats. *J. Nutr.* **119**, 181–188.

Coburn, S. P., Ziegler, P. J., Costill, D. L., Mahuren, J. D., Fink, W. J., Schaltenbrand, W. E., Pauly, T. A., Pearson, D. R., Conn, P. S., and Guilarte, T. R. (1991). Response of vitamin-B-6 content of muscle to changes in vitamin B-6 intake in men. *Am. J. Clin. Nutr.* **53**, 1436–1442.

Coburn, S. P., Mahuren, J. D., Kennedy, M. S., Schaltenbrand, W. E., and Townsend, D. W. (1992a). Metabolism of [^{14}C]- and [^{32}P]pyridoxal 5'-phosphate and [^{3}H]pyridoxal administered intravenously to pigs and goats. *J. Nutr.* **122**, 393–401.

Coburn, S. P., Mahuren, J. D., Pauly, T. A., Ericson, K. L., and Townsend, D. W. (1992b). Alkaline phosphatase activity and pyridoxal phosphate concentrations in the milk of various species. *J. Nutr.* **122**, 2348–2353.

Daft, F. S., McDaniel, E. C., Herman, L. G., Romine, M. K., and Hegner, J. R. (1963). Role of coprophagy in utilization of B vitamins synthesized by intestinal bacteria. *Fed. Proc.* **22**, 129–133.

Deshimaru, O., and Kuroki, K. (1979). Requirements of prawn for dietary thiamin, pyridoxine, and choline chloride. *Bull. Jpn. Soc. Sci. Fish.* **45**, 363–367.

Driskell, J. A., Strickland, L. A., Poon, C. H., and Foshee, D. P. (1973). The vitamin B-6 requirement of the male rat as determined by behavioral patterns, brain pyridoxal phosphate and nucleic acid composition and erythrocyte alanine aminotransferase activity. *J. Nutr.* **103**, 670–680.

Dubeski, P. L., Owens, F. N., d'Offay, J. M., and Song, W. O. (1993). Plasma B-vitamins in stressed weaned beef calves infected with bovine herpesvirus-1. *FASEB J.* **7**, A524.

Dzhelieva, Z. N., Kyulyan, G. M., Soloveva, G. A., Kotrikadze, N. G., and Bazilevskaya, L. G. (1974). Study of the requirement for some vitamins during their combined use in feed pellets for monkeys. *Ref. Nauchn. Soobsheh-Vses Biokhim. Sezd.* **2**, 187.

Easter, R. A., Anderson, P. A., Michel, E. J., and Corley, J. R. (1983). Response of gestating gilts and starter, grower and finisher swine to biotin, pyridoxine, folacin and thiamine additions to a corn-soybean meal diet. *Nutr. Rep. Int.* **28**, 945–954.

Emerson, G. A., Walker, J. B., and Ganapathy, S. N. (1960). Vitamin B-6 and lipid metabolism in monkeys. *Am. J. Clin. Nutr.* **8**, 424–433.

Fuller, H. L. (1964). Vitamin B-6 in farm animal nutrition and pets. *Vit. Horm.* **22**, 659–676.

Gershoff, S. N., Faragalla, F. F., Nelson, D. A., and Andrus, S. B. (1959). Vitamin B-6 deficiency and oxalate nephrocalcinosis in the cat. *Am. J. Med.* **27**, 72–80.

Grajal, A., Strahl, S. D., Parra, R., Dominguez, M. J., and Neher, A. (1989). Foregut fermentation in the Hoatzin, a tropical leaf-eating bird. *Science* **245**, 1236–1238.

Greenberg, L. D., and Peng, C. T. (1965). Metabolism of tritium labeled pyridoxine in the monkey. *Fed. Proc.* **24**, 625.

Greenberg, L. D., and Rinehart, J. F. (1949). Blood pyridoxine of vitamin B-6 deficient monkeys. *Proc. Soc. Exp. Biol. Med.* **70**, 20–29.

Gregory, J. F., and Ink, S. L. (1987). Identification and quantification of pyridoxine-beta-glucoside as a major form of vitamin B-6 in plant-derived foods. *J. Agric. Food Chem.* **35**, 76–82.

Gregory, J. F., Trumbo, P. R., Bailey, L. B., Toth, J. P., Baumgartner, T. G., and Cerda, J. J. (1991). Bioavailability of pyridoxine-5′-beta-D-glucoside determined in humans by stable-isotopic methods. *J. Nutr.* **121**, 177–186.

Gries, C. L., and Scott, M. L. (1972). The pathology of pyridoxine deficiency in chicks. *J. Nutr.* **102**, 1259–1267.

Groziak, S. M. and Kirksey, A. (1987). Effects of maternal dietary restriction in vitamin B-6 on neocortex development in rats: B-6 vitamer concentrations, volume and cell estimates. *J. Nutr.* **117**, 1045–1052.

Guilarte, T. R. (1986). Radiometric microbiological assay of vitamin B-6 and derivatives. *In* "Vitamin B-6" (D. Dolphin, R. Paulson, and O. Avramovic, eds.), pp. 595–627. Wiley, New York.

Halver, J. E. (1957). Nutrition of salmonoid fishes. III. Water-soluble vitamin requirements of chinook salmon. *J. Nutr.* **62**, 225–243.

Halver, J. E. (1972). The vitamins. *In* "Fish Nutrition" (J. E. Halver, ed.), pp. 29–103. Academic Press, New York.

Hanning, R. M., Paes, B., and Atkinson, S. A. (1992). Protein metabolism and growth of term infants in response to a reduced-protein, 40:60 whey: Casein formula with added tryptophan. *Am. J. Clin. Nutr.* **56**, 1004–1011.

Hardy, R. W., Casillas, E., and Masumoto, T. (1987). Determination of vitamin B-6 deficiency in rainbow trout by liver enzyme assay and HPLC analysis. *Can. J. Fish Aquatic Sci.* **44**, 219–222.

Heard, G. S., and Annison, E. F. (1986). Gastrointestinal absorption of vitamin B-6 in the chicken (*Gallus domesticus*). *J. Nutr.* **116,** 107–120.

Helgebostad, A., Svenkerud, R. R., and Ender, F. (1968). Sterility in mink induced experimentally by deficiency of vitamin B-6. *Acta Vet. Scand.* **4,** 228–237.

Helmreich, E.J.M. (1992). How pyridoxal 5'-phosphate could function in glycogen phosphorylase catalysis. *BioFactors* **3,** 159–172.

Hemilä, H. (1991). Is there a biochemical basis for nutrient need? *Trends Food Sci. Technol.* **2,** 73.

Herman, R. L. (1985). Histopathology associated with pyridoxine deficiency in Atlantic salmon (*Salmo salar*). *Aquaculture* **46,** 173–177.

Hove, E. L., and Herndon, J. F. (1957). Vitamin B-6 deficiency in rabbits. *J. Nutr.* **61,** 127–136.

Huff, J. W., and Perlzweig, W. A. (1944). A product of oxidative metabolism of pyridoxine, 2-methyl-3-hydroxy-4-carboxy-5-hydroxymethylpyridine(4-pyridoxic acid). *J. Biol. Chem.* **155,** 345–355.

Hughes, E. H., and Squibb, R. L. (1942). Vitamin B-6 in the nutrition of the pig. *J. Anim. Sci.* **1,** 320–325.

Ikeda, M., Hosotani, T., Kurimoto, K., Mori, T., Ueda, T., Kotake, Y., and Sakakibara, B. (1979a). The differences of the metabolism related to vitamin B-6 dependent enzymes among vitamin B-6 deficient germ-free and conventional rats. *J. Nutr. Sci. Vitaminol.* **25,** 131–139.

Ikeda, M., Hosotani, T., Ueda, T., Kotake, Y., and Sakakibara, B. (1979b). Effect of vitamin B-6 deficiency on levels of several water soluble vitamins in tissues of germfree and conventional rats. *J. Nutr. Sci. Vitaminol.* **25,** 141–149.

Johansson, S., Lindstedt, S., Register, U., and Wadstrom, L. (1966). Studies on the metabolism of labelled pyridoxine in man. *Am. J. Clin. Nutr.* **18,** 185–196.

Johnson, B. C., Pinkos, J. A., and Burke, K. A. (1950). Pyridoxine deficiency in the calf. *J. Nutr.* **40,** 309–322.

Jurss, K., Bastrop, R., and Vokler, T. (1988). Effects of salinity on vitamin B-6 deficiency in rainbow trout (*Salmo gairdneri* Richardson). *Comp. Biochem. Physiol.* **90B,** 891–895.

Kelsay, J., Baysal, A., and Linkswiler, H. (1968). Effect of vitamin B-6 depletion on the pyridoxal, pyridoxamine, and pyridoxine content of the blood and urine of men. *J. Nutr.* **94,** 490–494.

Keyhani, M., Guilaiani, D., and Morse, B. S. (1974). Erythropoiesis in pyridoxine-deficient mice. *Proc. Soc. Exp. Biol. Med.* **146,** 114–119.

Kirksey, A., Pang, R. L., and Lin, W. J. (1975). Effects of different levels of pyridoxine fed during pregnancy superimposed upon growth in the rat. *J. Nutr.* **105,** 607–615.

Kissil, G. W., Cowey, C. B., Adron, J. W., and Richards, R. H. (1981). Pyridoxine requirement of gilthead bream, Sparus aurata. *Aquaculture* **23,** 243–255.

Klosterman, H. J. (1979). Vitamin B-6 antagonists of natural origin. *Meth. Enzymol.* **62,** 483–495.

Kösters, W. W., and Kirchgessner, M. (1976). Effect of varying vitamin B-6 intake of early-weaned piglets on urinary xanthurenic and kynurenic acid excretion, serum transaminase activity and urea concentration. *Int. J. Vit. Nutr. Res.* **46,** 373–380.

Kösters, W. W., and Kirchgessner, M. (1977). Optimum vitamin B-6 requirement of early-weaned piglets. *Arch. Tierernähr.* **27,** 629–634.

Kretsch, M. J., Sauberlich, H. E., and Newbrun, E. (1991). Electroencephalographic changes and periodontal status during short-term vitamin-B-6 depletion of young, nonpregnant women. *Am. J. Clin. Nutr.* **53,** 1266–1274.

Krishnasawamy, K., and Rao, S. B. (1977). Failure to produce atherosclerosis in *Macaca radiata* on a high methionine, high fat pyridoxine deficient diet. *Atherosclerosis* **27**, 253–258.

Lall, S. P., and Weerakoon, D.E.M. (1990). Vitamin B-6 requirement of Atlantic salmon (*Salmo salar*). *FASEB J.* **4**, A912.

Leklem, J. (1990a). Vitamin B-6: A status report. *J. Nutr.* **120**, 1503–1507.

Leklem, J. E. (1990b). Vitamin B-6. *Food Sci. Technol.* **40**, 341–392.

Leoschke, W. L. (1960). "Mink Nutrition Research at the University of Wisconsin." Research Bulletin 222, University of Wisconsin, Madison, Wisconsin.

Linkswiler, H. M. (1981). Methionine metabolite excretion as affected by a vitamin B-6 deficiency. *In* "Methods in Vitamin B-6 Nutrition" (J. E. Leklem and R. D. Reynolds, eds.), pp. 373–381. Plenum Press, New York.

Lumeng, L., Brashear, R. E., and Li, T. K. (1974). Pyridoxal 5′-phosphate in plasma: Source, protein-binding, and cellular transport. *J. Lab. Clin. Med.* **84**, 334–343.

Lumeng, L., Ryan, M. P., and Li, T. K. (1978). Validation of the diagnostic value of plasma pyridoxal 5′-phosphate measurements in vitamin B-6 nutrition of the rat. *J. Nutr.* **108**, 545–553.

Lyon, J. B., Williams, H. L., and Arnold, E. A. (1958). The pyridoxine deficient state in two strains of inbred mice. *J. Nutr.* **66**, 261–275.

Mahuren, J. D., Pauly, T. A., and Coburn, S. P. (1991). Identification of 5-pyridoxic acid and 5-pyridoxic acid lactone as metabolites of vitamin B-6 in humans. *J. Nutr. Biochem.* **2**, 449–453.

Mak, T. K., and Vohra, P. (1982). Thiamin, riboflavin, pyridoxine and niacin requirements of growing Japanese quail fed purified diets. *Pertanika* **5**, 66–71.

Manore, M. M., Leklem, J. E., and Walter, M. C. (1987). Vitamin B-6 metabolism as affected by exercise in trained and untrained women fed diets differing in carbohydrate and vitamin B-6 content. *Am. J. Clin. Nutr.* **46**, 995–1004.

Marsh, M. E., Greenberg, L. D., and Rinehart, J. F. (1955). The relationship between pyridoxine ingestion and transaminase activity I. Blood hemolysates. *J. Nutr.* **56**, 115–127.

Massé, P. G., Vuilleumier, J. P., and Weiser, H. (1989). Pyridoxine status as assessed by the concentration of B-6 aldehyde vitamers. *Int. J. Vit. Nutr. Res.* **59**, 344–352.

Massé, P. G., Colombo, V. E., Gerber, F., Howell, D. S., and Weiser, H. (1990). Morphological abnormalities in vitamin B-6 deficient tarsometatarsal chick cartilage. *Scan. Microsc.* **4**, 667–674.

McCormick, D. B. (1989). Two interconnected B vitamins: riboflavin and pyridoxine. *Physiol. Rev.* **69**, 1170–1198.

McLaren, B. A., Keller, E., O'Donnell, D. J., and Elvehjem, C. J. (1947). The nutrition of rainbow trout I. Studies of vitamin requirements. *Arch. Biochem. Biophys.* **15**, 169–178.

Michaud, J., and Elvehjem, C. A. (1944). The nutritional requirements of the dog. *Nutr. Abst. Rev.* **13**, 324–331.

Mickelsen, O. (1956). Intestinal synthesis of vitamins in the nonruminant. *Vit. Horm.* **14**, 1–95.

Miller, E. C., and Baumann, C. A. (1945). Relative effects of casein and tryptophan on the health and xanthurenic acid excretion of pyridoxine deficient mice. *J. Biol. Chem.* **157**, 551–562.

Morris, H. P. (1947). Vitamin requirements of the mouse. *Vit. Horm.* **5**, 175–195.

Munger, J. C., and Holmes, J. C. (1988). Benefits of parasitic infection. A test using a ground squirrel trypanosome system. *Can. J. Zool.* **66**, 222–227.

Murray, J., and Stein, N. (1970). The effect of achylia gastrica in rats on the absorption of dietary iron. *Proc. Soc. Exp. Biol. Med.* **133**, 183–184.

Nomenclature Committee of the International Union of Biochemistry and Molecular Biology (1992). "Enzyme Nomenclature." Academic Press, New York.

Odynets, R. M., Aituganov, M. D., and Korokhova, V. V. (1975). Effect of vitamin B-6 and ammonium sulfate on the use of feed by growing sheep. *Fiziol. -Biokhim. Obosnovanie Normirovaniya Energ. Pitan. Vysokoproduktivnykh. Zhivotn. Tezisy. Dokl. Vses. Konf.* 73–74.

Ogino, C. (1965). B vitamin requirements of carp, *Cyprinus carpio*. I. Deficiency symptoms and requirements of vitamin B-6. *Bull. Jpn. Soc. Sci. Fish.* **31**, 546–551.

Orr, M. L. (1969). "Pantothenic Acid, Vitamin B-6 and Vitamin B-12 in Foods." Home Economics Research Report No. 36. U.S. Government Printing Office, Washington, D.C.

Pauly, T. A., Szadkowska, Z., Coburn, S. P., Mahuren, J. D., Schaltenbrand, W. E., Booth, L., Hachey, D. L., Ziegler, P. J., Costill, D. L., Fink, W. J., Pearson, D. R., Townsend, D. W., Micelli, R., and Guilarte, T. R. (1991). Kinetics of deuterated vitamin B-6 metabolism in men on a marginal vitamin B-6 intake. *FASEB J.* **5**, A1660.

Pavel, J., Zak, J., and Tylecek, J. (1968). Dynamics of the synthesis of several water soluble vitamins in the rumen of ruminants. *Arch. Tierernähr.* **18**, 264–273.

Petri, S., and Petri, C. (1970). Effect of vitamin B-6 upon gastroprival central nervous system degeneration. *Acta Med. Sci.* **187**, 129–131.

Petri, S., Rasmussen, F., and Petri, C. (1980). Anemia and growth retardation in totally gastrectomized swine. *Acta Vet. Scand.* **21**, 197–208.

Polansky, M. (1981). Microbiological assay of vitamin B-6 in foods. *In* "Methods in Vitamin B-6 Nutrition" (J. E. Leklem and R. D. Reynolds, eds.), pp. 21–44. Plenum Press, New York.

Poppen, K. J., Greenberg, L. D., and Rinehart, J. F. (1952). The blood picture of pyridoxine deficiency in the monkey. *Blood* **7**, 436–444.

Raiten, D. J., Reynolds, R. D., Andon, M. B., Robbins, S. T., and Fletcher, A. B. (1991). Vitamin B-6 metabolism in premature infants. *Am. J. Clin. Nutr.* **53**, 78–83.

Reid, M. E. (1954). Nutritional studies with the guinea pig. B-vitamins other than pantothenic acid. *Proc. Soc. Exp. Biol. Med.* **85**, 547–550.

Reid, M. E. (1964). Nutritional studies in the guinea pig. XI. Pyridoxine. *Proc. Soc. Exp. Biol. Med.* **116**, 289–290.

Reid, M. E., and Briggs, G. M. (1953). Development of a semi-synthetic diet for young guinea pigs. *J. Nutr.* **51**, 341–354.

Reynolds, R. D. (1983). Nationwide assay of vitamin B-6 in human plasma by different methods. *Fed. Proc.* **42**, 665.

Ribaya-Mercado, J. D., Otradovec, C. L., Russell, R. M., and Samloff, I. M. (1987). Atrophic gastritis does not impair vitamin B-6 status in the elderly. *Gastroenterology* **93**, 222.

Rinehart, J. F., and Greenberg, L. d. (1949). Arteriosclerotic lesions in pyridoxine-deficient monkeys. *Am. J. Pathol.* **25**, 481–491.

Rinehart, J. F., and Greenberg, L. D. (1956). Vitamin B-6 deficiency in the rhesus monkey with particular reference to the occurrence of atherosclerosis, dental caries, and hepatic cirrhosis. *Am. J. Clin. Nutr.* **4**, 318–328.

Ritchie, H. D., Miller, E. R., Ullrey, D. E., Hoefer, J. A., and Luecke, R. W. (1960). Supplementation of the swine gestation diet with pyridoxine. *J. Nutr.* **70**, 491–496.

Rogerson, G., and Singsen, E. P. (1976a). Effects of magnesium and high dietary intakes of pyridoxine on the chick. *Poultry Sci.* **55**, 883–891.

Rogerson, G., and Singsen, E. P. (1976b). Mineral metabolism in chicks on high dietary pyridoxine and magnesium. *Poultry Sci.* **55,** 1187–1194.
Roth-Maier, D. A., and Kirchgessner, M. (1977). Selected problems of B vitamins in animal nutrition. *Livestock Prod. Sci.* **4,** 177–189.
Roth-Maier, D. A., and Kirchgessner, M. (1981). Homoeostasis and requirement of vitamin B-6 of the growing rat. *Z. Tierphysiol. Tierernährg. Futtermittelkde.* **46,** 247–254.
Russell, L. E., Bechtel, P. J., and Easter, R. A. (1985a). Effect of deficient and excess dietary vitamin B-6 on aminotransferase and glycogen phosphorylase activity and pyridoxal phosphate content in two muscles from post pubertal gilts. *J. Nutr.* **115,** 1124–1135.
Russell, L. E., Easter, R. A., and Bechtel, P. J. (1985b). Evaluation of the erythrocyte aspartate aminotransferase activity coefficient as an indicator of the vitamin B-6 status of postpubertal gilts. *J. Nutr.* **115,** 1117–1123.
Sakaguchi, H., Takeda, F., and Tange, K. (1969). Vitamin requirements of yellowtails I. Vitamin B-6 and vitamin C deficiency. *Bull. Jpn. Soc. Sci. Fish.* **35,** 1201–1206.
Sakurai, T., Asakura, T., and Matsuda, M. (1988). Transport and metabolism of pyridoxine in the intestine of the mouse. *J. Nutr. Sci. Vitaminol.* **43,** 179–187.
Sakurai, T., Asakura, T., Mizuno, A., and Matsuda, M. (1992). Absorption and metabolism of pyridoxamine in mice. 2. Transformation of pyridoxamine to pyridoxal in intestinal tissues. *J. Nutr. Sci. Vitaminol.* **38,** 227–233.
Sampson, D. A., and O'Connor, D. K. (1989). Response of B-6 vitamers in plasma, erythrocytes and tissues to vitamin B-6 depletion and repletion in the rat. *J. Nutr.* **119,** 1940–1948.
Sampson, D. A., Young, L. A., and Kretsch, M. J. (1988). Marginal intake of vitamin B-6. Effects on protein sythesis in liver, kidney and muscle of the rat. *Nutr. Res.* **8,** 309–319.
Saroka, J. M., and Combs, G. F. (1986). The lack of effect of a pyridoxine deficiency on the utilization of the hydroxyl analog of methionine by the chick. *Poultry Sci.* **65,** 764–768.
Schaeffer, M. C., and Kretsch, M. J. (1987). Quantitative assessment of motor and sensory function in vitamin B-6 deficient rats. *Nutr. Res.* **7,** 851–863.
Schaefer, A. E., Whitehair, C. K., and Elvehjem, C. A. (1947). The importance of riboflavin, pantothenic acid, niacin and pyridoxine in the nutrition of foxes. *J. Nutr.* **34,** 131–139.
Schenker, S., Johnson, R. F., Mahuren, J. D., Henderson, G. I., and Coburn, S. P. (1992). Human placental vitamin B-6 (pyridoxal) transport—Normal characteristics and effects of ethanol. *Am. J. Physiol.* **262,** R966–R974.
Schwartzman, G., and Strauss, L. (1949). Vitamin B-6 deficiency in the Syrian hamster. *J. Nutr.* **38,** 131–153.
Shenk, J. S. (1976). The meadow vole as an experimental animal. *Lab. Anim. Sci.* **26,** 644–669.
Shenk, J. S., Elliott, F. C., and Thomas, J. W. (1970). Meadow vole nutrition studies with semi-synthetic diets. *J. Nutr.* **100,** 1437–1446.
Sherman, H. (1954). Pyridoxine and related compounds. In "The Vitamins" (W. H. Sebrell and R. S. Harris, eds.), Vol. 3, pp. 265–276. Academic Press, New York.
Shultz, T. D., and Leklem, J. E. (1981). Urinary 4-pyridoxic acid, urinary vitamin B-6 and plasma pyridoxal phosphate as measures of vitamin B-6 status and dietary intake in adults. *In* "Methods in Vitamin B-6 Nutrition" (J. E. Leklem and R. D. Reynolds, eds.), pp. 297–320. Plenum Press, New York.

Skala, J. H., Waring, P. P., Lyons, M. F., Rusnak, M. G., and Alletto, J. S. (1981). Methodology for determination of blood aminotransferases. In "Methods in Vitamin B-6 Nutrition" (J. E. Leklem and R. D. Reynolds, eds.), pp. 171–202. Plenum Press, New York.

Smith, C. E., Brin, M., and Halver, J. E. (1974). Biochemical, physiological and pathological changes in pyridoxine deficient rainbow trout (*Salmo gairdneri*). *J. Fish Res. Bd. Can.* **31,** 1893–1898.

Snyder, D. L., and Wostmann, B. S. (1989). The design of the Lobund aging study and the growth and survival of the Lobund-Wistar rat. In "Dietary Restriction and Aging" (D. L. Snyder, ed.), pp. 39–49. Liss, New York.

Subcommittee on Beef Cattle Nutrition (1984). "Nutrient Requirements of Beef Cattle." National Academy Press, Washington, D.C.

Subcommittee on Dairy Cattle Nutrition (1988). "Nutrient Requirements of Dairy Cattle." National Academy Press, Washington, D.C.

Subcommittee on Dog Nutrition (1985). "Nutrient Requirements of Dogs." National Academy Press, Washington, D.C.

Subcommittee on Horse Nutrition (1989). "Nutrient Requirements of Horses." National Academy Press, Washington, D.C.

Subcommittee on Laboratory Animal Nutrition (1978). "Nutrient Requirements of Laboratory Animals." National Academy Press, Washington, D.C.

Subcommittee on Poultry Nutrition (1984). "Nutrient Requirements of Poultry." National Academy Press, Washington, D.C.

Subcommittee on Rabbit Nutrition (1966). "Nutrient Requirements of Rabbits." National Academy Press, Washington, D.C.

Subcommittee on Rabbit Nutrition (1977). "Nutrient Requirements of Rabbits." National Academy Press, Washington, D.C.

Subcommittee on Sheep Nutrition (1985). "Nutrient Requirements of Sheep." National Academy Press, Washington, D.C.

Subcommittee on Swine Nutrition (1988). "Nutrient Requirements of Swine," 9th Ed. National Academy Press, Washington, D.C.

Sullivan, T. W., Heil, H. M., and Armintront, M. E. (1967). Dietary thiamin and pyridoxine requirements of young turkeys. *Poultry Sci.* **46,** 1560–1564.

Szadkowska, Z., Coburn, S. P., Hachey, D. L., Booth, L., Schaltenbrand, W. E., and Townsend, D. W. (1993). Urinary excretion of labelled pyridoxic acid after simultaneous oral administration of equimolar doses of D2-pyridoxamine, D3-pyridoxal, and D5-pyridoxine. *FASEB J.* **7,** A743.

Takeda, T., and Yone, Y. (1971). Studies on nutrition of red sea bream II. Comparison of vitamin B-6 requirement level between fish fed a synthetic diet and fish fed beef liver during pre-feeding period. *Rep. Fish Res. Lab. Kyushu Univ.* **1,** 37–47.

Tillotson, J. A., Sauberlich, H. E., Baker, E. M., and Canham, J. E. (1966). Uses of ^{14}C-labeled vitamins in human nutrition studies. Pyridoxine. *Proc. 7th Int. Congr. Nutr.* **5,** 554–557.

Toukairin-Oda, T., Sakamoto, E., Hirose, N., Mori, M., Itoh, T., and Tsuge, H. (1989). Determination of vitamin B-6 derivatives in foods and biological materials by reversed phase HPLC. *J. Nutr. Sci. Vitaminol.* **35,** 171–180.

Troelsen, J. E., and Bell, J. M. (1963). A comparison of nutritional affects in swine and mice. Response in feed intake, feed efficiency and carcass characteristics to similar diets. *Can. J. Anim. Sci.* **43,** 294–302.

Trumbo, P. R., and Gregory, J. F. (1989). The fate of dietary pyridoxine-beta-glucoside in the lactating rat. *J. Nutr.* **119,** 36–39.

Turkki, P. R., Ingerman, L., Schroeder, L. A., Chung, R. S., Chen, M., Russo-McGraw, M. A., and Dearlove, J. (1992). Thiamin and vitamin-B(6) intakes and erythrocyte transketolase and aminotransferase activities in morbidly obese females before and after gastroplasty. *J. Am. Coll. Nutr.* **11**, 272–282.

van den Berg, H., Bogaards, J.J.P., Sinkeldam, E. J., and Schreurs, W.H.P. (1982). Effect of different levels of vitamin B-6 in the diet of rats on the content of pyridoxamine 5'-phosphate and pyridoxal 5'-phosphate in the liver. *Int. J. Vit. Nutr. Res.* **52**, 407–416.

van Wagenen, G., and Catchpole, H. R. (1956). Physical growth of the rhesus monkey (*Macaca mulatta*). *Am. J. Phys. Anthropol.* **14**, 245–273.

Virtanen, A. I. (1963). Production of cow milk without protein with urea and ammonium salts as a nitrogen source and purified carbohydrates as an energy source. *Biochem. Z.* **338**, 443–453.

Waibel, P. E., Cravens, W. W., and Snell, E. E. (1952). The effect of diet on the comparative activities of pyridoxal, pyridoxamine and pyridoxine for chicks. *J. Nutr.* **48**, 533–538.

Waldroup, P. W., Maxey, J. F., Luther, L. W., Jones, B. D., and Meshew, M. L. (1976). Factors affecting the response of turkeys to biotin and pyridoxine supplementation. *Univ. Arkansas Bull.* **805**, 1–26.

Wintrobe, M. W., Follis, R. H., Miller, M. H., Stein, H. J., Alcayaga, R., Humphreys, S., Suksta, A., and Cartwright, G. E. (1943). Pyridoxine deficiency in swine. *Bull. Johns Hopkins Hosp.* **72**, 1–25.

Woodward, B. (1990). Dietary vitamin B-6 requirements of young rainbow trout (*Oncorhynchus mykiss*). *FASEB J.* **4**, A912.

Wostmann, B. S., Knight, P. L., and Kan, D. F. (1962). Thiamin in germfree and conventional animals: Effect of the intestinal microflora on thiamin metabolism of the rat. *Ann. N.Y. Acad. Sci.* **98**, 516–527.

Wozenski, J. R., Leklem, J. E., and Miller, L. T. (1980). The metabolism of small doses of vitamin B-6 in men. *J. Nutr.* **110**, 275–285.

Yen, J. T., Jensen, A. H., and Baker, D. H. (1976). Assessment of the concentration of biologically available vitamin B-6 in corn and soybean meal. *J. Anim. Sci.* **42**, 866–870.

Zinn, R. A., Owens, F. N., Stuart, R. L., Dunbar, J. R., and Norman, B. B. (1987). B-vitamin supplementation of diets for feedlot calves. *J. Anim. Sci.* **65**, 267–277.

Zubay, G. (1988). "Biochemistry," 2d Ed. Macmillan, New York.

Index

A

ADP-ribosyl cyclase, 208–212
 amino acid sequence, 209–211
 occurrence in animal tissues, 259–260
 soluble and membrane-bound forms, 209
Amino acids, sequence conservation
 TGF-β type II receptors, 126–127
 TGF-β type III receptors, 122
Anemia, in swine, 277
Antibodies, G protein-selective, 79
Antigens, CD38, 212–216
Antisense cDNA, 83–85
Antisense knockouts, 81–85
Antisense oligonucleotides, 81–83
Aplysia, ADP-ribosyl cyclase, 209–211
Aquatic species, vitamin B_6 requirements for growth, 283–287
Autophosphorylation, IGF-I receptor, 11–12

B

β cells, pancreatic
 cyclic ADP-ribose
  dependent Ca^{2+} release, 234–235
  role in insulin secretion, 245–248
  role in stimulus–secretion coupling, 247–248
Betaglycan
 structural features, 120–123
 TGF-β presentation to type II receptor, 147–150
Brain, microsomes, cyclic ADP-ribose-dependent Ca^{2+} release, 233–234
Bream, vitamin B_6 requirements for growth, 286
Breast cancer, cell lines, IGF expression, 35

C

Calcium
 release
  cGMP effects, 219–220
  cADP-ribose-dependent, 203, 221–236
   Ca^{2+} stores, 224–227
   mammalian systems, 231–236
   relationship with Ca^{2+}-induced release, 227–231
   sea urchin egg as model system, 221–224
  induced Ca^{2+} release, 227–231, 241
   caffeine effects, 229–230
   cyclic ADP-ribose-sensitive, 229–230
   ryanodine receptors, 228–229
  NAD^+-dependent, 201–202
 stores
  cADP-ribose-sensitive, 225–227
  IP_3- and ryanodine-sensitive, 224–225
Calcium channel, classes, 200
Caffeine, effects on Ca^{2+} channel, 229–230
Cancer, IGF in, 34–38
Carp, vitamin B_6 requirements for growth, 285
Cat, vitamin B_6 requirements for growth, 272–274

Catfish, vitamin B_6 requirements for growth, 283–286
Cattle, vitamin B_6 requirements for growth, 278–279
CD38, 212–216
ecto-expression, 215
Cell proliferation, cell-specific actions, fibroblast cells, 89–91
Chicken, vitamin B_6 requirements for growth, 281–283
Complementary DNA
antisense, 83–85
endothelin receptor, 167–168
G-protein specificity, 81–83
Crustaceans, vitamin B_6 requirements for growth, 286–287
Cyclic ADP-ribose, 199–251
antagonists, 239–243
binding sites, photoaffinity labeling, 243–244
Ca^{2+} mobilization, 229
model, 249–250
Ca^{2+} release, 203
discovery, 201–204
endogenous tissue levels, 206–207
hydrolytic products, 205
metabolism, enzymes involved in, 207–221
ADP-ribosyl cyclase, 208–212
cyclic ADP-ribose hydrolase, 212
lymphocyte CD38, 212–216
regulation by cGMP, 218–221
relationship with NAD^+ glycohydrolase, 216–218
physiological roles, 244–248
fertilization, 244–245
insulin secretion, 245–248
sensitive Ca^{2+} stores, 225–227
structure, 202, 204–205
8-substituted analogs, 239–243
Cyclic ADP-ribose hydrolase, 212
occurrence in animal tissues, 259–260
Cyclic ADP-ribose receptors, 236–244
antagonists, 239–243
binding to sea urchin egg microsomes, 237–239
photoaffinity labeling of binding sites, 243–244
Cyclic GMP, regulation of cyclic ADP-ribose, 218–221

D

4′-Deoxypyridoxine, 260–261
Desensitization, G protein linked receptors, redefinition, 97
Dimerization, TGF-β receptor activation, 140–142
DNA, complementary, see Complementary DNA
Dog, vitamin B_6 requirements for growth, 274
Duck, vitamin B_6 requirements for growth, 283
Dwarfism, Laron-type, 21

E

Eel, vitamin B_6 requirements for growth, 286
Embryo, insulin-like growth factor effects
postimplantation, 23–25
preimplantation, 22
Endoglin
role in signaling, 150–151
structural features, 124–125
Endothelin, 157–183
activation of transmembrane signaling systems, 170–172
biological actions, 172–175
discovery, 158–159
genes, structural organization, 164–165
in liver, 177–182
effect on hepatic portal pressure, 179–180
glycogenolytic activation mediation, 179
signaling mechanisms, 177
localization, 169–170
pathophysiology, 175–176
processing, 162–164
stimulation of production, 166
structure, 159–161
structure–activity relationships, 161–162
Endothelin receptors, 166–169
antagonists, 168–169
nomenclature, 167
regulation studies, 182
N-terminal region, 168
tissue distribution, 172–173
Endothelium-derived relaxing factor, 158, 173

INDEX 303

Enzymes
  bifunctional, 213–214
  CD38-like, 215–216, 249–250
  pyridoxal phosphate-containing, 208–209
Eukaryotic cells, heterologous expression of proteins, 62–64
Exotic animals, vitamin B_6 requirements for growth, 287–288

F

Fertilization, cyclic-ribose role, 244–245
Fibroblast cells, as models for signal transduction, 85–96
  cellular milieu, 93–94
  $G_{i/o}$-coupled receptors, cell specific signaling, 85–91
    cell proliferation, 89–91
    gene transcription, 89
    multiple signals, 85–89
  G_s-coupled receptors, cell-specific signaling, 91–93
  pathway-selective modulation, 94–96
Fox, vitamin B_6 requirements for growth, 274

G

Genes
  endothelin, 164
    structural organization, 164–165
  c-jun, 12
  transcription, cell-specific signaling properties, 89
Gerbil, vitamin B_6 requirements for growth, 270–271
$G_{i/o}$-coupled receptors
  cell specific signaling, 85–91
  in growth hormone pituitary cells, 73–75
  heterologous expression, 75–78
  multiple signals and receptor-specific efficacies, 77–78
  transfection method specificity, 75–77
G_s-coupled receptors, cell-specific signaling, 91–93
Glucocorticoid response element, 63
β-Glycan, see Betaglycan
Goat, vitamin B_6 requirements for growth, 280

G protein-linked receptors, 59–98; see also $G_{i/o}$-coupled receptors; G_s-coupled receptors
  characteristics, 59–60
  desensitization, 97
  induced signal initiation, essential components, 60
  reconstitution, antibody, and mutational approaches, 78–81
  stable transfection of antisense G-protein subunit cDNA constructs, 81–88
G proteins
  $G_{i/o}$, coupling specificities, 80
  specificity, 78–85
Growth hormone
  deficiency, 27
  effect on longitudinal growth, 21
Growth hormone pituitary cells
  as models for signal transduction, 64–85
  $G_{i/o}$-coupled receptors, 73–75
  G protein specificity, 78–85
  hormone secretion, 64–66
  ion channels and transporters, 66–68
  as models of hormone action, 68–73
  multiple signals and receptor-specific efficacies, 77–78
  receptors, 66
  transfection method specificity, 75–77
  hormone secretion, somatostatin-induced inhibition, 74
Guinea pig, vitamin B_6 requirements for growth, 271

H

Hamster, vitamin B_6 requirements for growth, 271
Herring, vitamin B_6 requirements for growth, 285
Heterologous expression systems, 61–64
  advantages and disadvantages, 61–62
  vectors for eukaryotic expression, 62–64
Hoatzin, vitamin B_6 requirements for growth, 283
Horse, vitamin B_6 requirements for growth, 277–278
Hypoglycemia, non-islet cell tumor-induced, 37–38

I

IGF, see Insulin-like growth factors
Inositol trisphosphate, 224–225
Insulin
  α subunit, 10
  biological actions, in nervous system, 34–35
  secretion, cyclic ADP-ribose role, 245–248
Insulin-like growth factors, 1–39
  binding to IGF-binding proteins, 17
  in cancer, 34–38
    IGF-binding proteins, 36–37
    ligands, 35–36
    receptors, 36
    tumor hypoglycemia, 37–38
  mRNAs, in situ hybridization, 23–25
  in nervous tissue
    function, 33–34
    ligands, 31–32
    receptors, 32–33
  physiological roles, 20–27
    cell cycle, 20
    development, 22–25
    in vivo biological actions, 21–22
    in vitro effects, 25–27
    as progression factor, 20
  in reproductive system, 27–31
    ovarian physiology, 28–29
    testes, 30–31
    uterus, 29–30
  subunits, 19
  tertiary structures, 2–3
Insulin-like growth factor-I, 2–5
  gene expression, 4–5
  mRNA, 4–5
Insulin-like growth factor-II, 5–6
Insulin-like growth factor-binding proteins, 15–19
  in cancer, 36–37
  expression
    in ovarian tissue, 28–29
    regulation, 17–18
  function, 19
  IGF binding, 17
  structure, 15–17
Insulin-like growth factor-I receptors, 6, 8–14
  αβ hemireceptors, 13

α and β subunits, autophosphorylation, 10–12
  binding characteristics, 10–11
  in cancer, 36
  expression, 8–10
  mRNAs, in situ hybridization, 23–25
  precursor, 8
  receptor heterogeneity, 13–14
  signal transduction, 10–13
Insulin-like growth factor-II/M-6-P receptors, 14–15
Ion channels, in growth hormone pituitary cells, 66–68

K

Kupffer cells, endothelin conversion, 179–180

L

Leydig cells, IGF expression, 30
Liver, endothelin in, 177–182

M

Mammalian systems, cyclic ADP-ribose-dependent Ca^{2+} release, 231–236
Messenger RNA
  endothelin, 169
  IGF-I, 4–5
  IGF-II, 6
  in situ hybridization, 23–25
  IGF-I receptor
Microsomes, sea urchin egg, cyclic ADP-ribose receptor binding, 237–239
Mink, vitamin B_6 requirements for growth, 274–275
Mitogen-activated protein 2 kinase, 12
Mouse, vitamin B_6 requirements for growth, 270

N

NAD^+, Ca^{2+} release, 201–202
NADase, relationship with cyclic ADP-ribose, 216–218
NAD^+ glycohydrolase, relationship with cyclic ADP-ribose, 216–218
Nervous tissue, IGF in, 31–34
Nitric oxide, synthesis, 220–221

O

Oligonucleotides, antisense, 81–83
Ovary, IGF expression, 28–29

P

Pancreas, β cells, see β cells, pancreatic
Peptide hormones, TGF-β family, 114–115
Peptides
  endothelin family, 159–160
  sequence conservation, TGF-β type III receptors, 121–123
Phosphoinositidase, 27
Phospholipases, activation by endothelin, 171–172, 180
Phosphoproteins
  IRS-I, 12
  pp70, 12
Photoaffinity labeling, cyclic ADP-ribose, binding sites, 243–244
Pituitary, growth hormone cells, see Growth hormone pituitary cells
Placenta, IGF-II effects, 23
Plants, vitamin B_6 derivatives, 290
Prawn, vitamin B_6 requirements for growth, 286–287
Preproendothelin, 162–163
  genes, 164–165
Primates, vitamin B_6 requirements for growth, 280–281
Protein kinase C, mediation of thyrotropin releasing hormone action, 72
Purkinje cells, IGF, 31
Pyridoxal 5'-phosphate, 259–260
4-Pyridoxic acid, urinary excretion, 261–262
Pyridoxine β-glucoside, 290

Q

Quail, vitamin B_6 requirements for growth, 283

R

Rabbit, vitamin B_6 requirements for growth, 272
Rat, vitamin B_6 requirements for growth, 266–269

Receptor signaling
  detailed mapping, 97–98
  pathway-selective modulation, 94–96
Receptor tyrosine kinases, dimerization, TGF-β receptors activation, 140–142
Reproductive system, IGF in, 27–31
Retinal ganglion cells, IGF, 31
RNA, messenger, see Messenger RNA
Ryanodine, sensitive Ca^{2+} stores, 224–225
Ryanodine receptors
  cardiac sarcoplasmic reticulum vesicles, 236
  types, 228–229

S

Salmon, vitamin B_6 requirements for growth, 284–285
Sarcoplasmic reticulum, cardiac, cyclic ADP-ribose-dependent Ca^{2+} release, 235–236
Sea urchin eggs
  as Ca^{2+} release model system, 221–224
  cGMP effects on Ca^{2+} release, 219–220
  cADP-ribose-sensitive Ca^{2+} stores, 225–227
  fertilization, cyclic ADP-ribose role, 244–245
  microsomes, cyclic ADP-ribose receptor binding, 237–239
Sertoli cells, IGF expression, 30
Sheep, vitamin B_6 requirements for growth, 279–280
Signal transduction, transmembrane, 180
Somatostatin, actions, 73
  pertussis toxin sensitivity, 74–75
Somatostatin receptors, 66
Squirrel, vitamin B_6 requirements for growth, 271–272
Swine, vitamin B_6 requirements for growth, 275–277

T

Testes, IGF expression, 30–31
TGF-β, see Transforming growth factor β
Thyrotropin releasing hormone, 65
  actions, time course, 68–69
  first phase of action, 70–71

receptor, 66, 68
second phase of action, 71–73
Transforming growth factor β
  definition, 111
  family, 113–114
  importance, 111–112
  signaling, gene responses, 145
  structure, 115–116
    functional relationship, 116–117
  superfamily, 114–115
Transforming growth factor-β binding proteins, cell surface, identification, 117–119
Transforming growth factor-β receptors
  activation mechanism, 140–143
  structure and function, 142–143
  pathways and signaling thresholds, 143–146
  signaling, 131–132
    endoglin role, 150–151
    complex, genetic evidence for, 132–134
    functional type I receptor, identification, 137–140
    other mechanisms, 146–147
    threshold, 143–146
    through heteromeric receptor complex, 134–136
  structural features, 119–131
    endoglin, 124–125
    type I, 129–131
    type II, 125–128
    type III, 120–123
    type V, 128–129
  type III, TGF-β presentation to type II receptor, 147–150
Transporters, in growth hormone pituitary cells, 68
Trout, vitamin B_6 requirements for growth, 284–285
Tumor, hypoglycemia, 37–38
Turbot, vitamin B_6 requirements for growth, 286
Turkey, vitamin B_6 requirements for growth, 283
Tyrosine kinase receptors
  growth factor-dependent activation, 143–144
  truncated, overexpression, 144

U

Uterus, IGF expression, 29–30

V

Vitamin B_6, 259–291
  deficiency
    birds, 281–283
    carnivores, 272–275
    horse, 277–278
    interspecies variation, 287–289
    primates, 280–281
    rodents, 266–272
    ruminants, 278–280
    swine, 275, 277
  forms, 259–260, 289–290
  low intake diet, 262–263
  measurement in biological samples, 262
  metabolism, 259–265
    liver role, 263–264
    in mammary tissue, 264
    protein binding role, 263
    urinary excretion, 261–262
  requirements
    assessment, 261
    for growth, 266–288
      aquatic species, 283–287
      birds, 281–283
      carnivores, 272–275
      exotic animals, 287–288
      horse, 277–278
      primates, 280–281
      rabbit, 272
      rodents, 266–272
      ruminants, 278–280
      swine, 275–277
    minimal and optimal, 264
    during pregnancy, 289
Vole, vitamin B_6 requirements for growth, 272

Y

Yellowtail, vitamin B_6 requirements for growth, 286

ISBN 0-12-709848-8